T0064995

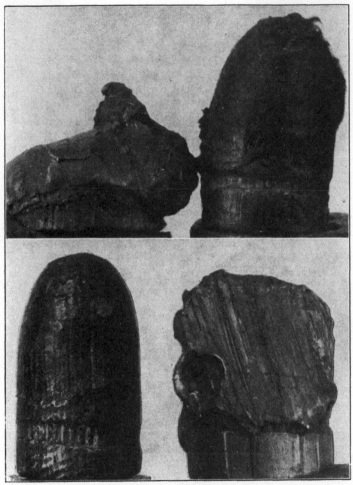

Signed Bullets.

THE IDENTIFICATION OF FIREARMS

From Ammunition Fired Therein
With an Analysis of Legal Authorities

BY

JACK DISBROW GUNTHER, A.B., LL.B.

Member of the New York State Bar

AND

CHARLES O. GUNTHER, M.E.

Professor of Mathematics, Stevens Institute of Technology; Lieutenant Colonel, Ordnance Department, The Army of the United States

Skyhorse Publishing

Visit our website at www.skyhorsepublishing.com.

10 9 8 7 6 5 4 3 2 1

Library of Congress Cataloging-in-Publication Data is available on file.

Cover design by Richard Rossiter

Print ISBN: 978-1-63220-276-5
Ebook ISBN: 978-1-63220-781-4

Printed in the United States of America

FOREWORD

It is only within the present generation that the science of identifying disputed documents has achieved judicial respectability. The ascent from the Avernus of deserved disrepute has been steep and difficult, obstructed by ignorance and slippery with fraud. Handwriting experts were long content to use the crudest of instrumentalities; they were quick to generalize from inadequate experience obtained in the most haphazard fashion; they made no real attempt to distinguish between the process of skilled observation and that of drawing deductions from observed data. In a word, they were of no more practical worth to a judge or jury than a graphologist would be. It took the patience, industry, skill, intelligence and expository power of an Albert S. Osborn to demonstrate that by the proper use of available aids in chemistry, microscopy and photography the pertinent data not only can be accurately observed by the expert, but also can be made understandable to the ordinary trier of fact, and that by long and carefully acquired experience genuine skill in interpreting such data can be attained. Today the real expert can be of invaluable assistance to the court. At the same time there will be no undue tendency to overrate his testimony. Judges and juries are too well acquainted with the variability of human reactions to fail to realize the great chance of error in attributing any specimen of handwriting to a particular individual.

The skepticism which expert testimony as to disputed documents encountered has not in general been opposed to expert testimony on firearms identification. Indeed, there has been too much of a tendency to take firearms experts at their own none too modest evaluation. There has been too great readiness to accept the notion that they deal with easily ascertainable data of exceptional exactness. It is easy to believe that the effect which one inanimate substance will produce upon another by a given interaction will always be the same. Hence, if a bullet of given size, weight and material is forced through a given rifle barrel, it will emerge with the same marks upon it as any other bullet of the same size, weight and material. For a quarter of a century a firearms expert for a great Commonwealth proceeded upon that theory, and assumed that the result would be the same whether the bullet were driven through with a rod and mallet or by any explosive. It is easy to credit the assertion that no two rifled barrels can be exactly the same because the cutting tool will necessarily be worn in the process of making each groove. From this the deduction is inevitable that a bullet shot from one gun cannot have the same markings as a bullet shot from any other gun. And so a widely known expert could confidently testify that a bullet was fired from a barrel which was indisputably shown not to have been in existence at the time of the shooting. Starting with plausible premises too many self-styled ballistics experts have found in their inadequate experience the results they were looking for; they have employed neither the approach nor the technique of the true scientist. They have confined their experiments to too few weapons, to too few types of ammunition, to too few conditions. They have emphasized similarities and minimized differences, or have relied upon casual dissimilarities and disregarded characteristics. They have failed to take into account the inherent deficiencies in their instrumentalities. Above all, they have been content to deduce universals from much too small a body of data.

It is the purpose of this book to make an honest exposition of the science of firearms identification in its relation to the judicial process. It is the result of long-continued and accurately controlled experiments with numerous specimens of various types and manufacture both of weapons and of ammunition. It neither exaggerates nor underrates the value of the function which the honest expert can perform. It does not disguise the evils that dishonest or ill-equipped experts may do. It should be an effective aid to court and counsel.

E. M. MORGAN

CAMBRIDGE, MASS.
HARVARD LAW SCHOOL
August 7, 1934

PREFACE

THE professional criminal evades the damning evidence of finger prints by the use of gloves and oiled surfaces. He prevents the tracing of firearms by filing off all such distinguishing marks as serial numbers. But when once a firearm is traced to his possession he will have much difficulty in removing therefrom the surfaces which have left their characteristic markings upon the surfaces of the bullets and cartridge cases fired therein. These markings furnish evidence more reliable than the testimony of eye-witnesses. The purpose of this book is to present such a study of the subject matter of the identification of firearms from ammunition fired therein as to demonstrate it as a science and to show its application in the courts by a discussion of the available authorities. If this purpose be achieved, the authors feel that they will have rendered a real service to the police, the trial lawyer, and the courts.

The authors wish to express their deep appreciation to the following for their invaluable assistance:

Edmund M. Morgan, Bussey Professor of Law, Harvard Law School;

Albert S. Osborn, author of "Questioned Documents," "The Problem of Proof";

F. T. Llewellyn, Research Engineer, United States Steel Corporation;

The Engineering Foundation;

Ordnance Department, U. S. A.;

and to extend their thanks to the following for their kind cooperation:

Carl L. Bausch, Bausch & Lomb Optical Co.

Stuart B. Campbell, Esq.

J. Victor D'Aloia, former Assistant Prosecutor of the Pleas, Essex County, New Jersey.

John Drewen, former Prosecutor of the Pleas, Hudson County, New Jersey.

Egbert C. Hadley, Remington Arms Co., Inc.

Hon. Roscoe T. Mauck, Judge of the Court of Appeals of Ohio.

E. Pugsley, Winchester Repeating Arms Co.

A. W. Schenck, Savage Arms Corporation.

Hon. Samuel E. Shull, President Judge, Court of Common Pleas, Pennsylvania.

Major D. B. Wesson, Smith & Wesson, Inc.

Colt's Patent Fire Arms Mfg. Co.

Mauser-Werke Aktiengesellschaft.

Peters Cartridge Co.

Webley & Scott, Ltd.

Western Cartridge Co.

Harry F. Butts, Commanding Officer, Ballistic Bureau, New York Police Dept.

JACK D. GUNTHER

CHARLES O. GUNTHER

GRAND VIEW-ON-HUDSON
NYACK, NEW YORK.
August 7, 1934.

CONTENTS

TABLE OF CASES

TABLE OF CASES

LIST OF ILLUSTRATIONS

INTRODUCTION

THERE are many natural laws whose cause and effect relationships have been long recognized by the courts, e.g., dry grass will ignite upon coming in contact with sparks from a locomotive; a charred substance indicates that it has been subjected to a high temperature; illuminating gas will ignite when brought in contact with a flame; a ship may suffer damage in passing through a hurricane. When the effect takes the form of making or marking a tangible object, its uniform operation furnishes a means for determining the identity of its product. This has been recognized with increasing clarity and effectiveness in tracing the origin of printed or typewritten material. The principles of typewriter identification are in many respects analogous to those of firearms identification, and a brief discussion of the former principles is a proper introduction to the relatively newer and possibly more complicated science of tracing the particular firearm employed to discharge a given bullet or cartridge case.

All typewriters constructed in accordance with the same specifications possess certain common physical attributes which may be called class characteristics. By means of these class characteristics it is possible to distinguish between typewriters of different manufacture. Class characteristics include the design, size, and proportions of each of the characters.

Theoretically all typewriters of common design will produce identical writings, but by experiment it is found to be highly improbable that any two machines will write exactly alike. Accidental variations occur despite the utmost care and skill in manufacture. When a typewriter is used it inevitably develops through wear and accident certain defects, such as scars or bruises on the type face and irregularities in the alignments of the characters. These accidental characteristics give each machine a true individuality and make possible a reliable basis of distinction between the matter that is typed on two machines of identical construction.

Osborn, in "Questioned Documents," second edition, pages 589 and 598, says:

Typewriting individuality in many cases is of the most unmistakable and convincing character and reaches a degree of certainty that can properly be described as almost absolute proof. The identification of a typewritten document in many cases is exactly parallel to the identification of an individual who exactly answers a general description as to features, complexion, size, etc., and *in addition* matches a detailed list of scars, birth-marks, deformities and individual peculiarities.

.   .   .   .   .

The identification in either case is based upon a definite combination of common or class qualities and features in connection with a second group of characteristics made up of divergences from class qualities which then become individual peculiarities.[1]

The procedure in establishing the individuality of a typewriter is first to compile a list of its accidental characteristics and then to determine by experiment with a large number of machines (if statistics are not available) the probable occurrence of each of these items separately. The probable coincidence of all the accidental characteristics is computed by the application of the following mathematical formula: the probability of the concurrence of all the items is equal to the continued product of the probabilities of all the separate items —thus if one item probably occurs once in twenty instances and another item probably occurs once in forty instances, the chance of their both occurring together is one in eight hundred. As the number of items is increased the improbability of an exact reproduction of a particular combination of accidental characteristics becomes greater, and the chance of duplication grows so negligible that it may be entirely disregarded and treated as an impossibility. A combination of class and accidental characteristics whose coexistence in more than one typewriter is proved improbable is called an individual peculiarity of that typewriter and differentiates it from all others.

The foregoing principles of identification have been successfully applied in courts for many years. As early as 1812[2] a plaintiff, attempting to prove that the defendant had printed a libelous article, was permitted to introduce in evidence for the scrutiny of the jury the alleged libelous copy and a specimen of printing from the defendant's shop. The jury was competent to determine whether the

[1] See also, Wigmore, "Principles of Judicial Proof," second edition, page 258, in which it is pointed out that the building up of an inference of identification is in accordance with the general principles of probative value.

[2] M'Corkle v. Binns, 5 Binney (Pa.) 340.

defendant was responsible for the printing. At that time there were relatively few presses in operation, and the jurors from their common experience were aware of the differences in the class characteristics of the printed material emanating from the small number of printing establishments.

A New York court in 1813[3] allowed the plaintiff to prove that the defendant had printed a libelous paper by the testimony of a former employee in the defendant's shop. The witness stated that in his opinion the libelous paper had originated at the defendant's press. He justified his conclusion on the theory that printers were able to identify a particular printing through an examination of its class characteristics.

The class characteristics of the type which served as the early basis of distinction between two or more specimens became inadequate with the extensive growth in the use of mechanized writing. The great number of printing presses and of typewriters of common design necessitated a scheme to identify a particular press or machine from others possessing the same class characteristics. As a consequence, a New Jersey court of 1893[4] permitted an expert to testify with respect to the accidental characteristics of a certain typewriter in the identification of a forged document. The witness was a man trained to notice defects as he was sent around the country to examine machines and ascertain if they were in good running order. He favorably impressed Pitney, V. C., by pointing out three peculiar irregularities: the period mark was always too low; the letter U was always off to the left; and the top section of the letter S was consistently more visible than the bottom.

A Utah court in 1906[5] admitted an expert to testify on the identification of a typewritten document. In establishing the relationship between a specimen proved to have been typed by the defendant's machine and the questioned document, the witness pointed out the following items: common class characteristics; the effect of certain defective letters which were broken and out of repair; the misalignment of particular letters; and the excessive spacing between certain letters. The expert testified that two machines out of repair might have precisely the same defects and produce the same faulty printing but it would be highly improbable for such a coincidence to occur.

[3] Southwick v. Stevens, 10 Johnson (N. Y.) 443. See also, Commonwealth v. Smith (1819) 6 Sergeant and Rawles (Pa.) 568.

[4] Levy v. Rust, 49 Atl. 1017.

[5] State v. Freshwater, 30 Utah 442, 85 Pac. 447.

A Maine court in 1917 [6] indicated its position on the question of typewriter identification when King, J., said:

We think the fact is patent and well recognized, requiring no expert testimony to establish it, that typewriting machines do develop by use some defects or irregularities in the alignment or position of its type, or in other features, and that such defects or irregularities are inevitably disclosed by the work produced upon such machines. If a proven specimen of work produced upon a certain typewriter corresponds identically with a disputed specimen in all of several defects, irregularities, and imperfections of the work, that fact would be pertinent and material to the question whether the disputed specimen was produced upon the same typewriter.

In Kerr v. United States [7] the defendant mailed a box of poisoned candy to one L. F. Kerr and was convicted for a violation of the postal laws by the trial court. The circuit court found no error in allowing an expert to identify the typewriter used to address the package of poisoned candy, and the general admission of this kind of evidence is the established law of today.[8]

In People v. Risley [9] the defendant was convicted for fraudulently inserting in an affidavit the words "the same" with his typewriter. In the lower court the trial judge permitted experts on typewriter identification to explain the defects existing in the defendant's machine and to show their agreement with the defects found in the inserted words. Subsequently the People called a professor of mathematics who applied the law of probability and indicated the impossibility of the same combination of defects existing in any other machine. The upper court held that the testimony of the mathematician was prejudicial to the defendant. Justice Hogan pointed out that the professor was not acquainted with the nature, causes, visibility, or permanency of the defects through personal observation, and that he should not be allowed to speculate in an abstract field. Justice Miller stated that the existence of the defects had been assumed in the questions put to the mathematician, and that the jury had not distinguished between existing and assumed defects. He

[6] Grant v. Jack, 116 Me. 342, 102 Atl. 38.

[7] 11 Fed. (2nd) 227 (1926).

[8] State v. Uhls (1926) 121 Kans. 377, 247 Pac. 1050; General Motors Acceptance Corporation v. Talbott (1924) 39 Idaho 707, 230 Pac. 30; Rudy v. State (Tex. 1917) 81 Crim. R. 272, 195 S. W. 187; Western Bottle Mfg. Co. v. Dufner (1914) 186 Ill. App. 235; People v. Storrs (1912) 207 N. Y. 147, 100 N. E. 730.

[9] 214 N. Y. 75, 108 N. E. 200 (1915).

likewise believed that the jury had not understood the precise relation between the defects in the defendant's machine and those appearing in the disputed writing, and that the jury relied largely upon the conclusions of the mathematician. ''The vice of the testimony consisted in its being purely an abstract theory having no relation to actual experience.''

However, it should be noted that the court was aware that an agreement between the defects in the defendant's machine and the defects appearing in the disputed words would be strong evidence of the defendant's guilt. The court was likewise aware of the important function of the law of probability in judicial proof, and it is common knowledge that the correct estimate of a probability involves the theory of probability. Therefore, it appears that the court might have ruled otherwise had the pertinent evidence been accurate and reliable and properly presented. For example, competent experts on typewriter identification should have clearly explained and pointed out to the jury the defects common to the standard of comparison and the disputed writing. They should likewise have testified as to the probable occurrence of each individual defect (based upon constructive experience with a large number of machines). With this foundation it would seem that a competent mathematician should be allowed to compute the probable concurrence of all the defects in order to show the extent of the probability. Of course, the questions would have to be put in hypothetical form, that is, if these individual probabilities exist, what is the probability of a combined occurrence? The mathematician would testify with respect to data observed by others, and it is in the province of the jury to determine as facts the correct data. Accordingly the jury would adopt the opinions consistent with the data as they found them.[10]

Generally speaking, the courts commonly recognize that the theory of probability must be applied in judicial proof when the fact to be proved is the probability of the happening of a future event, such as the expectancy of life of a particular individual. The same necessity is present in the field of identification. For example, it is impossible to examine all of the finger prints, typewriters, or firearms in existence. Therefore, the identification in any of these cases is predicated upon the results of research involving a large number of objects

[10] Typewriter identification and handwriting identification are fully presented in an excellent manner by Albert S. Osborn in ''Questioned Documents,'' second edition. See also ''The Problem of Proof,'' second edition, by Albert S. Osborn; ''The Principles of Judicial Proof,'' second edition, by John H. Wigmore.

which represent a cross-section of all the objects in existence. From research it is possible to determine the probable duplication of particular characteristics; and when once the individual probabilities are established, it is possible to determine the probability of the coincidence of any group of characteristics. All identifications depend upon this principle.

THE IDENTIFICATION OF FIREARMS

CHAPTER I

THE PRINCIPLES OF FIREARMS IDENTIFICATION FROM AMMUNITION FIRED THEREIN

TYPES OF PROBLEMS. DEFINITIONS

THE science of identification of firearms from the ammunition fired therein [1] concerns itself primarily with the development of methods by whose application it may be possible to solve six types of problems:

Type I. Given a bullet to determine the type and make of firearm from which it was fired.

Type II. Given a fired cartridge case to determine the type and make of firearm in which it was fired.

Type III. Given a bullet and a suspected firearm to determine whether or not the bullet was fired from the suspected firearm.

Type IV. Given a fired cartridge case and a suspected firearm to determine whether or not the cartridge case was fired in the suspected firearm.

Type V. Given two or more bullets to determine whether or not they were fired from the same firearm.

Type VI. Given two or more fired cartridge cases to determine whether or not they were fired in the same firearm.

The first steps in a logical development of these methods are to define terms which are pertinent to the subject matter, and to establish the basic principles involved.

[1] "Principles of Firearms Identification," by Charles O. Gunther. *Army Ordnance*, March-April and July-August, 1932.

"Markings on Bullets and Shells Fired From Small Arms," by Charles O. Gunther. *Mechanical Engineering*, February and December, 1930; May, 1932.

A **firearm** [2] may be defined as any instrument or device with which it is possible to propel a projectile by the expansive force of the gases generated by the combustion of an explosive substance. In its simplest form it consists of a **tube** or **barrel** containing a cylindrical passage, called the **bore,** through which the projectile is propelled by the expansive force of the **gases**; a **chamber** at one end of the barrel to receive the explosive substance and the projectile; and a means for igniting the explosive substance. The end of the barrel from which the projectile is discharged is called the **muzzle,** and the opposite end the **breech.** The **breechblock** is that part of a firearm which closes the breech and prevents the escape of the gases generated by the combustion of the explosive substance.

A firearm may have one or more barrels each with its own chamber, or it may have a number of chambers in a cylinder which can be rotated about an axis, thus bringing the chambers into successive alignment with a single barrel.

In ordnance, firearms which propel projectiles of less than one inch in diameter are generally classed as **small arms.** The science of identification of firearms from the ammunition fired therein deals primarily with small arms, particularly those which are capable of being concealed upon the person.

PROPELLANTS

Explosive substances [3] which can be used in a firearm to propel a projectile are classed as **propellants.** The various propellants in use

[2] No detailed descriptions of various types of firearms or ammunition are contemplated as there are a number of excellent books which cover these subjects, e.g.:

"American Small Arms," by Edward S. Farrow. The Bradford Co., New York, 1904.

"Firearms in American History," by Chas. W. Sawyer, Boston.

"Pistol and Revolver Shooting," by A. L. A. Himmelwright. The Macmillan Co., New York, 1928.

"Rifle and Pistol Ammunition Hand Book," published by the Western Cartridge Company, East Alton, Illinois.

"The Ideal Hand Book," published by the Lyman Gun Sight Corporation, Middlefield, Conn.

"The Modern Gunsmith," by James V. Howe. Funk & Wagnalls Co., New York, 1934.

"Textbook for Small Arms." London: His Majesty's Stationery Office, 1929.

[3] "Military Explosives," War Department Document No. 947. Government Printing Office, Washington, 1924.

today are termed **propellent powders**. The quantity of a propellent powder used in a firearm to propel the projectile through the bore is referred to as the **powder charge**.

Black powder is the oldest form of propellent powder used in firearms. It is a mechanical mixture of potassium nitrate (niter), charcoal, and sulphur approximately in the proportions of 75, 15, and 10.

Berchtold Schwarz was the first (A.D. 1313) recorded user of black powder in the propelling of stones from a gun. In the early days of black powder, or **gunpowder,** as it was called, it was used in the form of a fine powder or dust. Later developments led to powder grains of various sizes and shapes, obtained by compressing the finely divided powder into larger grains of greater density. At the present time black powder is usually made up in the form of small black grains which are polished by glazing with graphite.

Brown powder contains a larger percentage of potassium nitrate than black powder and a smaller percentage of sulphur. Its color is caused by an underburned charcoal.

Both black powder and brown powder produce a considerable volume of smoke. These powders contain inorganic substances and therefore leave a large quantity of solid residue in the bore of a firearm after the ignition of a charge.

Smokeless powders were introduced in about the year 1886. These powders are colloidal mixtures of organic compounds. Two general classes of smokeless powders are used in small arms: **nitrocellulose** and **nitroglycerin.**

Nitrocellulose powders are colloided masses of nitrocellulose containing some volatile solvent and diphenylamine which acts as a stabilizer. They are generally made in the form of cylindrical single-perforated grains or round flakes which are usually coated with a small percentage of graphite.

Nitroglycerin powders are mixtures of nitrocellulose with nitroglycerin. They usually appear in the form of cylindrical single-perforated grains or round or square flakes.

"Technical Regulations No. 1370-A." War Department, March 24, 1930. U. S. Government Printing Office, Washington.

"Notes on Military Explosives," by E. M. Weaver. Fourth edition. John Wiley & Sons, Inc., New York, 1917.

"The Manufacture and Testing of Military Explosives," by John A. Marshall. McGraw-Hill Book Co., Inc., New York, 1919.

Many varieties of smokeless powder are used in small arms in this country and abroad. A few of them will be mentioned here.

Ballistite, a typical nitroglycerin powder, is obtained by gelatinizing a low nitrated nitrocotton with nitroglycerin.

Cordite, a nitroglycerin-nitrocellulose powder, is a modification of ballistite. It derives its name from its cord-like appearance.

Bull's-eye powder is another nitroglycerin-nitrocellulose powder. It is granulated in solid cylindrical disks.

Pistol powder No. 5 is a nitrocellulose powder.

E. C. powder and **Kynoch** are both mixtures of nitrocellulose with the nitrates of potassium and barium.

Smokeless powders are not entirely smokeless. Smokeless powders which contain only organic compounds do not leave any solid residue in the bore of a firearm after the ignition of a charge. Because of the inorganic compounds they contain, E. C. powder and Kynoch leave some solid residue in the bore of a firearm after the ignition of a charge.

Semi-smokeless powders are a mechanical mixture of nitrocellulose, potassium nitrate, charcoal, and sulphur. These powders have an advantage over black powder in that they develop less smoke and leave a smaller solid residue in the bore of a firearm after the ignition of a charge.

In Figs. 1 to 6 are shown photomicrographs (photographs made with a microscope) of the following powder grains:

Fig. 1. Black powder.

Fig. 2. Semi-smokeless powder.

Fig. 3. Bull's-eye powder.

Fig. 4. Single perforated disks of smokeless powder.

Fig. 5. Single perforated cylinders of smokeless powder.

Fig. 6. German smokeless powder, green in color.

Types of Firearms

In the early types of small arms the bore had a smooth surface. The projectile consisted of a lead ball, and the powder charge and the projectile were introduced into the chamber at the breech from the muzzle end of the bore, hence the name **muzzle-loader.** Instead of a single lead ball of approximately the diameter of the bore, it was also possible to use a number of lead pellets of smaller diameter, called **shot** or **buckshot,** according to their size. The means for igniting the powder charge were found in the matchlock, wheel lock, flintlock, and percussion lock.

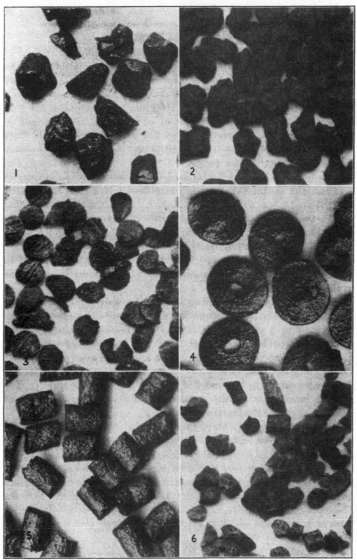

FIGS. 1 TO 6.—Photomicrographs (× 10) of Powder Grains.

Later developments produced small arms which had a number of **helical** (spiral) **grooves** cut in the smooth surface of the bore. Such arms are referred to as having **rifled barrels**.

Small arms are made with bore diameters of different size. The various sizes are indicated by the **gage** or **caliber**. Originally the term "gage" as applied to the now obsolete types of smooth-bore firearms indicated the number in a pound of lead balls of the size adapted to the arm. As applied to shotguns it indicates that the bore diameter is equal to the diameter of a lead ball whose weight in pounds is equal to the reciprocal of the gage index; e.g., the bore diameter of a 12-gage shotgun is equal to the diameter of a lead sphere weighing one-twelfth of a pound.

The term "caliber" was also used to indicate the bore diameter of firearms with smooth-bore barrels which fired a lead ball; thus caliber .50 indicated a bore diameter of 0.50 inch. With the advent of rifling barrels the term was retained, but today it only approximately indicates the bore diameter of a firearm; e.g., a caliber .38 revolver of a certain make has a bore diameter of 0.36 inch.

In countries using the metric system the caliber is expressed in millimeters, e.g., the metric caliber 6.35 (millimeters) is equivalent to the nonmetric caliber .25.

The **percussion cap** is used to ignite the powder charge in muzzle-loading firearms with percussion locks. It consists of a small metallic cup containing a **priming mixture**. It is placed on a nipple located at the breech end of the barrel. A blow from the hammer of the firearm, when released by a pull on the trigger, crushes and explodes the priming mixture. The flame thus produced is communicated to the powder charge in the chamber through a vent in the nipple. The **priming mixture** is usually a composition containing fulminate of mercury as one of the ingredients.

Muzzle-loading was ultimately superseded by **breech-loading**. In about 1815 a breech-loader was developed in which the powder charge and projectile were assembled in a paper case and introduced at the breech, but the paper case was soon replaced by a copper case. The powder charge was ignited by means of a percussion cap. Further development led to the present-day type of small arms in which the projectile, powder charge, and priming mixture are assembled in the form of a **cartridge** which is introduced as a unit into a chamber at the breech end of the barrel.

AMMUNITION

A **cartridge** consists of a **cartridge case** containing the powder charge, a bullet (projectile) rigidly fixed in the mouth of the case, and the priming mixture introduced in the base of the cartridge case. The base of the cartridge case is commonly termed the **head,** although the term **base** would seem to be the more appropriate. The priming mixture is exploded by the impact of a hammer or plunger, and the flame thus produced is communicated to the powder charge. Ammunition assembled in the form of cartridges is termed **fixed ammunition.** Cartridges can be obtained which are loaded with shot or buckshot instead of a single bullet, and shotgun cartridges can be obtained loaded with a single ball. Three types of fixed ammunition are used

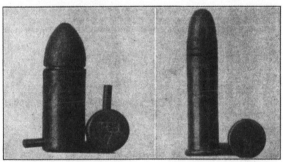

Fig. 7.—Pin-fire Cartridge.   Fig. 8.—Rim-fire Cartridge.

in small arms: **pin-fire, rim-fire,** and **center-fire.** Each type has its particular means for introducing and exploding the priming mixture.

In **pin-fire** ammunition, Fig. 7, a primer consisting of a small cylindrical cup containing the priming mixture is placed in a cavity on the inside of the head of the cartridge case. The priming mixture is exploded by the impact of the hammer on a pin which extends radially through the head of the cartridge case into the primer.

In **rim-fire** ammunition, Fig. 8, the priming mixture is placed in the cavity formed in the rim of the head of the cartridge case. The priming mixture is crushed and exploded either by a direct blow from the hammer on the rim or by a blow from the hammer on one end of a plunger, called the **firing pin,** driving the other end of the plunger into the rim of the head of the cartridge case. The flame so produced is in direct communication with the powder charge.

In **center-fire** ammunition, Fig. 9, the primer is forced into a small cylindrical chamber in the head of the cartridge case and the priming mixture is exploded by the impact of the firing pin. The flame is communicated to the powder charge through vents leading into the powder chamber. An early form of center-fire ammunition called "centre-primed, metallic cartridges," Fig. 10, resembled the present rim-fire ammunition in appearance.

The term **metallic ammunition,** as used in this discussion, applies to cartridges with metallic cases which are normally loaded with a

FIG. 9.—Center-fire Cartridge.   FIG. 10.—"Centre-primed" Cartridge.

single bullet, and the term **shotgun cartridges** applies to ammunition designed for use in shotguns.

Pin-fire ammunition is manufactured abroad and is used in Lefaucheux and other revolvers, carbines, and shotguns.[4] American manufacturers are large producers of both rim-fire and center-fire

[4] In the catalogues of the Imperial Chemical Industries Limited of London, England, dated 1925, and currently distributed as late as December, 1933, three pages are devoted to percussion caps, one page to pin-fire shotgun ammunition, and one page to pin-fire metallic ammunition. The metallic ammunition is made in caliber 5, 7, 9, 12, and 15 mm. Shotgun cartridges are made in 12, 16, and 20 gage.

ammunition, and both of these types of ammunition are also manu- factured abroad.

Both **brass** and **gilding metal** are alloys of copper and zinc, the gilding metal having the higher copper content. In each, the percent- ages of copper and zinc are governed by the degree of hardness de- sired in the alloy. Both brass and gilding metal are used extensively in the manufacture of cartridge cases for metallic ammunition. Shot-

FIG. 11.—Primer, Cup, and Anvil.

gun cartridge cases are made either of brass or of paper with brass heads, and some paper cases are metal lined.

One type of primer used in center-fire ammunition, Fig. 11, con- sists of a cup made of gilding metal or some other metal that is softer than the brass of the cartridge case. The cup contains the primer composition against which a paper disk is tightly pressed, and over which an anvil is forced into the cup. The anvil is made of brass and resists the blow of the firing pin, which crushes the composition between the cup and the anvil; the flame thus produced is communi- cated to the charge by the two vents in the anvil through a hole in the head at the base of the powder chamber in the case. In another type of primer used by some foreign manufacturers, Fig. 12, the anvil is formed in the head of the cartridge case in the cylindrical chamber which receives the primer cup containing the primer composition.

In metallic ammunition, cartridges of the same type are made in different calibers according to the firearms adapted to their use. Cartridges of the same type and caliber may be made in different sizes and with various types and weights of bullet. Some of the types of bullet used are the following: lead, full metal case, metal point, soft point, flat point, and hollow point. Blank cartridges are also available in certain types, calibers, and sizes of ammunition.

Lead bullets and the cores for metal case bullets are usually made of lead which has been hardened by the addition of a small percentage of either antimony or tin, or both. The jackets of metal case bullets are usually made of gilding metal or **cupro nickel,** the latter being an alloy of copper and nickel, high in copper content, the percentage of nickel depending upon the degree of hardness desired in the alloy.

The cylindrical portion of a bullet is generally provided with one or more circumferential grooves called **cannelures.** These cannelures

FIG. 12.—Anvil Formed in Head of Cartridge Case.

are usually knurled and may be used to hold the lubricant or to receive the **crimp** formed at the mouth of the cartridge case. The original purpose of the crimp was to prevent the bullet in the cartridge case from moving forward, as in firing a revolver it occasionally happened that, when one or more shots were fired, the bullets of the unfired cartridges moved or jumped forward so that their points jammed against the side of the barrel under the frame, thereby preventing the cylinder from revolving. A bullet may also be secured to the cartridge case by indenting the case into the surface of the bullet at two or more points.

Cartridge cases are made with either rimmed or rimless heads. In center-fire ammunition, the rimless cartridge case has a groove turned into the head for engaging the extractor.

The **extractor** is that mechanism in a firearm by which a cartridge or fired cartridge case is withdrawn from the chamber.

The **ejector** in a firearm is that mechanism which throws the cartridge or fired cartridge case from the firearm.

In some firearms one mechanism serves as both extractor and ejector.

Center-fire revolver cartridges have rimmed heads whereas cartridges for use in automatic (auto-loading) pistols are rimless. By using a clip as shown in Fig. 13, it is possible to use rimless cartridges in certain revolvers.

Cartridge cases of cartridges loaded with smokeless powder usually have a circumferential groove (Fig. 13) to prevent the bullet from being forced into the case beyond this groove, as such backward movement would be dangerous in that it would reduce the volume of the powder chamber and result in developing excessive pressure.

FIG. 13.—Clip.

The calibers .38 Smith and Wesson and .38 Smith and Wesson Special are examples of cartridges of different sizes of the same caliber of center-fire ammunition. The caliber .38 S. & W. cartridge will not enter the chamber of a revolver chambered for caliber .38 S. & W. Sp'l because the diameter of the cartridge case of the former is slightly larger than that of the latter. The caliber .38 S. & W. Sp'l cartridge is longer than the cylinder of a revolver chambered for the caliber .38 S. & W. cartridge and it can not be inserted to its full length in a chamber fitted for the caliber .38 S. & W. cartridge. In some makes of revolvers the cylinder chambers are made of uniform diameter. If a caliber .38 S. & W. Sp'l cartridge were introduced into such a chamber of a revolver adapted to the caliber .38 S. & W. the cylinder could not rotate. The caliber .38 S. & W. Sp'l lead bullet is the heavier of the two and has two grease cannelures whereas the caliber .38 S. & W. lead bullet has one.

The same cartridge cases may be used for cartridges of different sizes of the same caliber, e.g., the same cartridge cases are used in both the caliber .22 long and the caliber .22 long rifle rim-fire cartridges. The caliber .22 long rifle cartridge has a larger powder charge and a heavier and longer bullet than the caliber .22 long.

It is found that there is sufficient variation in the weights of bullets of the same caliber, type, size, and manufacture, so that in general it is only necessary to express the weight of a bullet to the nearest 0.5 grain. If the metric system of weights is used, the conversion from grams to grains for the purpose in hand, can be made by using 0.0648 gram as the equivalent of 1 grain avoirdupois.

Variations are found in the dimensions of the chambers of firearms adapted to cartridges of the same caliber and size. Manufacturers of ammunition must therefore control the dimensions of their cartridges so that the largest cartridges will fit the chambers of the firearms with the smallest dimensions, with the result that in many instances the cartridges fit the chambers loosely.

Shotgun cartridges are made in different sizes according to the gage of the shotgun adapted to their use.

The majority of manufacturers of ammunition stamp the heads of the cartridge cases, and some manufacturers stamp the primer cups in center-fire ammunition. In Fig. 7, the head of the cartridge case is stamped $\frac{"BB"}{7}$ This is a caliber 7 mm. pin-fire cartridge with a lead bullet manufactured by Braun & Bloem, Düsseldorf, Germany. In Fig. 8, the head is stamped with an "H." This is a caliber .22 long rifle rim-fire cartridge with a "Spatter Proof" bullet manufactured by the Winchester Repeating Arms Co. In Fig. 9, the head is stamped "PETERS .38 S. & W. SP'L." This is a caliber .38 Smith and Wesson Special center-fire cartridge with a lead bullet manufactured by the Peters Cartridge Co. Fig. 10 is a "calibre .50, centre-primed, metallic cartridge" manufactured at Frankford Arsenal, April, 1873.

In center-fire ammunition, in which the primer can be removed from the fired cartridge case, cartridge cases which have been fired may be reloaded with home-made bullets, bullets made with molds of standard makes, or with bullets purchased directly from the manufacturers of ammunition.

The term **shell** is popularly applied to the cartridge case. This is an undesirable practice, inasmuch as the same term is thus used for two entirely different objects, for in firearms adapted to fixed ammunition of calibers larger than one inch the cartridges have the same components as cartridges for small arms, and the projectile is a **shell** containing a high explosive, gas, or shrapnel.

Shotgun cartridges are popularly called "shotgun shells" in this country. Perhaps some justification for this practice is found in the similarity which exists between a shotgun cartridge and a projectile (shell) loaded with shrapnel.

RIFLING

Rifling consists of a number of helical (spiral) grooves cut in the surface of the bore. The raised helical surfaces thus formed are called the **lands.** The breech end of the lands are chamfered to form the **forcing cone** through which the bullet is led into the bore.

The purpose of the rifling is to impart to an elongated projectile a motion of rotation about its longer axis (axis of symmetry) and thus insure the necessary stability in its flight.

Rifling is of two kinds: **uniform twist,** in which the twist is constant throughout the bore; and **increasing twist,** in which the twist increases from the breech toward the muzzle end of the bore.

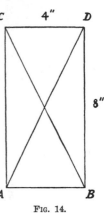

FIG. 14.

With a very few exceptions, the barrels of small arms are rifled with a uniform twist, and this discussion will be confined to rifling with a uniform twist.

The **twist of rifling** may be either **right-handed** or **left-handed.** In small arms it is expressed in the number of units of length (inches or millimeters) of bore in which it makes one complete turn.

The **tangent of the angle of twist** is equal to the ratio of the circumference of the bore to the distance to make one complete turn. The **bore diameter** is the diameter of the original smooth bore. The **groove diameter** is equal to the bore diameter increased by twice the depth of a groove.

The **angle of twist** is analogous to the angle between a tangent to a helix of uniform pitch (twist) at any point and the axis of the cylinder upon which the helix is described.

If a sheet of paper 4 inches by 8 inches in size, as represented by *ABCD* in Fig. 14, upon which the diagonals *AD* and *BC* have been drawn, be rolled into a cylinder bringing the edge *BD* in contact with the edge *AC*, keeping the diagonals *AD* and *BC* on the inside of the cylinder thus formed, then the diagonal *AD* will represent a helix of uniform left-handed pitch (twist) and the diagonal *BC* a helix of uniform right-handed pitch, each making one complete turn in 8 inches. The circumference of the cylinder, which corresponds to the circumference of the bore, is 4 inches, and the tangent of the angle

FIG. 15.

of twist in both helices is 4/8 or 1/2. The angle of twist for the helix *AD* is the angle *CAD,* and the angle of twist for the helix *BC*

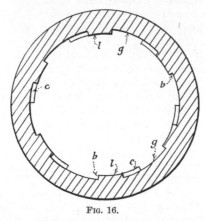

FIG. 16.

is the angle *DBC*. Obviously the angle *CAD* is equal to the angle *DBC*. The angle of twist in either case is approximately 26.5 degrees.

Fig. 15 is a view looking into the muzzle of a barrel, and Fig. 16 is a diagram of a cross-section of the barrel. The rifling consists of six helical grooves *g* with a uniform left-handed twist. The raised portions *l* are the lands. The sides of the lands, *b* and *c*, are called the **land shoulders.** The left-hand side *b* of the bottom land, or the corresponding side of any other land, is the pressure side and is called the **carry shoulder** or **driving edge** of the land. On account of the reflection of light this side of the land appears in the photo-

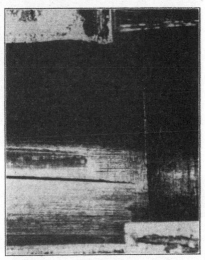

Fig. 17.

graph to be beveled, whereas it is actually the same as the right-hand side. Fig. 17 is a view of a longitudinal section of an old barrel showing the bullet seat and the forcing cone. Fig. 18 is a photomicrograph of a longitudinal section of a barrel like that of Fig. 15.

Fig. 19 is a view looking into the muzzle of a barrel in which the rifling consists of five helical grooves with a uniform right-handed twist. In the case of rifling with a right-handed twist the right-hand side of the bottom land, or the corresponding side of any other land, is the pressure side and is called the driving edge.

Fig. 20 is a photomicrograph of a longitudinal section of a barrel in which the rifling consists of six helical grooves with a uniform right-handed twist.

FIG. 18.

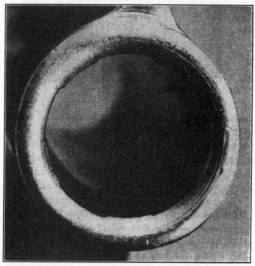

FIG. 19.

THE MICROSCOPE. PHOTOMICROGRAPHY

The **microscope** is the most important scientific instrument used in the identification of firearms from the ammunition fired therein. Professor Gage gives the following definitions:

A **simple microscope** or **magnifier** is a lens or a combination of lenses to use with the eye. But one image is formed and that is upon the retina. The enlarged image has all its parts in the same

FIG. 20.

position as they are in the object itself, that is, the image appears exactly as with the naked eye, except that it is larger.

A **compound microscope** is one in which a lens, or combination of lenses, called an **objective**, forms a real image, and this real image is looked at by the eye and a magnifier, or **ocular**. The image seen has the object and its parts inverted. In the compound microscope then, two images are formed, one by the objective independent of the eye, and the other on the retina by the action of the eyelens of the ocular and the cornea and crystalline lens of the eye.[5]

[5] "The Microscope," by Simon Henry Gage. Fifteenth edition, page 8. The Comstock Publishing Company, Ithaca, N. Y., 1932.

The **field** or **field of view** of a microscope is the area visible through a microscope when it is in focus.[6]

The **magnification, amplification,** or **magnifying power** of a simple or compound microscope is the ratio between the apparent and real size of the object examined.[7]

Magnification is expressed in diameters or times linear; that is, but one dimension is considered.[8]

Thus if a circle be viewed through a microscope and the diameter of the apparent size of the circle is found to be five times the diameter of the real circle, then the microscope has a magnification of five diameters, or x5, the word "magnification" being usually indicated by the sign of multiplication. The area of the apparent circle would of course be 25 times the area of the real circle.

In the examination of fired cartridge cases or bullets a magnification of 15 to 20 diameters is sufficient for ordinary purposes. In special cases it may be necessary to use a magnification of 30 diameters.

The microscope may be used for measuring objects. One method of making such measurements is by means of an ocular or eyepiece micrometer. The **eyepiece micrometer** consists of a glass disk, with a graduated scale, which is placed upon a diaphragm of the ocular or eyepiece of the microscope and brought into focus so that the scale appears sharply defined to the observer. This scale is then used to measure the microscopic image in the field of the microscope. The divisions of the eyepiece micrometer are calibrated by replacing the object by a stage micrometer, the scale of which is graduated in known units. For example, if the stage micrometer has a scale of which each interval measures one millimeter and it is found that five divisions of the scale of the eyepiece micrometer correspond to one division of the microscopic image of the scale of the stage micrometer, then each division of the eyepiece micrometer corresponds to 1/5 or 0.2 millimeter.

More refined measurements can be made with a micrometer eyepiece with a movable scale or a Filar micrometer eyepiece, which usually consists of an ocular with fixed cross lines and a movable line. The movable line is controlled by rotating a graduated drum, the circumference of which is generally divided into one hundred parts, one complete turn of the drum translating the movable line through one division of the scale in the eyepiece.

6 *Id.,* page 66.   7 *Id.,* page 282.   8 *Id.,* page 283.

Measurements of great precision are possible only by those skilled and trained in the art. The value of precision measurements made with metallic objects involving curved surfaces which have been more or less distorted is very questionable, and needless to say, precision measurements of such objects must be made at constant temperature. When consideration is given to the differences in the coefficients of expansion of steel, lead, copper, zinc, tin, antimony; the plasticity and elasticity of metals and alloys; the shock effect produced when the bullet comes in contact with a resisting medium; one recognizes the futility of attempting to establish the fact that a bullet was fired from a particular firearm by a comparison of measurements obtained from the bullet and the firearm.

In the identification of firearms from the ammunition fired therein it is of paramount importance to be able to compare the microscopic images of two objects. This is accomplished by using a **comparison ocular** or **comparison eyepiece** in combination with two compound microscopes of identical magnifying power with matched objectives. The comparison eyepiece is an optical instrument which consists of a series of prisms in combination with an ocular arranged to place half the fields of two microscopes side by side. The comparison eyepiece has been known in one form or another since about 1886. Some comparison eyepieces are made to fit into the tubes of the two microscopes, thus replacing the oculars of the two microscopes; others are made to fit over the oculars. In the early forms of comparison eyepiece, the fields of the two microscopes were divided by a horizontal diameter so that the rear half of the field of one microscope was combined with the front half of the field of the other. In 1920, Carl Zeiss, Inc., brought out a comparison eyepiece in which the fields of the two microscopes were divided by a vertical diameter thus combining the left-hand half of the field of one microscope with the right-hand half of the field of the other. This latter form of comparison eyepiece is best adapted for use in the identification of firearms. The assembly of two compound microscopes and a comparison eyepiece is popularly called a "comparison microscope," and it seems probable that Philip O. Gravelle, F.R.P.S., F.R.M.S., was the first recorded user of the comparison eyepiece in the identification of firearms.[9] The intelligent use of the comparison microscope requires training and experience.

[9] The following is quoted from a letter from Mr. Philip O. Gravelle to Colonel Gunther under date of September 10, 1933:

"I first met C. E. Waite in May, 1924, at which time he invited me to visit

In the apparatus shown in Fig. 21, the illumination is furnished by a 40-watt T-8 lamp in combination with a reflector made from a sheet of thin white cardboard. This source of illumination has proved very satisfactory for use within the range of magnification required.

The bullets are attached to the holders by the use of beeswax. When comparing bullets fired from firearms with a right-handed twist of rifling, they should be placed under the microscopes with the nose pointing to the right of the observer so as to throw the light into the shoulders formed on the cylindrical surfaces of the bullets by the driving edges of the lands. For the same reason bullets fired from firearms with a left-handed twist of rifling should be compared with the nose pointing to the left of the observer.

A number of types of comparison microscopes designed especially for comparing bullets and cartridge cases are now on the market in this country and abroad.

---

him at the Hotel Latham to outline his ideas regarding the markings left on bullets as a means of identifying them with a particular firearm. He had been successful in the Stielow Case and felt the work should be developed to a more scientific method, but had few ideas as to how it could be accomplished. Fortunately or otherwise I fell in line and spent considerable time and effort in working out photographic procedure which later was followed with the comparison microscope assembly.

"This was shown for the first time to Mr. Wesley W. Stout, associate editor of the *Saturday Evening Post,* April 2, 1925, while he and Mr. Waite were visiting my laboratory in South Orange. Waite did not grasp the significance of the microscope assembly until several days later. Then followed the *Saturday Evening Post* article of June, 1925, which you know about. This was the first published description of the use of the comparison microscope in matching striae on bullets, etc. . . .

"My work covering the optical and photographic procedure in relation to small arms was published in May, 1931, in the Symposium Number of the *Journal of the Photomicrographic Society* (London)."

Waite's ideas are set forth in an article by Calvin H. Goddard entitled "Scientific Identification of Firearms and Bullets" which appeared in the *Journal of Criminal Law and Criminology,* Vol. XVII, No. 2, August, 1926.

John H. Fisher developed another instrument for Mr. Waite. This instrument, which Mr. Fisher called the "helixometer," is an adaptation of the optical system of the cystoscope. With this instrument it is possible to examine visually the interior surfaces of the bore of a firearm as well as to measure the pitch of the rifling. This instrument which is now on the market is of value to the student in that it enables him to inspect visually the interior surfaces of the bore to study the effects of corrosion and erosion in conjunction with his study of the effects produced by corrosion and erosion on the surfaces of bullets which have passed through the bore.

A **photomicrograph** is a photograph of a small or microscopic object usually made with a camera in which the optical system of a microscope constitutes the lens of the camera.

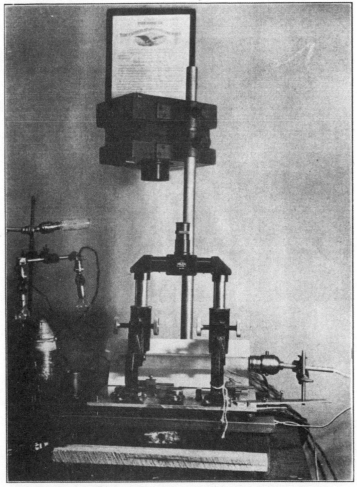

FIG. 21.—Comparison Microscope with Camera Body Swung Back.

Photomicrographs can also be made with the comparison ocular by connecting the camera body with the ocular as shown in Fig. 22.

FIG. 22.—Comparison Microscope with Camera Body in Position.

With the camera body swung back (Fig. 21) the apparatus is available for visual use. The whole apparatus rests on sponge rubber pads to eliminate vibration. All photomicrographs used in illustrating this book were made by the authors with the apparatus shown in Fig. 22.

Low-power photomicrographs can be made with a long-extension camera fitted with a photographic lens of short focus.

An **enlargement** of a photograph may be made by passing the rays of light from the illuminant (daylight or artificial light) through the negative (plate or film) and then through a lens which forms the image on the sensitized paper or material.

For these reasons it should be noted that an enlargement, as its name implies, is only an enlargement of the detail in the negative from which it is obtained. If, for example, a picture were printed on a sheet of thin rubber, the picture would be enlarged by stretching the rubber, but no new detail would appear.

A photomicrograph combines magnification (enlargement) with the resolving power of the microscope. In the enlargement the resolving power of the microscope is lacking.

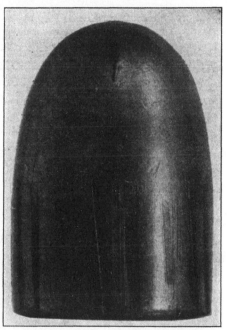

FIG. 23.—Photomicrograph (× 5.6) of a Cal. .45 Bullet.

Fig. 23 is a photomicrograph (x5.6) of a bullet. Fig. 24 is a portion of the photomicrograph (Fig. 23) enlarged about 5.5 times. In the latter photograph, the area is presumably shown with a magnification of approximately 30 diameters, but in reality it has only the microscopic detail of the 5.6 magnification. Fig. 25 is a photomicrograph with a magnification of about 30 diameters of the same area on the bullet which appears in Fig. 24. The difference in the

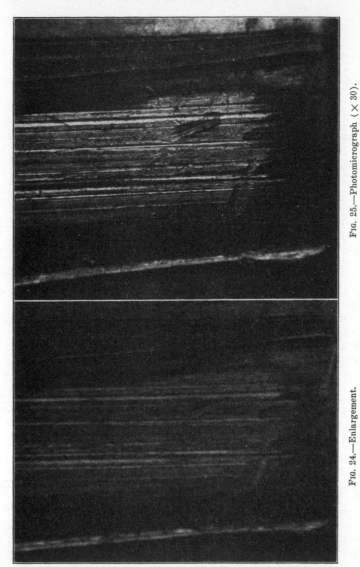

FIG. 25.—Photomicrograph (× 30).

FIG. 24.—Enlargement.

microscopic detail is quite evident from a comparison of these two photographs.

An enlargement may also be made from the positive print by photographing it, using a long-extension camera fitted with a short-focus lens, as in the case of a low-power photomicrograph. In the latter case the enlargement is in reality a photomicrograph of a photograph, magnifying all the details visible on the photograph, including the grain of the paper.

When taking photomicrographs with the same intensity of illumination on the object the time of exposure must be increased in the proper proportion as the magnification is increased. Taking a photomicrograph with the comparison microscope also requires a longer exposure than using either microscope alone.

Tool-Marks

A study under the microscope of the surfaces resulting from various hand-tool and machining operations enables one to recognize a particular operation by an examination of the tool-mark pattern on the surface. In Fig. 26 are photomicrographs (x5.6) of the surfaces of pieces of cold-rolled steel showing the tool-mark patterns made by:

A. A power-driven hack saw.
B. Fine and coarse files.
C. A grinding wheel.
D. A lathe tool.
E. A milling-machine cutter.
F. A shaper tool.

If the surface of a softer metal such as brass, lead, or copper is brought in contact under pressure with one of the surfaces of the pieces of cold-rolled steel shown in Fig. 26, it is quite evident that the tool-mark pattern on the surface of the cold-rolled steel will form its impression on the surface of the softer metal. "Ridges" on the surface of the cold-rolled steel will produce "valleys" on the surface of the softer metal, and *vice versa*. This fact furnishes one reason why a photomicrograph of a tool-mark pattern should not be compared with a photomicrograph of an impression when it is desired to determine whether or not the impression was formed by the tool-

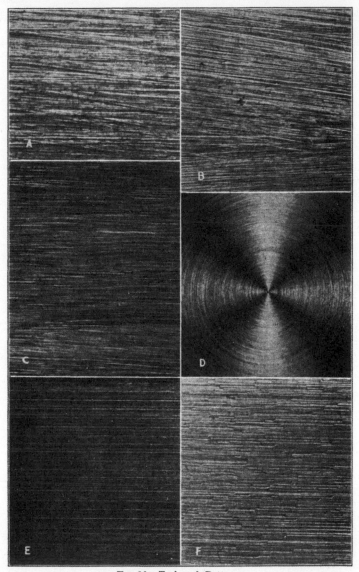

Fig. 26.—Tool-mark Patterns.

mark pattern. Another reason lies in the fact that the images in the two photomicrographs are reversed.

The underlying principle may be stated as follows:

If a surface is brought in contact under pressure with another harder surface, the resultant effect upon the softer surface will depend upon the relative hardness of the two surfaces, the character of the harder surface, the magnitude of the pressure, and the relative motion of one surface with respect to the other.[10]

The principle as thus formulated is expressed in simple terms. It includes four factors: relative hardness, character of the harder surface, magnitude of the pressure, and relative motion. The term relative hardness is intended to embrace the physical properties of the metal or alloy, which in turn depend upon its grain structure. In many of the applications of this principle its four factors may not remain constant at all times, in which case the principle should be qualified as follows:

If a surface is brought in contact under pressure with another harder surface, the resultant effect upon the softer surface *at any instant* will depend upon the relative hardness of the two surfaces, the character of the harder surface, the magnitude of the pressure, and the relative motion of one surface with respect to the other.

The tool-mark patterns on surfaces formed by machining operations are all illustrative of this principle. That the four factors do not remain constant at all times is quite evident from the following considerations: metals and alloys are not perfectly homogeneous; the cutting speed varies; the operation of all types of metal-cutting machines is accompanied by more or less vibration; all metal-cutting operations are accompanied by the generation of heat; and the tool becomes dull with wear. In some machines provisions are made for controlling the temperature.

In any metal-cutting operation the metal in contact with the cutter is stressed beyond its ultimate strength, and the result of any metal-cutting operation is a tearing rather than a true shearing action, the fineness or coarseness of the tear varying with the properties of the metal, the depth of the cut, and the properties and shape of the cutter. The tool-marks are produced by the *tearing of the metal* and *not by the blunting of the cutter*. The contour of the edge of the cutter governs the distribution of the stresses in the metal.

10 "Markings on Bullets and Shells Fired From Small Arms," By Charles O. Gunther. *Mechanical Engineering*, December, 1930, page 1065.

The same edge of a cutter will produce tool-mark patterns on soft
metals or alloys which are different from those it will produce on
harder metals or alloys.

Fig. 27 is a photomicrograph (x30) of the cutting edge, 1/8 inch
wide, of a tool bit made of high-speed steel. Fig. 28 is a photomicrograph (x30) of the same tool bit after it was treated in an abusive

FIG. 27.

manner in a shaper to cut a groove in a piece of cold-rolled steel
to a depth of 1/16 inch. Fig. 29 is a photomicrograph (x30) of a
portion of the groove cut in the cold-rolled steel by this same tool,
and the result of the tearing action is well illustrated. The white
spots are caused by the reflection of light on polished irregularities
in the surface. A comparison of the negatives of Figs. 27 and 28

FIG. 28.

indicates no appreciable change in the contour of the cutting edge
of the tool, even though the edge is shown with a magnification of
30 diameters.

The importance of this principle becomes evident when consideration is given to the fact that, in small arms adapted to fixed ammunition, there are certain component parts, depending on the type of firearm, which are brought in contact under pressure with the softer
metallic components of the cartridges fired in them.

CLASS AND ACCIDENTAL CHARACTERISTICS

There is an important analogy between the tracing of the origin of printed or typewritten material and the identification of a firearm from the ammunition fired therein. In both cases class characteristics and accidental characteristics play a controlling part.[11]

Class characteristics are those which are determinable *prior* to manufacture. Before a particular type of firearm goes into mass pro-

FIG. 29.

duction there will have been prepared the drawings, specifications, schedules of manufacturing operations; and since interchangeability of parts is essential, provision will have been made for the necessary gages, jigs, and fixtures. All information relating to the firearm which can be obtained or formulated from an examination and study of these drawings, specifications, and provisions will be grouped under the head of class characteristics. Such information may relate to the dimensions of component parts; peculiar features of design, operation, or construction; or the machining and other operations by which certain parts are produced. Class characteristics are controlled by man.

[11] See Introduction.

Accidental characteristics are those which are determinable only *after* manufacture. They are characteristics whose existence is beyond the control of man and which have a random distribution. Their existence in a firearm is brought about through the failure of a tool in its normal operation, through wear, abuse, mutilation, corrosion, erosion, and other fortuitous causes. The tool-mark patterns found on the various surfaces of a firearm are accidental characteristics. Dimensional variations in class characteristics, whether within or outside of the tolerances allowed by the specifications, are accidental

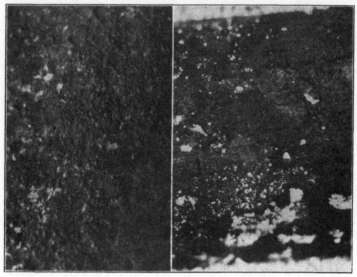

Fig. 30.   Fig. 31.

characteristics, and these become apparent only when instruments of precision are employed. Accidental characteristics exist in all firearms and are independent of class characteristics.

Corrosion is the eating away of the surface of the metal by the formation of rust or other chemical action.

Erosion is the result of firing; it is the breaking down and wearing away of the metal at the surface of the bore and the rifling. When a powder charge is ignited in a firearm, the gases generated by the combustion of the powder develop a high temperature and great pressure. When these gases, moving at high velocity, escape past the projectile, they attack the walls of the bore and cut irregular channels in the metal, destroying the surface of the bore and the rifling.

Fig. 30 is a photomicrograph showing the effect of corrosion on the nickel-plated surface of a revolver barrel. Fig. 31 is a photomicrograph of a longitudinal section of a revolver barrel showing the condition produced in the bore by permitting the accumulation of black powder residue to remain for some years without cleaning.

In the identification of firearms from the ammunition fired therein, the only useful class and accidental characteristics are those whose existence in a firearm become evident from an examination of the effects they produce on the metallic components of cartridges fired in it. Such characteristics will be termed **pertinent characteristics.**

CLASSIFICATION

Identification and classification are so closely associated that a person can not think of one without, consciously or unconsciously, involving the other. There are several ways in which characteristics may be classified, each serving a special purpose, each having its particular mode of grouping, but all with the same purpose in view— identification.

A classification is necessarily based on a similarity of some sort, and it is therefore possible to group firearms according to pertinent class characteristics. For example, small arms can be classified under two general groups:

Group I. Small arms with smooth-bore barrels. This group includes the various types of shotguns, the now obsolete types of smooth-bore firearms, and freakish devices.[12]

Group II. Small arms with rifled barrels.[13] This group includes rifles, carbines, single-shot pistols, automatic pistols, revolvers, derringers, machine guns, automatic and semiautomatic rifles, and freakish devices.

It should be noted that combination shotguns have one or two smooth-bore barrels in combination with another rifled barrel for use with a ball cartridge.

[12] The term ''freakish devices'' applies to firearms made in a form intended to disguise their real nature. Firearms which are made to resemble such forms as pocket knives, fountain pens, walking sticks, or flashlights are all classed as freakish devices.

[13] Small arms have been made with rifled barrels in which the bore is tapered. See ''The Gerlich Rifle and Bullet,'' by Glenn P. Wilhelm. *Army Ordnance,* March-April, 1933.

Each of these groups can next be classified under four subgroups:

Subgroup A. Small arms adapted to center-fire ammunition.
Subgroup B. Small arms adapted to rim-fire ammunition.
Subgroup C. Small arms adapted to pin-fire ammunition.
Subgroup D. Small arms with flintlocks, matchlocks, wheel locks, or percussion locks.

Firearms which may be manufactured so that they can be adapted to both rim-fire and center-fire ammunition would necessarily be included in both *Subgroup A* and *Subgroup B*.

The process of classification may be continued by grouping according to pertinent class characteristics. After pertinent class characteristics are exhausted for the purpose of grouping, the further classification of a group of firearms with common pertinent class characteristics can be accomplished only by a grouping according to pertinent accidental characteristics.

In such a classification, as the number of groups is increased the number of firearms in each group is decreased and the detail in the description of the firearms in each group is increased. The more minute the detail becomes in the description of an object the fewer will be the objects which answer the description.

In the following description of a person it is quite evident that, as each item is added to the specification, the number of persons answering the description will be decreased:

1. Human being. 2. Male. 3. White. 4. Age, 35 years. 5. Height, 6 ft. 6. Weight, 185 lb. 7. Brown hair. 8. Blue eyes. 9. Birthmark on left cheek. 10. Mole on right shoulder. 11. Scar on left portion of neck. 12. Right thumb amputated.

Pertinent accidental characteristics are of vital importance in determining the identity of a particular firearm. Obviously not all firearms in existence can be examined for accidental characteristics, and therefore the science of identification becomes fundamentally a mathematical science in that it must determine from an examination of relatively small groups of firearms the probable distribution of accidental characteristics in the larger groups. The science of identification consequently has recourse to the laws of permutations and combinations, the theory of probability, and other mathematical principles—the same mathematical fundamentals which, in one form or another, find their application in practically every field of human endeavor. Faith in the use of finger prints as a means of identification has been built up even though there have been examined and classified the finger prints of but relatively few persons when compared with the total population of the earth.

Thus if research develops the fact that in a particular type and make of firearm a certain accidental characteristic is likely to occur once in 100 firearms and another accidental characteristic once in 75 firearms, then the probability of the coexistence of these two accidental characteristics in one firearm is once in 75×100 or once in 7500 firearms.

If a firearm possesses a combination of pertinent class and accidental characteristics not found in any other firearm, then such a combination becomes an **individual peculiarity** of the firearm by which it is differentiated from all other firearms in existence. There may be more than one combination of pertinent class and accidental characteristics which establish an individual peculiarity of a firearm.

For the purpose of establishing the identity of the manufacturer of a particular cartridge case or bullet, ammunition may be classified in the same manner as firearms by grouping cartridge cases and bullets according to class characteristics. Further classification by grouping according to acidental characteristics is manifestly unnecessary.

By way of illustration, photomicrographs of lead bullets and their bases used in the caliber .44 Smith & Wesson Special cartridge by the following American manufacturers are shown in Figs. 32 to 35:

Fig. 32. Peters Cartridge Co.

Fig. 33. Remington Arms Company-Union Metallic Cartridge Co.

Fig. 34. Western Cartridge Co.

Fig. 35. Winchester Repeating Arms Co. and the United States Cartridge Co.

Class characteristics of these bullets are found in:

1. Size and shape of bullet as indicated by the contour of a longitudinal cross-section.
2. Width of cannelures.
3. Location of cannelures.
4. Spacing of the teeth in the knurling in the cannelures.
5. Design of the base of the bullet.

The Winchester Repeating Arms Co. has for some years been manufacturing ammunition for the United States Cartridge Co. At

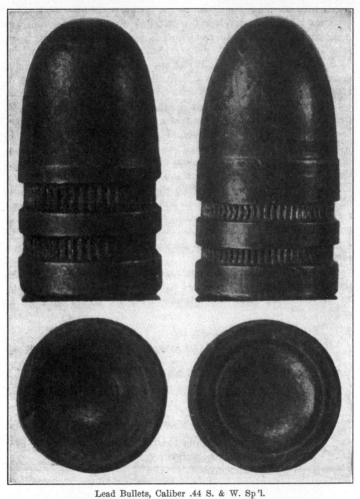

Lead Bullets, Caliber .44 S. & W. Sp'l.

FIG. 32.—Peters Cartridge Co.   FIG. 33.—Remington Arms Co.-Union
                                                    Metallic Cartridge Co.

Lead Bullets, Caliber .44 S. & W. Sp'l.

FIG. 34.—Western Cartridge Co.　　FIG. 35.—Winchester Repeating Arms
Co., and United States Cartridge Co.

the time of this writing, for the sake of economy the same bullets are used in certain calibers and sizes of ammunition for both Winchester Repeating Arms Co. and United States Cartridge Co. Given such a bullet, it would be impossible to tell whether it was from a cartridge put out under the name of Winchester Repeating Arms Co. or of United States Cartridge Co. Recently the Western Cartridge Co. took over the Winchester Repeating Arms Co., and it is not improbable that in the near future the same bullets will be used in certain calibers and sizes of ammunition put out under the names of Winchester Repeating Arms Co., United States Cartridge Co., and Western Cartridge Co.

Considering the types of problems in connection with the identification of firearms, it is found that every firearm adapted to fixed ammunition engraves its **signature** on the ammunition fired in it, one part of the signature appearing on the cartridge case and the other part on the bullet. In the case of a shotgun cartridge, part of the signature appears on the surface of those pellets which have come in contact with the surface of the bore. In the case of a muzzle-loading firearm, all the signature appears on the bullets fired therefrom. If the firearm is loaded with shot or buckshot, the signature is distributed over the surfaces of those pellets which come in contact with the walls of the bore.

The signature is engraved in accordance with the principle stated on page 27.

If a surface (cartridge case, primer, bullet, or pellet of shot) *is brought in contact under pressure with another harder surface* (surface of component part of a firearm), *the resultant effect upon the softer surface* (cartridge case, primer, bullet, or pellet of shot) *at any instant will depend upon the relative hardness of the two surfaces, the character of the harder surface, the magnitude of the pressure, and the relative motion of one surface with respect to the other.*

The component parts, depending on the type of firearm, with which a cartridge case may come in contact are: chamber, breechblock, recoil plate, firing pin, hammer, extractor, ejector, magazine, and those with which the cartridge comes in contact from the time it reaches the top of the magazine until it is in the chamber of the firearm.

In a firearm with a rifled barrel the component parts with which a bullet may come in contact are: chamber, forcing cone, lands and grooves.

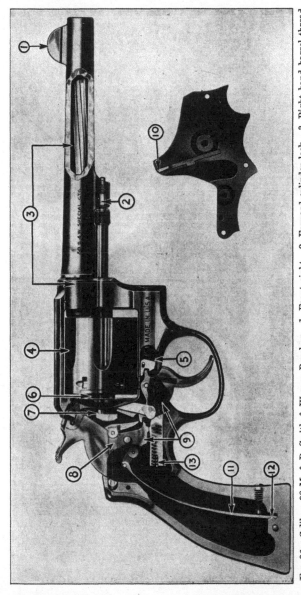

FIG. 36.—Caliber .38 M. & P. Smith & Wesson Revolver. 1. Front sight. 2. Forward cylinder lock. 3. Right hand barrel thread and right hand rifling. 4. Cylinder. 5. Cylinder stop. 6. Ratchet. 7. Rear cylinder lock. 8. Thumb piece. 9. Hammer and trigger. 10. Hammer block. 11. Main spring. 12. Main spring anchorage. 13. Trigger spring and hammer rebound.

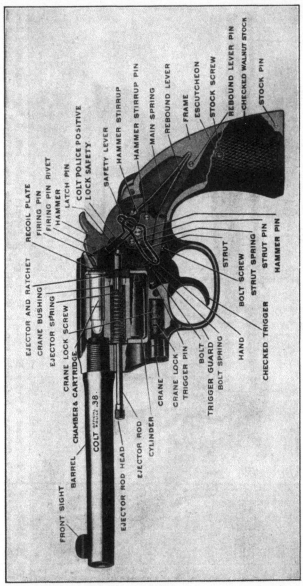

FIG. 37.—Phantom View of Colt Official Police Model Revolver.

For the nomenclature of the component parts of a revolver see Figs. 36 and 37. For the nomenclature of the component parts of an automatic pistol see Fig. 38.

The signature plays the same part in the identification of a firearm that the typewritten material does in the identification of a typewriter. Obviously **all conclusions as to the identity of a particular firearm must be based solely on an analysis of the signature.**

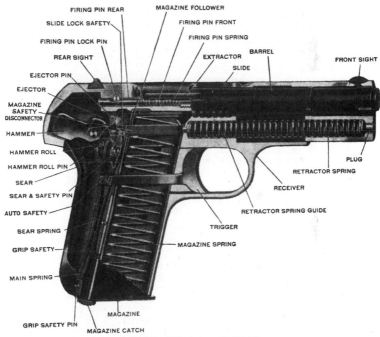

FIRING PIN REAR   MAGAZINE FOLLOWER

SLIDE LOCK SAFETY   FIRING PIN FRONT

FIRING PIN LOCK PIN   FIRING PIN SPRING

REAR SIGHT   BARREL   FRONT SIGHT

EXTRACTOR

EJECTOR PIN   SLIDE

EJECTOR

MAGAZINE SAFETY DISCONNECTOR

HAMMER

HAMMER ROLL

HAMMER ROLL PIN   PLUG

SEAR   RETRACTOR SPRING

SEAR & SAFETY PIN   RECEIVER

AUTO SAFETY   RETRACTOR SPRING GUIDE

SEAR SPRING   TRIGGER

GRIP SAFETY   MAGAZINE SPRING

MAIN SPRING

GRIP SAFETY PIN   MAGAZINE

MAGAZINE CATCH

FIG. 38.—Colt Automatic Pistol.

An individual peculiarity of a firearm may be established from a combination of pertinent class and accidental characteristics which are recognizable from an examination of that part of the firearm's signature which appears on the bullets fired from it, or the individual peculiarity may be established from a combination of pertinent class and accidental characteristics which are recognizable from an examination of that part of the firearm's signature which appears on the cartridge cases fired therein.

ANALYSIS OF MOTION OF BULLET THROUGH BORE OF AUTOMATIC PISTOL

The way in which a firearm engraves its signature on the ammunition fired in it is well illustrated by the automatic pistol, caliber .45, model of 1911, U. S. Army.[14]

In passing through the bore of a firearm, the lands of the bore form grooves on the cylindrical portion of the bullet, and the engravings found in these grooves will be referred to as **land engravings**.

The engravings on the raised parts of the cylindrical portion of the bullet (lands) are the result of contact with the grooves of the bore, and these engravings will be termed **groove engravings**.

FIG. 39.

A land shoulder on a bullet which was in contact with a driving edge of a land in the bore of a firearm will be called the "driven edge of the land."

The barrels of this pistol are interchangeable. Fig. 39 is a view of a section of a barrel with a ball cartridge (center-fire) in the chamber. The total length of the ball cartridge is 1.261 inches,[15] and its components consist of cartridge case, primer, powder charge, and bullet.

The cartridge case, Fig. 40, is 0.895 inch long and is made of brass. In the head of the case there is a small cylindrical chamber to

[14] A full description of this pistol will be found in "Training Regulations, No. 320-15," issued by the War Department under date of March 3, 1924.

[15] All dimensions given are subject to variations within limits set by the tolerances.

receive the primer. This chamber is provided with a small hole which communicates with the powder chamber. The primer, Fig. 41, is of the same type as that described on page 9 and shown in Fig. 11.

The bullet consists of a jacket made of gilding metal, enclosing a core of lead and antimony composition. The bullet weighs 230 ± 2 grains and has a length of 0.662 inch. The cylindrical part of the bullet has a diameter of 0.45015 inch. The bullet is seated in the cartridge case to a depth of 0.296 inch. In the manufacture of ammunition for this pistol, present methods of the Ordnance Department provide for seating the bullet in the cartridge case without crimping the case to the bullet and without the use of indents.

FIG. 40.—Cartridge Case.

The powder is smokeless; the charge varies with the powder and is usually about 5 grains.

The cartridge, although rimless, is intended also for use, when clipped, in both the Colt and Smith & Wesson army revolvers, M-1917. (See Fig. 13.)

The barrel is 5.025 inches long. The rifling consists of six helical grooves cut in the surface of the bore, which is 4.130 inches long

FIG. 41.—Primer.

and has a diameter of 0.445 inch. (See Figs. 15 and 16.) The rifling is of uniform left-handed twist, making one complete turn in 16 inches. The tangent of the angle of twist is equal to the ratio of the circumference of the bore to the distance to make one complete turn, or $0.445 \ \pi/16 = 0.0875$, and therefore the angle of twist is approximately 5 degrees. The grooves are 0.1522 inch wide and 0.003 inch deep. The lands are 0.072 inch wide. The forcing cone (Fig. 17) is formed by chamfering the breech ends of the lands, and the lands rise to their full height of 0.003 inch in a distance of 0.086 inch from the beginning of the bullet seat. The lands therefore rise at an angle of 2 degrees.

Before it becomes possible for the bullet to take up the motion of rotation about its longer axis, its cylindrical part must be provided with grooves into which the lands of the bore will fit. The formation of these grooves when the bullet is driven through the bore by mechanical means will be first considered.

The diameter of the cylindrical part of the bullet (0.45015 inch) is larger than the bore diameter (0.445 inch) and smaller than the groove diameter (0.451 inch) of the bore; therefore, when a bullet is forced into the bore for a short distance by mechanical means, it will become wedged in the forcing cone and be subjected to compressive forces which tend to imbed the chamfered ends of the lands in the surface of the jacket of the bullet, decrease the diameter of the bullet across opposite lands of the bore, and increase its diameter across opposite grooves of the bore. Fig. 42 is a diagram showing the wedge formed by the chamfered ends of two opposite lands. When the force acting on the base of the bullet is removed there is also a tendency for the bullet to increase in length on account of the radial compression.

Driving a bullet through the forcing cone is in the nature of a drawing operation, and in some types of firearms the metal of the jacket in the grooves formed on the cylindrical portion of the bullet is drawn so that it extends below the base of the bullet.

Under ideal conditions the radial compressive forces will so deform the bullet that in its deformed state it will conform to the contour of the cross-section of the bore and thus seal the bore and prevent the escape of powder gases. In the automatic pistol the pressure of the powder gases expands the cartridge case; this expansion tends to prevent the escape of gases to the rear. There is some escape of powder gases past the bullet as it moves through the bore; but the pressure of the gases exerted on the base of the bullet tends to give obturation, as a bullet under normal conditions eventually expands so as practically to fill up the grooves of the barrel.[16]

[16] See "Spark Photography and Its Application to Some Problems in Ballistics," by Philip P. Quayle, in Scientific Papers of the Bureau of Standards,

In order to drive a bullet through the bore it is necessary that the force applied to the base of the bullet be greater than the opposing force of friction. While a bullet is being driven through the bore, its surface is subjected first to the abrasive action of the chamfered ends of the lands (forcing cone) and then to the plowing action of the driving edges of the lands and the abrasive action of the surfaces of both lands and grooves. The engraving of the surface of the bullet is therefore the combined result of all these actions.

In the diagram, Fig. 43, let *abcd* represent the surface of a land of the bore; *ab*, the end of the land, or the intersection of the cylin-

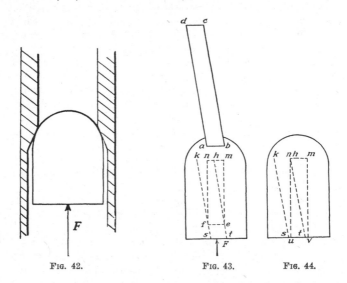

Fɪɢ. 42.   Fɪɢ. 43.   Fɪɢ. 44.

drical surface of the land with the conical surface of the forcing cone; and *F* a force acting in a direction normal to the base of the bullet. The force *F* in pounds is equal to the product of the intensity of the pressure per square inch and the area of the base of the bullet, the latter being approximately 0.16 square inch. Initially the motion of the bullet is one of pure translation along the axis of the bore. As the bullet moves forward in the bore under the action of the force *F,* the end *ab* of the land first comes in contact with the surface of the bullet at *nm.* When the end of the land *ab* has reached the position *fe,* the point *a* will have described the line *nf* and the point *b* the line *me,* and the end of the land *ab* will have scraped the surface of the bullet over the area *nmef.* The lines *nf* and *me* will be parallel to the axis of the bore. The driving edge of the land will be in the position *kf,* and the area *nfk* will have been subjected to the plowing action of the driving edge *ad.* If now the driving edges of the lands have imbedded themselves in the bullet to such a depth that the metal in contact with them is sufficient to withstand the tangential force exerted by them through the force *F,* the bullet will take up a motion of rotation and the end of the land *ab* will scrape the area *fets* as the bullet continues its motion. The surface of the bullet is therefore no longer subjected to the scraping action of the ends of the lands after the base of the bullet has passed into the bore, but only to the abrasive action of the surface of the lands. When the radial compression has deformed the bullet so that its surface also comes in contact with the surface of the grooves of the bore, then such surface of the bullet in contact with the surface of the grooves of the bore will also be subjected to an abrasive action.

If the bullet does not take up a motion of rotation until after the the base of the bullet has passed into the bore, the driving edge of the land may occupy the position *ks* in the diagram, Fig. 44, and the land shoulder the position *ht.* While there is always a tendency for the driving edge of the land to imbed itself in the surface of the bullet and thus prevent the escape of powder gases along *ad,* Fig. 43, the same is not true for the land shoulder. If the bullet does not take up a motion of rotation until after the base of the bullet has entered the bore as shown in Fig. 44, powder gases will escape at *tv* along the area *thmv,* and it is this escape of powder gases under high pressure and temperature that causes erosion along the land shoulder.

It is of utmost importance to note that the relative positions of the areas *htsk* and *metsfn* in Fig. 43, and the areas *htsk* and *mvun* in Fig. 44, depend upon the position of the end of the land *ab* with reference to the bullet at the instant that the latter takes up the motion of rotation. The area *htsk* represents the area of the surface of the groove formed in the surface of the bullet to receive the corresponding land of the bore, and the area *metsfn* in Fig. 43, or the area *mvun* in Fig. 44, represents the area of the surface of the bullet that has been subjected to the scraping action of the ends of the lands *ab*.

Fig. 45 is an enlargement of a photograph of the surface of a bullet corresponding to the areas referred to in Figs. 43 and 44. The breech end of one of the lands of the barrel from which this bullet was fired has a number of irregularities which produce the lines shown in the photograph that are parallel to the axis of the bore. This can be verified without firing a bullet through the barrel; it is only necessary to force a bullet into the breech end of the barrel until the base of the bullet is flush with the breech end of the lands and then force it out of the same end. An examination of the land engraving on the surface of this bullet will reveal all striae made by the irregularities which exist at the breech end of the lands or in the edge *ab* in Fig. 43.

To obtain some idea of the magnitude of the force *F*, experiments were made at the Carnegie Laboratory of Engineering, Stevens Institute of Technology, in which bullets were slowly driven through the bore and the resistance encountered was measured in a testing machine. In one of these tests a bullet was slowly driven through the bore of a

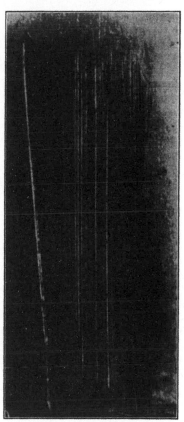

FIG. 45.—Land Engraving.

barrel, a short distance at a time. The bullet was of Frankford Arsenal manufacture with a jacket of gilding metal. The barrel was one that had been used considerably, and the bore showed signs of erosion. The force necessary to drive the bullet through the bore was initially 230 lb. (1437 lb. per sq. in.). From this it gradually increased to 350 lb. (2187 lb. per sq. in.), and then gradually decreased to 214 lb. (1337 lb. per sq. in.) when the point of the bullet was 1.5 in. from the muzzle. From this point the force again increased to 300 lb. (1875 lb. per sq. in.) at the muzzle. The greatest decrease in the driving force took place while the bullet was in the area of the bore which had been subject to erosion. The results of the tests also indicate that the resistance decreases somewhat as the speed of the bullet increases.

Considering now what occurs when a bullet is driven through the bore of the pistol by the force of the expansion of the powder gases. After the charge is ignited the powder burns and produces a large volume of gas. As soon as the expansive force of the powder gases is sufficient to overcome the friction of the wall of the cartridge case, the bullet moves forward in the bore, and after it has moved forward about 0.07 in. it comes in contact with the chamfered ends of the lands forming the forcing cone. The maximum pressure developed in the barrel is between 12,000 and 14,000 lb. per sq. in., and hence the maximum force is between 1920 and 2240 lb. The pressure probably reaches a maximum at the instant the bullet strikes the forcing cone. The bullet passes through the bore in approximately 0.001 sec. and leaves the muzzle with a velocity of 802 ft. per sec.; its kinetic energy upon leaving the muzzle is therefore 329 ft-lb. Since the twist of the rifling makes one complete turn in 16 in., the bullet will rotate at the rate of $802 \times 12/16 = 601.5$ turns per second or 36,090 turns per minute at the instant it leaves the muzzle.

The resistance encountered by a bullet in its passage through the bore is called the **passive resistance.** The accelerating force at any instant is therefore the difference between the force acting on the base of the bullet and the resisting force or passive resistance.

Chronographs are employed to determine the time it takes a projectile to travel a known distance between two fixed points at some distance from the muzzle. This gives the average velocity for the distance measured and this average velocity is called the **instrumental velocity** and is the actual velocity the projectile will attain at a point approximately midway between the two fixed points. From this instrumental velocity the muzzle velocity is calculated. In certain types of firearms the projectile is further accelerated by the blast of the powder gases after it leaves the muzzle. In such cases the calculated muzzle velocity will be slightly higher than the actual muzzle velocity because the calculated muzzle velocity is determined on the assumption that there is no muzzle blast.

A complete analysis of the deformation that the bullet undergoes as it passes through the bore of the pistol may therefore be divided into two parts: the first dealing with the formation of grooves in the surface of the bullet, and the second with marks and striae that the surface of the bullet reveals on account of its contact under pressure with the surface of the walls of the bore and the driving edges of the lands.

When a bullet is fired from a pistol the formation of the grooves is largely dependent upon variations in the dimensions of both the bore and the bullet, to which reference has already been made. A bullet of normal size fired in a bore whose diameter is larger than normal would show an effect similar to that of an undersized bullet fired in a bore of normal diameter.

The formation of the grooves is also dependent upon the hardness of the metal of the jacket of the bullet as well as upon the position of the axis of the bullet with respect to the axis of the bore. The average length of the grooves formed in the surface of the bullet is about 0.30 in. Under normal conditions it is found that the distance km in Figs. 43 and 44 is wider when a bullet is driven through the bore at high velocity by the force of the expansion of the powder gases than when driven slowly by mechanical means, which indicates that in the latter case there is less **stripping** and the bullet takes up the motion of rotation sooner. The distance km increases as the barrel becomes worn and eroded.

Stripping is said to take place when the bullet is moving with a motion of translation accompanied by a motion of rotation less than that provided for by the rifling.

The second part of the analysis resolves itself into an interpretation of the marks and striae that the surface of the bullet reveals on account of its contact under pressure with the surface of the walls of the bore and the driving edges of the lands.

If there is a slight protrusion at some point in the bore, then such protrusion will scratch the surface of the bullet upon coming in contact with it as it passes through the bore, and the shape of the scratch will depend upon the relative motion of the bullet with reference to the point at which the protrusion is located.

Consider the convex surface of the cylindrical part of the bullet as that of a right circular cylinder. The development of this surface will be a rectangle *ABCD,* Fig. 46, in which the length of the side *AB* is equal to the circumference of the cylinder, and that of the side *AD* to its altitude. A helix of uniform pitch described on the surface of the cylinder will appear as a straight line *ab* in the development. If *ac* be drawn perpendicular to *AB*, then *ca* represents the translation of the cylinder along its axis in the direction *c* to *a*, and *bc* represents the rotation of the cylinder about its axis in the direction *b* to *c*.

If X represents the angle between *ab* and *ac*, then $bc/ac = \tan X$, in which X is the angle of twist previously defined.

In the development, therefore, a straight line indicates a motion of pure translation of the cylinder along its axis when $X = 0$ degrees; a motion of pure rotation of the cylinder about its axis when

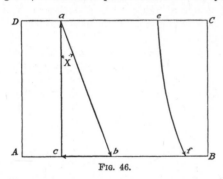

FIG. 46.

$X = 90$ degrees; and a helix of uniform pitch and hence a motion of translation and rotation in which the rotation is proportional to the translation, when X is greater than 0 degrees and less than 90 degrees.

If follows, then, that a curved line such as *ef* indicates a motion of translation and rotation in which the rotation varies with the translation, in which case the angle of twist is a variable angle.

Considering the surface of the bullet and assuming the axis of symmetry of the bullet coincident with the axis of the bore as the bullet passes through the bore, then a slight protrusion at some point in the bore will produce a scratch parallel to the axis of the bullet when the motion of the bullet is one of pure translation. In this case the protrusion must be located in the bore at or very close to the breech under normal conditions.

A scratch parallel to the impression made in the surface of the bullet by a driving edge of a land of the bore indicates that the bullet was moving with a motion of translation and the full rotation provided for by the rifling, and therefore the protrusion cannot be located in the bore in the vicinity of the breech under normal conditions.

The striae in the grooves of the bore may be regarded as tiny

ridges and valleys running parallel to the lands, and these can produce corresponding striae on the surface of the bullet only when
the latter has a motion of translation and the full rotation provided
for by the rifling; if the rotation is less than that provided for by the
rifling, then these groove striae will scrape the surface of the bullet
with which they come in contact. In the same way a tiny ridge in
the bore parallel to the axis can produce a scratch parallel to the axis
of the bore only when the bullet has a motion of translation alone.

A scratch starting out in a direction parallel to the axis of the
bore, and then gradually changing its direction until it is parallel
to the impression made by a driving edge of a land of the bore,
indicates a motion of translation with an accelerating rotation and
must therefore take place at the instant that the rifling takes effect;
the curve ef in Fig. 46 illustrates the development of such a scratch;
the protrusion producing it must therefore be located in the vicinity
of the breech under normal conditions.

The direction of a straight line or the direction of a curve at any
point is indicated by the angle X in the development, and the magnitude of this angle cannot exceed that of the angle of twist of the
rifling of the bore through which the bullet subjected to this analysis
has passed. In other words, if a scratch on the surface of a bullet
is developed and this development indicates a direction for which the
angle X exceeds the angle of twist of the bore through which the
bullet passed, then such scratch was *not* produced while the bullet
passed through the bore. This is of extreme importance in making
an analysis of the scratches on the surface of a bullet which has met
with an obstruction in its path and as a result of which it has been
much deformed and its surface scratched and mutilated.

In the case of a barrel in which the rifling is of uniform right-
handed twist, the point b is laid off on the line AB to the left of the
point c in Fig. 46.

Following the same line of reasoning outlined above, it becomes
possible to interpret the motion of a bullet at the instant that any
particular scratch was produced on its surface. If, however, a bullet
were fired through a barrel in the bore of which there happened to be
a loose particle of steel, then it would be a difficult matter to predict
how this would affect the engraving of the surface of the bullet. On
the other hand, in forming the grooves in the surface of the bullet
the driving edges of the lands cut away tiny particles of metal from
the jacket, and if one of these particles becomes lodged in the bore
it may produce a very pronounced scratch on the surface of the next

bullet fired; this scratch will probably not again appear on the surface of succeeding bullets.

MANUFACTURE OF PISTOL BARRELS

The operations incident to the manufacture of barrels, and more particularly, the methods adopted by the Ordnance Department in the manufacture of barrels for the automatic pistol, caliber .45, Model 1911, U. S. Army, are as follows:

The barrels are drop-forged in pairs from a rectangular bar of hot-rolled manganese steel. These forgings (Fig. 47) are then subjected to the following operations: (1) annealing, (2) pickling, (3) trimming, (4) inspection, (5) heat treating (quench in oil, draw), (6) test for hardness, (7) pickle, (8) straightening (when necessary), (9) inspection, (10) burring, (11) mill between lugs, (12) mill ends, (13) center—both ends, (14) rough turn ends, (15) rough turn between lugs, (16) inspection, (17) shouldering, (18) inspection, (19) drilling bore, (20) first reaming of bore (2 reamers), (21) inspection, (22) finish turning, (23) inspection, (24) grinding body, (25) inspection, (26) second reaming (rough—2 reamers), (27) third reaming (finish—3 reamers), (28) inspection.

The barrels are now ready for the rifling operation. The rifling consists of six helical grooves cut in the surface of the bore with a uniform left-handed twist, making one complete turn in 16 inches. The barrels are rifled with what is known as a **scrape** cutter. Fig. 48 shows a close-up view of one of these cutters in place in the rifling head. The length of the cutting edge is 0.45 inch and the width of the finished groove is 0.1522 inch, hence the ratio of groove width to length of cutting edge is approximately 1 to 3. Fig. 49 shows a longitudinal view of the rifling head with the two cutters and wedge in place. The cutters are elevated by means of the wedge shown entering the left end of the rifling head.

For the convenience of explanation, the consecutive grooves of the bore will be referred to by the numbers 1 to 6, inclusive, and the cutters will be designated by the letters A and B. Cutter A has one corner knocked off and cutter B has the opposite corner knocked off so that, when the rifling head is passed through the barrel the first time, cutter A cuts one corner of groove 1 and more than half the groove width while cutter B cuts the opposite corner of groove 4 and more than half the groove width. After the cutters have passed through the barrel and back in the same groove, the barrel

is rotated one-sixth of a turn and the rifling head passed through
the barrel and back for the second time making the initial cuts of
grooves 2 and 5. The barrel is again rotated one-sixth of a turn and
the initial cuts of grooves 3 and 6 are made on the third pass of the
rifling head.

The barrel is again rotated one-sixth of a turn and the rifling
head passed through the barrel and back for the fourth time, cutter

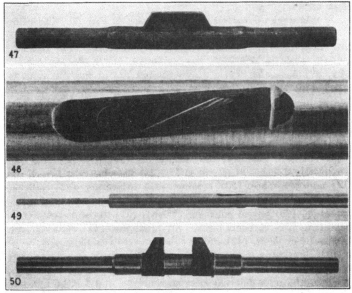

Fig. 47.—Single Forging for Pair of Barrels.
Fig. 48.—"Scrape" Cutter.
Fig. 49.—Longitudinal View of Rifling Head with Two Cutters and Wedge in
Place.
Fig. 50.—Rifled Barrels before Separation.

B completing the initial cut of groove 1 and cutter A the initial
cut of groove 4. After the barrel has been rotated one complete
turn and the rifling head has passed through the barrel and back six
times, the initial cuts of all six grooves have been completed. The
wedge is now fed in the proper distance to elevate the cutters for
the next cut, and the cycle as described is repeated until the grooves
have been cut to the proper depth. The wedge is fed in approxi-

mately seventy times during the rifling operation so that the rifling head passes through the barrel and back approximately 420 times. After the required depth of groove has been obtained, the rifling head with cutters remaining set at that position is allowed to pass through the barrel and back for approximately 15 minutes as a smoothing operation. About five barrels (double) can be rifled before the cutters require regrinding. In 1931 the use of double cutters was discontinued, and pistol barrels are now rifled with a single cutter.

At this stage the barrels appear as shown in Fig. 50. The rifling is followed by the following operations in the sequence given: (1) inspection, (2) cut in single lengths, (3) inspection, (4) face rear end, (5) chamfer both ends, (6) inspection, (7) mill right and left sides, (8) inspection, (9) rotary mill rear end of lug and taper on barrel, (10) burr rear corner of lug, (11) inspection, (12) handmill face of breech, (13) inspection, (14) drill and ream link pin hole, (15) burring, (16) form front of lug to shape, (17) burring, (18) mill radius on bottom of lug, (19) mill locking grooves, (20) inspection, (21) mill link slot, (22) grind top of breech, (23) burring, (24) inspection, (25) chamber, (26) inspection, (27) first polishing (approximately 2 inches of muzzle end), (28) chamber over breech end, (29) mill headspace, (30) burring, (31) inspection, (32) mill cartridge clearance cut, (33) inspection, (34) face muzzle to length, (35) inspection, (36) second polishing (finish), (37) file and corner, (38) inspection, (39) carbonia black, (40) proof fire, (41) cleaning, (42) stamp—proof mark, (43) grease (rust-preventive compound).

It is interesting to note that the interior surfaces of a pair of barrels manufactured in accordance with the operations enumerated above will be quite different from one another, and for the following specific reasons: [17]

First, in chambering the separated barrels, approximately 2 inches of the rifling at the center of the double barrel is removed and therefore there will be a difference in phase of about one-eighth of a turn or 45 degrees between the breech ends of the corresponding grooves in the two barrels when they are placed in the position occupied before separation as shown in Fig. 50.

Second, if the successive grooves at the right-hand end of the double barrel in Fig. 50 be numbered 1 to 6 in a *clockwise* direction, then these same grooves will carry through to the left-hand end of

[17] "Principles of Firearms Identification," by Charles O. Gunther. *Army Ordnance,* July-August, 1932.

the double barrel with the numbers 1 to 6 in a *counter-clockwise* direction. In other words bullets passing through these barrels when separated will *rotate in opposite directions with reference to the axis of the double barrel,* and the driving edges of corresponding lands in the two barrels will be reversed. In one of the separated barrels the driving edges of the lands will have been formed by cutter A while those of the second barrel will have been formed by cutter B.

Third, the various operations to which the barrels are subjected after separation introduce new accidental characteristics which will produce their resultant effect upon bullets fired from them and the probability of the accidental characteristics so introduced in one of the barrels being exactly reproduced in the other barrel is very remote.

Fourth, if any similarity in the signatures of such a pair of barrels did exist it could become plainly evident only from a comparison of the signatures on the bullets when they are placed base to base under the comparison microscope and not nose to base as is normally the case. Even if corresponding grooves of a pair of barrels did produce the same effect upon the surfaces of bullets fired from them, the pattern of the groove engraving on bullets fired from one barrel would necessarily be reversed on the bullets fired from the other barrel, unless such pattern was symmetrical about the center line of the groove engraving; however, the production of a pattern of perfect symmetry about the center line of the groove engraving is highly improbable.

Engravings on bullets fired from a pair of double barrels manufactured at Springfield Armory in 1930 are shown in Figs. 51 to 54. Fig. 51 is a comparison of land engravings produced by the same land of the double barrel with bullets compared nose to base, and in Fig. 52 the same engravings are shown with the bullets compared base to base. Fig. 53 is a comparison of the engravings produced by the same groove of the double barrel with the bullets compared nose to base, and in Fig. 54 the same engravings are shown with bullets compared base to base.

The automatic pistol is also manufactured by the Colt's Patent Fire Arms Manufacturing Co. and is known as their Government Model, Automatic Pistol, Caliber .45. The Colt barrels are each manufactured from special rolled steel which is made up to their own specifications. The rifling tool is drawn through one barrel at a time in the direction from breech to muzzle, and the barrels are not lead lapped after the rifling operation.

In some types of firearms the barrels are lead lapped or leaded as a finishing operation. This lapping is usually accomplished by means of a lead, tin, and antimony rod which has been cast, using the barrel as a mold and charging with a very fine abrasive and oil.

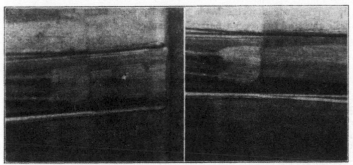

FIG. 51.   FIG. 52.

In general, barrels designed primarily for use with metal-jacketed bullets are not lapped.

Some manufacturers use a **hook** cutter in rifling barrels. The cutting edge of this type of cutter lies in a plane perpendicular

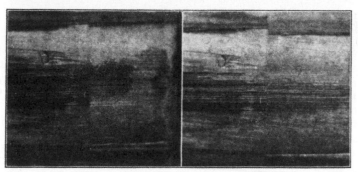

FIG. 53.   FIG. 54.

to the axis of the bore, and the ratio of length of cutting edge to groove width is unity. In barrels rifled with a single cutter, all grooves are formed by the same cutting edge unless the cutter is removed during the operation and reground.

If it were possible to make photomicrographs of the entire interior surfaces of the bores of barrels of common class characteristics, it would probably develop that no two of these photomicrographs would agree in all details. The tool-marks on the surfaces of the lands, grooves, land shoulders, and forcing cone form in any one barrel a pattern which will not be reproduced in its entirety in any other barrel with the same class characteristics; but not all variations in this pattern are recognizable from the land and groove engravings on bullets fired through these barrels.

RECOVERY OF BULLETS

Test bullets may be recovered by firing them into two or more one-pound rolls of absorbent cotton placed end to end in a container, or the container may be filled with a soft grade of cotton waste. When ammunition loaded with smokeless powder is used the muzzle of the firearm may be held close to the cotton. When ammunition loaded with black or semi-smokeless powder is used, the muzzle of the firearm must be held at some distance from the cotton, usually between 3 and 5 feet, to prevent the cotton from being ignited. In general when firing ammunition loaded with smokeless powder in firearms with short barrels a considerable number of unburned powder grains may be forcibly ejected and found adhering to the cotton. Before any test shots are fired a mark should be made at the nose of each bullet, using a small file for this purpose. Each cartridge should be placed in the chamber of the firearm so that this mark will always occupy a definite position at the instant of firing. By noting the position of the mark at the nose of the bullet in reference to some identifying letter or mark on the head of the cartridge case, the position of the cartridge case at the instant of firing will also be known.

The bullets shown in Figs. 55 to 61 were fired from different barrels in an automatic pistol, caliber .45, Model 1911, U. S. Army. The bullet shown in Fig. 55 was fired from a new barrel manufactured at Springfield Armory. The bore diameter of this barrel measures 0.4449 inch at the muzzle; at 1 inch from the muzzle the diameter is 0.4447 inch, and this dimension continues on to the breech.

The bullet shown in Fig. 56 was fired from a barrel manufactured by the Colt's Patent Fire Arms Manufacturing Co. As indicated by the land engraving on the bullet, the bore diameter of this barrel is

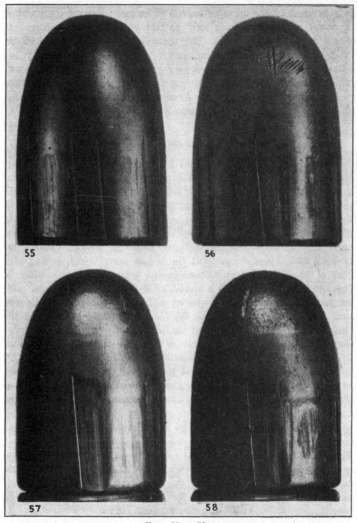

FIGS. 55 TO 58.

slightly larger at the breech than farther along the bore to the muzzle.

The bullet shown in Fig. 57 was fired from a barrel which has had considerable use, and the land engraving shows the effect of erosion; the bullet shown in Fig. 58 was fired from a barrel in which the erosion had reached the stage at which the barrel was no longer serviceable for accurate shooting.

The land engraving on the bullet shown in Fig. 59 shows the effect of corrosion; this barrel had not been given any attention for many years prior to which it had been used very little.

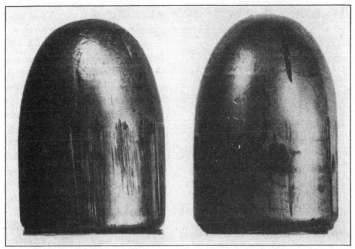

FIG. 59.   FIG. 60.

The bullet shown in Fig. 60 was undersize and made very little contact with the lands. The black area is the result of the action of the powder gases escaping past the bullet under high pressure and temperature.

The jacket of the bullet is drawn, and its surface reveals striae parallel to the axis of symmetry as a result of this operation. These striae must not be confused with scratches that the bullet receives in its passage through the bore. Fig. 61 shows clearly the striae resulting from the drawing operation.

Fig. 62 is a view of the breechblock of a pistol with the firing pin and the extractor removed. The tool-mark pattern on the **breech-**

ing face (surface of the breechblock) leaves its impression on the head of the cartridge case and the primer cup as these are forced against the breechblock by the expansive force of the powder gases, since the pressure developed by the powder gases is exerted in all directions.

The extractor, Fig. 63, will leave a mark on the side of the cartridge case near the base by which the position of the cartridge in the chamber of the barrel at the instant of firing can be determined. Extractor marks are also found on the cases of unfired cartridges which have been ejected from the pistol.

The end of the firing pin, Fig. 64, will generally show one or more circumferential toolmarks and irregularities which will leave their impression in the indentation formed in the primer cup when the firing pin is driven into the primer by the blow from the hammer. In this pistol the firing pin can be rotated, thus making it possible to change the relative position of the firing-pin impression.

The automatic pistol automatically ejects the exploded cartridge case after the first

FIG. 61.

shot is fired and reloads from a magazine; it is, however, necessary to pull the trigger to fire each shot. In the process of ejecting the fired cartridge case from the pistol, the case is brought into forcible contact with the ejector as the slide recoils to the rear, and the ejector leaves its mark on the head of the fired cartridge case.

In the automatic pistol and other firearms in which the cartridges are loaded into magazines, the sharp edges of the magazine will produce marks on the cartridge cases, but such marks will not appear on cartridge cases of cartridges which have been placed directly into the chamber of the barrel. A cartridge case may carry a series of these magazine marks, depending upon the number of times the cartridge was loaded in and removed from the magazine.

The lower part of the breechblock, Fig. 62, is brought in contact with the top of the head of the uppermost cartridge in the magazine in the process of forcing the cartridge out of the magazine into the chamber and leaves its impression on the head of the case. The relative position of this impression with reference to the marks pro-

FIG. 62.—View of Breechblock of Automatic Pistol with Firing Pin and Extractor Removed.

duced by the ejector and extractor furnishes an important pertinent class characteristic.

Figs. 65 to 67 are photomicrographs of the primers of cartridges fired in the same caliber .45 automatic pistol. The imprint of the tool-mark pattern of the breeching face (Fig. 62) and the firing-pin (Fig. 64) indentation are seen in Figs. 65 and 66. The firing-pin indentation in Fig. 65 shows a very pronounced lip, and this lip was not formed in the firing-pin indentation in Fig. 66 even though both of these cartridges were of the same manufacture and taken from the same box. On account of the relatively slow action of the

FIG. 63.—Extractor.

spring in returning the firing pin to its normal position in the rear end of the slide, it sometimes happens that the ejection of the case may take place while part of the firing pin is still protruding through the breechblock, with the result that the case in the grip of the extractor will pivot on the end of the firing pin during the process of ejection, and a more or less pronounced lip will be formed at the top of the firing-pin indentation in the primer cup, as shown in Fig. 65.

In Fig. 67 the metal in the primer cup was too soft. The pressure developed in the barrel forced the soft metal of the primer cup into the hole in the breechblock around the firing pin to practically the full depth of the indentation, and this metal was sheared off in the process of ejection. In the process of ejection the head of the cartridge case moves downward, while the mouth of the cartridge case moves upward and to the right as it encounters

FIG. 64.—Firing Pin.

the ejector and is thrown forcibly from the firearm. It is in this downward movement of the head of the cartridge case that the metal is sheared off and the tiny teeth in the lower half of the circumference of the hole through which the firing pin strikes (see Fig. 62) produce the marks shown on the primer cup in Fig. 67.

In general, with center-fire ammunition, if a primed cartridge case—a cartridge from which the powder and bullet have been removed—is fired in a firearm, the pressure developed in the primer chamber in the base of the cartridge case by the ignition of the primer composition will drive the primer partly out of the case and against the breeching face with a force sufficient to leave the impression of the breeching face on the primer cup. The indentation thus formed will not be as deep as when a fully loaded cartridge is fired.

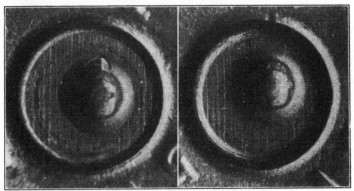

FIG. 65.   FIG. 66.

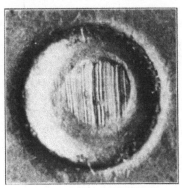

FIG. 67.

ANALYSIS OF SIGNATURES ON BULLETS FIRED FROM REVOLVERS

The analysis of the signatures on bullets fired from a revolver presents a more complex problem than in the case of bullets fired from an automatic pistol.

In the automatic pistol the chamber is an integral part of the barrel. When a cartridge is chambered, the mouth of the cartridge case seats against a square shoulder and the bullet lies in the bore in close proximity to the forcing cone. In the revolver the chambers are located in the cylinder, and when a chamber is in line with the barrel it is separated from the barrel by an air space. When a cartridge is chambered, the rim at the base of the cartridge case seats against the cylinder and the bullet lies in the chamber at some distance from the forcing cone in the barrel. In some revolvers the cylinder is chambered for a particular size of cartridge; in others the cylinder chamber is of uniform bore. In general the ammunition manufactured for use in revolvers has the mouth of the cartridge case crimped to the bullet.

In the automatic pistol the bullet is well in contact with the lands before the base of the bullet is out of the cartridge case; in the revolver the bullet does not as a rule come in contact with the lands until after the bullet is entirely free from the cartridge case; it has attained considerable velocity when it strikes the forcing cone, with the result that, because of inertia, in the case of a jacketed bullet the core is driven into the jacket, increasing its diameter, and in the case of a lead bullet, its length is shortened, with a consequent increase in diameter. The use of improper ammunition in a revolver may produce conditions which would result in splitting the barrel. In the revolver the bullet must jump from the cylinder chamber into the barrel, with an attendant escape of powder gases between the cylinder and the breech end of the barrel. This escape of gases also explains why a revolver cannot be silenced. The pressure of the powder gases expands the cartridge case, and this expansion tends to prevent the escape of gases to the rear; but there is also some escape of powder gases past the bullet as it moves through the bore of the barrel.

Fig. 19 is a view looking into the muzzle of the barrel of a Smith and Wesson revolver, caliber .38, Safety Hammerless, New Departure, loaned by Major D. B. Wesson. The barrel is 3.25 in. long and is lead lapped. The rifling consists of five helical grooves cut in the surface of the bore and is of uniform right-handed twist, making one complete turn in 18.75 in. The grooves are 0.114 in. wide and 0.005 in. deep. The lands are 0.1059 in. wide. The bore diameter has a small limit of 0.350 in. and a large limit of 0.351 in. The barrels are leaded as a finishing operation.

The cylinder is 1.215 in. long and has five chambers. The chamber diameter has a small limit of 0.388 in. and a large limit of 0.389 in. The diameter of the charge hole at the front end of the chamber is 0.362 in. The bullet is seated in the cartridge case to a depth of 0.25 in. and moves forward about 0.6 in. before it comes in contact with the forcing cone in the barrel.

Fig. 68 is a view of the frame back of the cylinder showing the recoil plate and the firing pin. Fig. 69 shows the firing-pin impression in the primer of a cartridge that was fired in this revolver.

Fig. 70 shows corresponding surfaces of three bullets fired from the same chamber in this revolver. These bullets are all from ammunition manufactured by the Western Cartridge Co. The one on the left is a 145-grain Lubaloy-coated bullet; the one in the center is a 145-grain lead bullet; and the one on the right is a full-metal-patch (jacketed) bullet from a cartridge loaded with smokeless powder. Each bullet was fired from a clean barrel. In each bullet the groove shown in the illustration was made by the same land of the bore;

Fig. 68.—Frame of Revolver Back of Cylinder Showing Recoil Plate and Firing Pin.

the groove width is narrowest in the jacketed bullet on the right, and widest in the lead bullet in the center. This indicates that there was less stripping with the jacketed bullet than with the other two types of bullets.

In a revolver, under ideal conditions, the chamber and barrel are in perfect line—the axis of the bore is coincident with the axis of the chamber in the cylinder—and the bullet moves through the chamber and bore with its axis coincident with the axis of the bore.

Under these conditions the land impressions will be of uniform
length.

FIG. 69.—Impression of Firing Pin Shown in Fig. 68.

FIG. 70.

If a bullet strikes the forcing cone with its axis parallel to but
not coincident with the axis of the bore, the land impressions will be

nonuniform in length as shown in Fig. 71. In a lead bullet some of the metal may be sheared off by the barrel.

If a bullet strikes the forcing cone with its axis inclined to the axis of the bore a double land impression, as shown in Fig. 72, is formed by the rotation of the axis of the bullet as it tends to align itself with the axis of the bore after striking the forcing cone.

Under normal conditions the

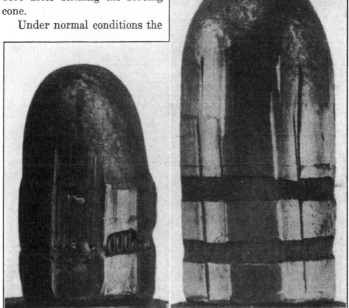

FIG. 71.   FIG. 72.

bullet should strike the forcing cone with its axis coincident with the axis of the bore. The following conditions, when present, are disturbing influences:

       Cylinder chamber is out of line with the barrel.
       Play in the axis of the cylinder.
       Variations in the head space and barrel clearance on the cylinder.
       Defective crimp in cartridge.

Use of a cartridge of shorter length than the cartridge for which the cylinder is chambered.

Firing pin with large **angle of percussion** in revolvers adapted to center-fire ammunition, particularly with loosely fitting cartridges.

In some firearms the firing pin strikes the primer in a direction parallel to the axis of the bore. In many revolvers the firing pin is an integral part of the hammer and may strike the primer in a direc-

FIG. 73.—Angle of Percussion.

tion considerably inclined to the axis of the bore. Let O (Fig. 73) be the center about which the hammer turns when released by the trigger pull, y, the distance from O to nose of firing pin, and z the normal distance from O to the plane of the head of the cartridge case in the chamber; then the firing pin will strike the primer in a direction inclined to the axis of the bore by the **angle of percussion,** x, whose sine is z/y. A microscopic examination of a cast of the firing-pin indentation in the base of the primer cup will enable one to determine the shape of the nose of the firing pin and its approximate diameter. Such a cast also offers possibilities for developing a method by which the angle of percussion may be determined.

A bullet fired from a cartridge loaded with black or semi-smokeless powder will show a black residue on its base, whereas the base of a bullet fired from a cartridge loaded with smokeless powder will be

FIG. 74.

comparatively clean. In Fig. 74 are shown the bases of two bullets. The bullet on the left was fired from a cartridge loaded with black powder; that on the right, from a cartridge loaded with smokeless powder. If a bullet is fired from a cartridge loaded with smokeless

FIG. 75.   FIG. 76.

powder in a firearm in which a number of cartridges loaded with black or semi-smokeless powder had previously been fired, the base of the bullet will show some signs of black deposit. Similarly the effect of the residue produced by the use of ammunition loaded with black or semi-smokeless powder is quite noticeable in the land and groove engravings on the surfaces of the bullets. Fig. 75 is a photo-

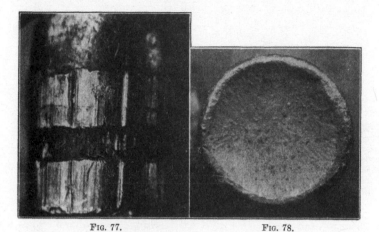

FIG. 77.   FIG. 78.

micrograph of a land engraving on the surface of a caliber .32 S. & W. lead bullet from a cartridge loaded with smokeless powder and fired from a Smith and Wesson revolver with a clean barrel, and Fig.

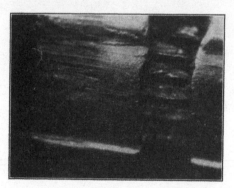

FIG. 79.

76 is a photomicrograph of the base of this bullet. Fig. 77 is a photomicrograph of the corresponding surface of a similar bullet fired in the same revolver after two smokeless and five black powder

cartridges had been previously fired, and Fig. 78 is a photomicrograph of the base of this bullet. Fig. 79 is a photomicrograph of a comparison of the land engravings shown in Figs. 75 and 77. Fig. 80 is a photomicrograph of the base of a bullet from a cartridge loaded with black powder and fired in the same revolver. These photomicrographs illustrate the result produced by the black powder fouling.

The irregular indentations in the base of the bullet (Fig. 80) are produced by the hard black powder grains being forced against the base of the bullet. The volume of a charge of black or semismokeless powder is considerably greater than the volume of the equivalent charge of smokeless powder.

Every reputable manufacturer stamps his name and serial number on each firearm he manufactures. Firearms, probably of foreign manufacture, have been found which are entirely devoid of any such identifying marks. When the serial number of a firearm is known it is possible for the

Fig. 80.

manufacturer thereof, if his records are complete, to furnish information relating to the firearm, including the date the firearm was shipped and the name and address of the consignee. For obvious reasons this information is available only to the proper authorities.

The manufacturer's records should show the range of the serial numbers of all firearms manufactured in accordance with a particular set of specifications. When changes in the specifications are made the record should indicate the serial number of the firearm with which such change was initiated. To credit a general statement that the approximate serial number and the approximate date of manufacture of any firearm can be determined from a bullet fired therein would require too great a stretch of the imagination even if one attributed a most remarkably elastic interpretation to the word "approximate." Fig. 81 is a photomicrograph of a comparison of the groove widths at the base of two bullets fired from two new caliber .38 S. & W. Sp'l revolvers of the same model and make with consecutive serial numbers. As indicated these two revolvers have different land widths, and the greater land width is found in revolvers bearing both higher and lower serial numbers than those of the revolvers with the consecutive serial numbers.

FIG. 81.

In many types and makes of firearms the parts are interchangeable, and therefore it might be advisable for manufacturers to stamp the serial number on the frame, cylinder, and barrel of a revolver, and on the frame, slide, and barrel of an automatic pistol. If this were done then the principal parts of the firearm with which the ammunition fired therein comes in contact would bear the same serial number when the firearm leaves the factory.

REPRODUCTION OF GROOVE ENGRAVINGS

In studying the groove engravings on metal-jacketed bullets fired in a Savage rifle,[18] caliber .32–20, the barrel of which was rifled with a hook cutter and not lapped, it was found that the tool-mark effect as indicated by the groove engravings on the bullet was reproduced by all six grooves. Photomicrographs were made with the apparatus shown in Fig. 22. One of the test bullets was placed under the microscope on the right showing the engraving on its surface made by one of the grooves in the barrel. The second test bullet was then placed under the microscope on the left and the engravings on its surface made by the six grooves of the barrel successively matched with the groove engraving on the surface of the bullet on the right. These photomicrographs are reproduced in Figs. 82 to 87.

If more than one barrel was rifled with this same cutter before it was reground, bullets fired from these barrels would probably show this same reproduction of groove engravings.

A second rifle of the same make manufactured about seven years later was next examined. A similar comparison of two bullets fired in this second rifle indicated considerable reproduction of groove engrav-

[18] Commonwealth v. Armbruster (Pa. 1931) County Court, Pike County.

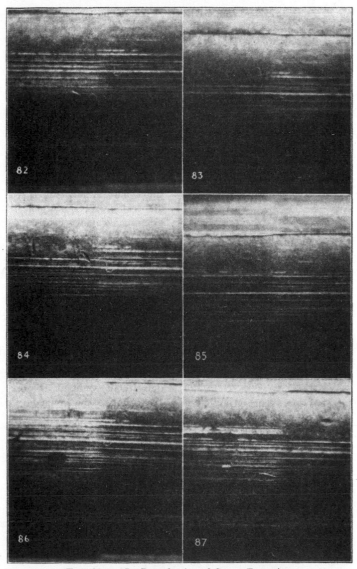

FIGS. 82 TO 87.—Reproduction of Groove Engravings.

ing in the six grooves, and, curiously, a photomicrograph (x15) of a comparison of the groove engraving on the surface of one of these test bullets shown on the left in Fig. 88 with the same engraving shown on the right in Figs. 82 to 87, demonstrates an interesting similarity.

The reproduction of groove engraving creates a situation which is not only serious but extremely significant and conveys a warning that may not be ignored. Lead lapping a barrel will to a large extent destroy the evidence of the reproduction of the groove engraving in barrels rifled with a hook cutter.

However, in certain high-grade revolvers in which the barrels are rifled with a scrape cutter, the lapping may produce perplexing similarities in the groove engravings on bullets fired from different revolvers when the groove engravings consist of a series of fine striae of practically uniform weight.

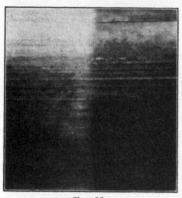

FIG. 88.

In the case of single barrels rifled consecutively, even if there were a reproduction of groove engravings on bullets fired from them, the operations incident to their manufacture subsequent to the rifling operation will introduce accidental characteristics which will produce their resultant effects in the land engravings on bullets fired from them. The probability of the coexistence of these accidental characteristics in the successive barrels is very remote.

In taking photomicrographs of bullets or cartridge cases, three dimensions are involved, and it is therefore necessary to illuminate the bullet or cartridge case properly in order to produce a true representation. Fig. 89 is a photomicrograph showing the impression of the tool-mark pattern of the breeching face on the primer cup. Fig. 90 is a photomicrograph of the same cartridge case rotated through an angle of 90 degrees. In the latter the tool-mark pattern has been eliminated, and this illustrates how photomicrographs can be made of an object in order to convey a false impression by the simple manipulation of the direction of the illumination.

Before considering the six types of problems in the identification of firearms from the ammunition fired therein, it should be noted that the problem of identification involves, first, a skilled observation and perception of the data; and second, a skilled interpretation of the data. The investigator must not only be able to ascertain the pertinent data, but he must also be capable of determining the significance and the comparative value of the characteristics relative to the identification. The art of identification concerns reasoning with respect to similarities and differences. It is found that there are differences in the signatures of a particular firearm and that similarities

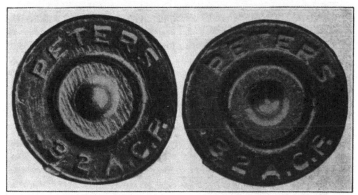

Fig. 89.   Fig. 90.

exist in the signatures of different firearms. Therefore an investigator can never intelligently ignore either similarities or differences—the identification must always be predicated upon a proper consideration of both. When identifying two signatures as having been made by the same firearm, he must be able to reconcile the differences found therein; and, likewise, when of the opinion that two signatures have not been made by the same firearm, he must be able to reconcile any existing similarities in the two signatures. It is quite obvious that a skilled investigator will necessarily have spent considerable time in acquiring the proper experience based upon training and controlled research.

In discussing the following problems the expression "type and make of firearm" refers to all arms of a certain type and make without reference to length of barrel, finish, stocks, grips, or other non-essentials.

SOLUTION OF PROBLEMS

First type of problem: Given a bullet, to determine the type and make of firearm from which it was fired.

The solution of problems of this type is limited to bullets which have not been subjected to excessive deformation or mutilation.

There are devices with which it is possible to fire a cartridge of the same caliber, but smaller in size than the cartridge to which the firearm is adapted, and other devices enable the firing of a cartridge of smaller caliber than the cartridge to which the firearm is adapted. The magazines of the Luger automatic pistols, caliber .30 (7.65 mm.)

FIG. 91.—Caliber .30 and Caliber 9 mm. Luger Cartridges.

and caliber 9 mm., are interchangeable and it is possible to fire a caliber .30 Luger cartridge (Fig. 91) in the caliber 9 mm. Luger automatic pistol with the result that the mouth of the cartridge case is ruptured. This discussion will be confined to bullets fired under normal conditions.

An examination of the bullet will indicate whether it was fired from a firearm with a smooth bore or rifled barrel. If the bullet was fired from a rifled barrel, grooves will be found in the cylindrical surface of the bullet which were formed by the lands in the barrel. The diameter of the bullet will indicate the caliber

of the firearm from which it was fired. The class characteristics of the bullet may indicate the size of the cartridge from which it was fired. In some cases it is possible to determine whether or not the bullet was fired from a revolver. The type of bullet may give information relating to the type and make of firearm from which it was fired. In the case of a bullet fired from a rifled barrel, it is possible to determine from a study of the bullet and the firearm's signature on the bullet, the following class characteristics of the firearm from which it was fired: bore diameter, groove diameter, depth of grooves, number of grooves, direction of twist of rifling, pitch of rifling, land widths, and groove widths. The problem is completely solved only when the class characteristics so determined lead to a group embracing but one type and make of firearm.

A study and comparison of the specifications of many manufacturers, both American and foreign, indicate sufficient variation in the items mentioned above to adopt them as class characteristics, and this adoption is made necessary by the fact that these items are the only ones in reference to which information can be obtained from the signature on the bullet. In this connection it should be noted that certain manufacturers have in the past made changes in their rifling specifications, and that manufacturers may make changes in the

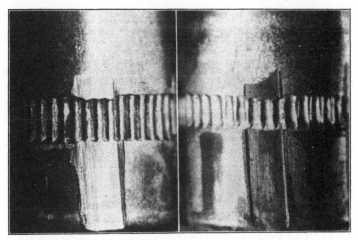

FIG. 92.—Buffalo.   FIG. 93.—Colt.

future. The dimensions given in the specifications are subject to variations governed by the tolerances. In some instances the rifling specifications of two different manufacturers are found to be very much alike. For instance a bullet (Fig. 92) fired from a "Buffalo" automatic pistol, caliber 7.65 mm., made in Spain, might easily be mistaken by an inexperienced observer for a bullet (Fig. 93) fired from a Colt automatic pistol, caliber .32, unless carefully compared with such a bullet under a comparison microscope.

Although it is possible to obtain by appropriate measurements made with instruments of precision the class characteristics of the firearm from which the bullet was fired, it is perhaps a better plan to compare the bullet with standards of comparison consisting of bullets fired from known types and makes of firearms, using the comparison microscope for this purpose. In this way it is possible to compare diameters, land and groove widths, pitch of rifling, and, in many cases, depth of grooves.

An extensive collection of standards of comparison is probably of far greater value to the investigator than a collection of rifling specifications. Especially is this true with reference to a variety of cheaply manufactured firearms which found their way into this

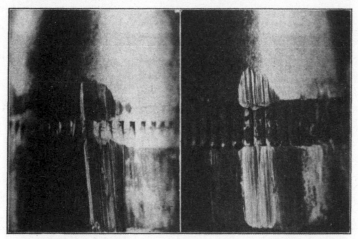

FIG. 94.—Bayard.   FIG. 95.—Colt.

country some years ago and for which it would probably be impossible to obtain rifling specifications.

Standards of comparison have a further advantage in that the investigator becomes familiar with certain characteristics of particular types and makes of firearms from a study of these standards, and furthermore, he becomes familiar with the variations *actually* found in the land and groove widths on bullets fired from firearms of a particular type and make. (See Fig. 81.)

In the case of shotguns, pellets which have come in contact with the surface of the bore will have parts of their surfaces flattened to conform to the curvature of the surface of the bore. Such pellets can be used to establish the gage of the shotgun from which they were fired by placing these flattened surfaces in contact with the surfaces of a series of gage standards and photographing them against a source of light.   The gage standard whose surface shows the most intimate contact with the flattened surface of the pellet indicates the gage of the shotgun in which the pellets were fired.   This method is described by Dr. Otto Mezger on page 755 of *Chemiker-Zeitung* (Köthen), September 27, 1930.

Bullets fired from various makes of automatic pistols of caliber .32 (7.65 mm.) are shown in Figs. 94 to 99.   The differences in the rifling specifications are quite apparent.

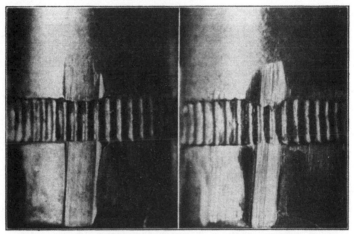

FIG. 96.—Savage.   FIG. 97.—Stenda.

Fig. 94. Bayard.
Fig. 95. Colt (old type).
Fig. 96. Savage.
Fig. 97. Stenda.
Fig. 98. Ortgies.
Fig. 99. Stosel.

Fig. 100 is a photograph of a jacketed bullet fired from a caliber .45 Colt revolver; Fig. 101 is a photograph of a lead bullet fired in the same revolver.   The jacketed bullet struck the forcing cone with its axis inclined to the axis of the bore.   The lead bullet is somewhat longer than the jacketed bullet and it struck the forcing cone true.

Second type of problem: Given a fired cartridge case, to determine the type and make of firearm in which it was fired.

An examination of the fired cartridge case develops information with regard to the following:

1. Type of ammunition: pin-fire, rim-fire, or center-fire.
2. Caliber or gage.
3. Paper or metal case.
4. Head: rimmed or rimless.

In the case of metallic ammunition, the size of the cartridge, if not indicated by the stamping on the head, must be determined from the size and shape of the cartridge case. This has its limitations in that the same case may be used in more than one size of cartridge.

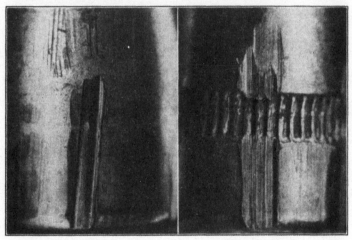

FIG. 98.—Ortgies.   FIG. 99.—Stosel.

If a fired cartridge case proves to be that of a cartridge adapted to but one type and make of firearm, the problem is solved; if not, then the class characteristics of the type and make of firearm in which the cartridge case was fired must be determined from an analysis of the firearm's signature on the cartridge case.

The solution of problems of this type therefore depends entirely upon the extent to which it is possible to establish pertinent class characteristics for the various types and makes of firearms in order to differentiate between them, keeping in mind that pertinent class characteristics are those recognizable from an examination of the signature on the fired cartridge case.

When a cartridge is fired, the pressure developed by the powder gases is exerted in all directions. The cartridge case is expanded in the chamber, bringing the convex surface of the cartridge case into intimate contact with the surface of the chamber, thus sealing the chamber and preventing the escape of gases to the rear. In some firearms the head of the cartridge case is forced against the breeching face; in some revolvers it is forced against the surface of the recoil plate.

An examination of a variety of types and makes of small arms indicates many possibilities for establishing pertinent class character-

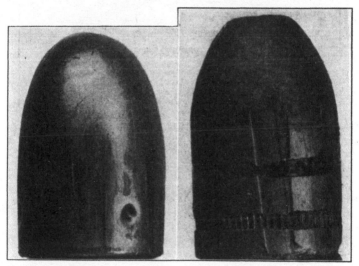

FIG. 100.   FIG. 101.

istics. For instance, differences are found to exist in the design and construction of breechblocks and recoil plates. Some manufacturers have adopted hand-tool or machining operations which are different from those used by other manufacturers in producing the breeching face; these become evident from an examination of the breeching-face impression on the head of a fired cartridge case. In center-fire ammunition, a part of the breeching-face impression normally appears on the base of the primer cup.

A study of various types of firing pins develops further information for establishing class characteristics. In firearms adapted to rim-fire ammunition, the size and shape of the impression made on the head of the case by the nose of the firing pin are class characteristics. In firearms adapted to center-fire ammunition, differences are found to exist in the type, diameter, and shape of nose of the firing pin, and in the hand-tool and machining operations incident to their manufacture.

FIG. 102.—Breech Mechanism, Savage Automatic Pistol, Caliber .32.

Further data for developing class characteristics are obtained from the size and shape of the impression made by the nose of the hammer in firearms adapted to rim-fire ammunition.

Other class characteristics are found in the size and shape of the hole in the breechblock or recoil plate through which the firing pin or hammer strikes, and in the angle of percussion.

When a criminal uses a revolver in the perpetration of a crime, he is not likely to eject the fired cartridge cases deliberately at the scene of the crime unless he is forced to reload. If he uses an automatic firearm, the fired cartridge cases will be automatically ejected with considerable force, and it is unlikely that he would attempt to hunt for and collect the fired cartridge cases. Therefore, when fired cartridge cases are recovered at the scene of the crime, usually they have been fired in an automatic firearm.

Cartridge cases fired in automatic firearms show more marks, deformations, and impressions than do those fired in revolvers, and therefore offer greater possibilities for determining the class characteristics of the firearm in which they were fired. Automatic (autoloading) pistols differ both in design and construction. Pertinent

FIG. 103.—Savage.

class characteristics are found in the relative position of the extractor and the ejector. The surface brought in contact with the top of the head of the uppermost cartridge in the magazine, in the process of forcing the cartridge out of the magazine into the chamber, leaves its mark on the head of the cartridge case. The relative position of this mark on the head of the fired cartridge case with reference to the marks produced by the ejector and extractor furnishes an important class characteristic.

Fig. 102 is a photomicrograph of the breech mechanism of a Savage automatic pistol, caliber .32. The concentric circles in the toolmark pattern on the breeching face indicate a turning operation. The extractor is shown on the left, diametrically opposite to the rectangular opening through which the ejector operates. The lowest part of the breech mechanism shown is that which comes in contact with the top of the head of the cartridge case of the uppermost cartridge in the magazine in the process of loading it into the chamber.

Fig. 103 is a photomicrograph of the head of a cartridge fired in a Savage automatic pistol, caliber .32, showing the impression of the tool-mark pattern of the breeching face (Fig. 102) and the indentation made in the primer by the blow from the firing pin which strikes through the hole shown in the breechblock.

Fig. 104 is a photomicrograph of the deformation produced by

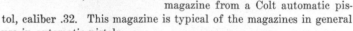

the ejector of a Savage automatic pistol, caliber .32.

Fig. 105 is a photomicrograph of the breech mechanism of a Steyr automatic pistol, caliber .25. The extractor shown on the left is an integral part of the breechblock. The firing pin is seen in the circular opening in the breechblock.

Fig. 106 is a photomicrograph of the deformation made by the extractor in a Colt automatic pistol, caliber .32.

FIG. 104.—Ejector Mark, Savage Automatic Pistol, Caliber .32.

Fig. 107 is a photograph of a magazine from a Colt automatic pistol, caliber .32. This magazine is typical of the magazines in general use in automatic pistols.

Photomicrographs of the primers of cartridges fired in various makes of automatic pistols caliber .32 (7.65 mm.) are shown in the following figures:

Fig. 89. Buffalo (Spanish).
Fig. 108. Bayard.
Fig. 109. Colt.
Fig. 110. Colt (old type).
Fig. 111. Stenda (German).
Fig. 112. Ortgies.
Fig. 113. Stosel.

Photomicrographs of primers of cartridges fired in different revolvers, caliber .32 S. & W., are shown in Figs. 114 and 115.

Fig. 114. Smith and Wesson.
Fig. 115. Iver Johnson.

Differences in the size and shape of the impression made either by the nose of the hammer or by the firing pin in rim-fire ammunition are illustrated in Figs. 116 to 119.

Fig. 116. J. M. Marlin revolver, caliber .32.

Fig. 117. Smith and Wesson revolver, caliber .32.

Fig. 118. Colt revolver, caliber .30.

Fig. 119. J. Stevens Arms Co., single shot target pistol, caliber .22.

Fig. 105.—Breech Mechanism, Steyr Automatic Pistol, Caliber .25.

In general then, pertinent class characteristics of the type and make of firearm in which the cartridge case was fired must be determined from the marks, impressions, and deformations on the cartridge case which constitute one part of the firearm's signature. Pertinent class characteristics may be found relating to the following component parts of the firearm (depending upon the type) with which the cartridge case may come in contact:

Breechblock or recoil plate: Design features and machining operations.

Firing pin: Size and shape of nose as indicated by the impression on base of cartridge case in rim-fire ammunition. Type, diameter, shape of nose, angle of percussion, machining operations, in center-fire ammunition.

Hammer: Size and shape of nose as indicated by the impression on base of cartridge case in rim-fire ammunition.

Extractor.

Ejector.

Magazine.

In automatic (auto-loading) pistols the relative positions of the marks made by the ejector, extractor, and the surface in contact with the head of the cartridge case in the process of loading furnish an important class characteristic.

The problem is completely solved only when the class characteristics so determined point to a group embracing but one type and make of firearm.

In the practical solution of this problem an extensive collection of standards of comparison consisting of cartridge cases fired from known types and makes of firearms will be found invaluable.

Third type of problem: Given a bullet and a suspected firearm, to determine whether or not the bullet was fired from the suspected firearm.

FIG. 106.—Extractor Mark, Colt Automatic Pistol, Caliber .32.

To solve the third type of problem test bullets fired from the suspected firearm are used as the standards of comparison. Before any test shots are fired the condition of the bore of the suspected firearm as well as the condition of the chambers in the cylinder of a revolver should be recorded for possible future reference. This is important for the reason that the firearm may have passed through various hands from the time the crime was committed until its delivery to the investigator, and the condition of the bore may be very much different from what it was at the time the crime was committed. For example, the fatal shot may have been fired from a dirty, rusty, and fouled barrel and yet the investigator may receive the firearm with a clean barrel.

The condition of the bore may indicate whether or not the firearm was recently fired as well as the kind of powder with which the ammunition last used was loaded. Sometimes one or two unburned grains of smokeless powder may be found in the bore of a firearm with a relatively short barrel.

As a general precaution the bore should always be inspected before any shots are fired in order to avoid a possible bursting of the barrel by an obstruction in the bore. The interior of the bore of a

FIG. 107.—Magazine, Colt Automatic Pistol, Caliber .32.

firearm can be viewed through the muzzle end of the barrel by reflecting light into the bore at the breech end of the barrel with a small mirror such as is commonly used by dentists.

In the principle as stated on page 36 there are four factors: relative hardness, character of the harder surface, magnitude of the pressure, and relative motion.

As to the relative hardness, it is found that there are variations in the relative hardness of the alloys used in the manufacture of bullets.

Initially the character of the surface of the bore of a firearm is that which it possesses by virtue of its contact with the surfaces of the various tools employed in its manufacture. The character of the surface will change with erosion, corrosion, and fouling. It is evident that, if black or smokeless powder ammunition is used, the character

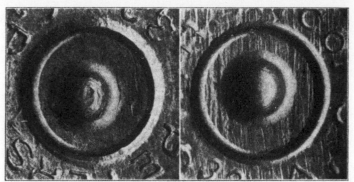

FIG. 108.—Bayard.   FIG. 109.—Colt.

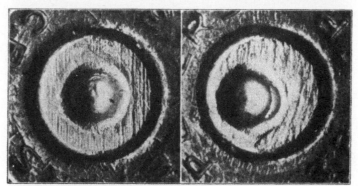

FIG. 110.—Colt.   FIG. 111.—Stenda.

FIG. 112.—Ortgies.   FIG. 113.—Stosel.

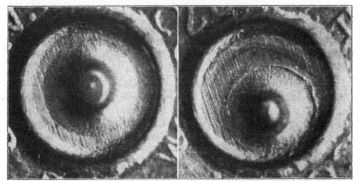

FIG. 114.—Smith & Wesson Revolver.   FIG. 115.—Iver Johnson Revolver.

FIG. 116.—J. M. Marlin Revolver.   FIG. 117.—Smith & Wesson Revolver.

FIG. 118.—Colt Revolver.   FIG. 119.—J. Stevens Single Shot
Target Pistol.

of the surface of the bore will change from shot to shot. It should also be noted that the accumulation of dirt, grit, or other foreign substances in the barrel will affect the character of the surface. **Metallic fouling** is caused by the deposit of particles of the bullet in the bore of the firearm.

The magnitude of the pressure is dependent upon the pressure developed by the powder gases. The pressure will vary throughout the travel of the bullet through the bore of a firearm, and this pressure distribution may vary from shot to shot even though

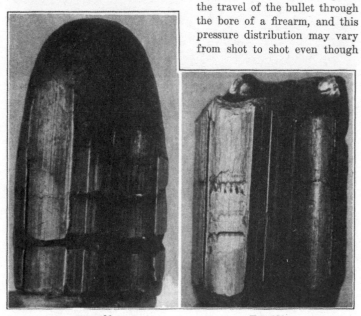

FIG. 120.   FIG. 121.

the cartridges are of the same manufacture and taken from the same original package.

Relative motion in general is also a variable factor and is affected by changes in the pressure, relative hardness, and the angle between the axis of symmetry of the bullet and the axis of the bore at the instant the bullet strikes the forcing cone.

From the aforesaid principle it is evident that a lead slug which has been forced through the bore of a firearm by mechanical means has but little probative value in identifying the signature on bullets fired from the firearm. Such a slug could be used to determine the class characteristics of the firearm. However, if a groove in the bore of a firearm has one or more pronounced tool-marks which run the full length of the bore in a direction parallel to a land shoulder of the groove, then such tool-marks will produce the same resultant effect with a lead slug forced through the bore as with a lead bullet fired from the firearm. This is illustrated in the corresponding groove engravings shown in Figs. 120 and 121. Fig. 120 is a lead bullet fired from a revolver, and Fig. 121 is a lead slug forced through the bore of the same revolver. By noting the variations in the passive resistance as a slug is forced through the bore it is possible to detect existing enlargements in the cross-section of the bore.

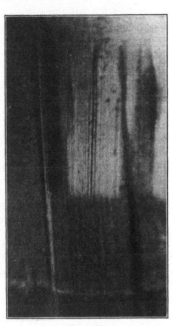

FIG. 122.

Fig. 122 is a photomicrograph of a comparison of the land engraving on a jacketed bullet fired in a caliber .45 automatic pistol and the corresponding land engraving on a lead bullet forced through the bore by mechanical means. The striae parallel to the axis of the bore which appear on the jacketed bullet in the central portion of the land engraving do not appear on the lead bullet, and they will not appear on any lead bullet fired from the pistol. With a jacketed bullet the firearm's signature is the cumulative effect of all contacts from breech to muzzle. With a lead bullet there is more stripping, and because striae parallel to the axis of the bore must be produced at or near the breech, such striae are entirely obliterated in the stripping.

In order that the four factors involved in the principle may as nearly as possible be reproduced in the firing of test shots, the given bullet should be examined to determine the manufacturer of the ammunition, and cartridges of the same manufacture as that of the given bullet should preferably be used for test purposes. If the given bullet shows a black deposit and irregular indentations on its base it is an indication that the bullet was from a cartridge loaded with black or semi-smokeless powder.

Unless there are good reasons to the contrary the first shot for test purposes should be fired from a clean barrel and the firing continued without further cleaning. A study of these test bullets will indicate the effect produced on the firearm's signature by the fouling which may accumulate from shot to shot.

Test shots should be fired under normal conditions. No attempt to improve the firearm's signature should be made by oiling the cartridge cases or by applying grease or oil to the bullets of the ammunition used for test purposes. The trained investigator has sufficient confidence in his own ability to be content to deal with realities.

No two oak leaves may be exactly alike, but the exact counterpart of a small area of one oak leaf can probably be found in other oak leaves.

It is probably true that no two firearms with the same class characteristics will produce the same signature, but it is likewise true that each element of a firearm's signature may be found in the signatures of other firearms. Each element of the signature produced by a firearm with a clean bore is the resultant effect of either a pertinent class or accidental characteristic of the firearm. A firearm's signature is therefore a combination of elements, and the coexistence of this same combination of elements in the signatures of other firearms is highly improbable. In some respects a firearm's signature is like a personal signature. If two personal signatures are found to coincide exactly in every detail when superimposed, it indicates that probably one of the signatures is a forgery.

Some variations in a firearm's signature which occur from shot to shot are caused by conditions existing in the firearm itself, and others are caused by the non-uniformity which exists in ammunition in general.

The problem of determining whether or not a given bullet was fired from a suspected firearm therefore resolves itself into the problem of determining whether or not the signature on the given bullet can be identified with the signatures on bullets fired from the suspected firearm.

Common sense dictates that this problem can be solved only if it is possible to establish: (a) that the signature on the given bullet was engraved by a firearm with the same class characteristics as those of the suspected firearm; (b) that the same combination of identifying elements exists in the signatures on all bullets (except those under size) fired from the suspected firearm at the time, and all variations found in these signatures can be reconciled; (c) that the same combination of identifying elements exists in the signature on the given bullet; (*d*) that all variations existing in the signature on the given bullet and the signatures of the suspected firearm can be reconciled; and (*e*) that the identifying elements as determined form a combination the coexistence of which is highly improbable in the signatures of other firearms with the same class characteristics.

When comparing two signatures engraved by the same firearm, **corresponding land engravings** are those produced by the same land, and **corresponding groove engravings** are those produced by the same groove in the bore of a firearm. **Corresponding driven edges of lands** are those which were in contact with the driving edge of the same land in the bore of the firearm. Two signatures engraved by the same firearm are said to be in **phase** when corresponding land or groove engravings are in perfect alignment in the field of the comparison microscope as indicated by the alignment of corresponding driven edges of lands. **Corresponding elements** are those elements which occupy the same phase relation in the signatures produced by the same firearm.

Those elements of a firearm's signature which are the resultant effects of accidental characteristics of the firearm, and which are reproduced on every bullet (except those under size) fired from the firearm at the time, will be called **elements of identity.** The elements of identity are necessarily corresponding elements and therefore will appear as coincident adjacent elements in the field of the comparison microscope when two signatures of the same firearm are compared in phase. Such coincidence or agreement of corresponding elements will be termed an **identity.** When no such coincidence or agreement exists in the corresponding elements of the signatures of the same firearm, it will be termed a **diversity.**

When comparing two signatures, it should be noted that agreement in the elements which are the resultant effects of class characteristics of a firearm only establishes the fact that the two signatures were engraved by firearms with the same class characteristics. On the other hand, the elements of identity furnish the means by which the identity of a particular firearm in a group of firearms with common class characteristics is established. An individual peculiarity of a firearm can therefore be established by elements of identity which form a combination the coexistence of which is highly improbable in the signatures of other firearms with the same class characteristics.

When two signatures of the same firearm on bullets with unequal stripping are compared in phase, corresponding elements of identity, whose relative position with respect to a driven edge of a land varies with the stripping, will be out of phase by an angle depending upon the difference in the amount of stripping in the two bullets. (See page 44.)

When comparing two signatures of different firearms with the same class characteristics, or two signatures of the same firearm out of phase (the driven edges of the lands in alignment are not corresponding), the term **congruence** will be used to indicate the coincidence of two adjacent elements in the field of the comparison microscope; and the two adjacent elements so coinciding will be called **congruent elements**. The term **incongruence** will be applied when there is no such coincidence or agreement in the adjacent elements.

The first step in the solution of this problem is to determine whether or not the given bullet was fired from a firearm with the same class characteristics as those of the suspected firearm. This is accomplished by comparing the signature on the given bullet with the signature on one of the test bullets. If the requisite agreement in class characteristics is found, the next step is to compare the signatures on two or more test bullets when these are in phase, to determine the identities and to reconcile the diversities. The proper phase relation is established by the file mark at the nose of each test bullet. (See page 55.)

Having completed the comparisons of the signatures on test bullets, the investigator is now in a position to compare the signature on the given bullet with the signature on a test bullet. These signatures are compared in various phase relations, always keeping the driven edges of adjacent lands in alignment, and the number of congruences noted. It will be found that when two signatures of the same firearm are compared in the various phase relations, the sum of the congruences will change with each phase relation, but the sum of the congruences will be a maximum when the signatures are in phase, in which case the congruences become identities and the incongruences become diversities.

If, then, in the phase relation for which the sum of the congruences is a maximum, it is found that the congruences are in agreement with the identities existing in the signatures on the test bullets, and all incongruences can be reconciled, the signature on the given bullet is identified with the signatures on the test bullets, provided that the congruences establish an individual peculiarity of the firearm. If no phase relation can be found in which the congruences are in agreement with the identities existing in the signatures on test bullets, then it is evident that the signature on the given bullet can not be identified with the signatures on the test bullets.

A bullet which is evidence of an act or deed, the nature of which makes it desirable to obtain information relating to the identity of the firearm from which it was fired, will be called the **evidence bullet.** In homicide cases the evidence bullet is often referred to as the ''fatal bullet,'' ''mortal bullet,'' ''crime bullet,'' or ''bullet in evidence.''

If the evidence bullet is so distorted and mutilated that only a small portion of the signature remains intact, it will often be impossible to reach any definite conclusions. In some cases it may be possible to determine definitely only whether or not the evidence bullet was fired from a firearm with the same class characteristics as those of the suspected firearm. If, however, sufficient of the signature remains intact, and it is possible to determine a few congruences which are in agreement with the identities found in the signatures on test bullets, and all incongruences can be reconciled, then the signature on the evidence bullet is identified with the signatures on the test bullets, provided that the congruences determined under these adverse conditions establish an individual peculiarity of the firearm.

An element which, by reason of contrast, stands out in such a way as to attract attention, will be called a **conspicuous** element. Conspicuous elements of identity are of more probative value than those which are more or less uniform in character.

If, in a signature, each groove engraving forms a pattern distinctly different from the others, then such groove engravings are of more probative value than those which are all of a common pattern.

In general, more importance should be attached to the elements in the land engravings because the portion of the cylindrical surface of the bullet which comes in contact with the lands in its motion through the bore is subjected to the greatest radial compressive forces.

No test bullets should be discarded in any investigation. The trained investigator retains, studies, and evaluates all observable data.

A comparison of the signature of a certain firearm with its signature after more than 100 shots had been fired over a period of time indicated that no change had occurred in the elements of identity. In a firearm in which the powder gases develop a relatively high pressure and the bullet attains a relatively high velocity, erosion may develop more rapidly, and the experienced investigator will be guided accordingly with regard to the number of shots fired for test purposes in order that no vital changes in the firearm's signature may be introduced by erosion.

Generally, if two bullets have passed through the same bore, and one bullet is heavily engraved and the other lightly engraved, the obvious inference is that the more heavily engraved bullet has the larger diameter of the two. On the other hand, it is quite conceivable that just the reverse might be true. If a lead bullet is fired from a firearm in the bore of which there is an accumulation of dirt, rust, and black powder fouling, and a second bullet fired from the same barrel after it has been thoroughly cleaned, it may develop that the first bullet will be more heavily engraved and of smaller diameter than the second.

When using ammunition with lead bullets which have been plated with a thin metallic coating, considerable variation in the signature of a firearm will be found. This variation is caused by the fact that

FIG. 123.

considerable of the metallic coating on the cylindrical surface of the bullet may be removed as it passes through the bore.

When consideration is given to the condition of evidence bullets generally it is doubtful whether the most skilled investigator will be able to reach definite conclusions in more than six out of ten cases.

Fig. 123 is a photomicrograph of a comparison of corresponding groove engravings and illustrates the coincidence of two corresponding conspicuous elements of identity.

Fig. 124 is a lead evidence bullet, caliber .38 S. & W. Sp'l, and Fig. 125 is a lead evidence bullet, caliber .38 S. & W. Fig. 126 is a photomicrograph of the comparison of land engravings found on these two bullets. The elements of the signatures consist of fine striae and are common to the land engravings of different revolvers.

In Fig. 127 are shown photomicrographs of the corresponding surfaces of two lead bullets from cartridges loaded with black powder and fired from different chambers in the same revolver. Fig. 128 is a photomicrograph of a comparison of the corresponding land engravings (Fig. 127), and illustrates both identities and diversities. The diversities

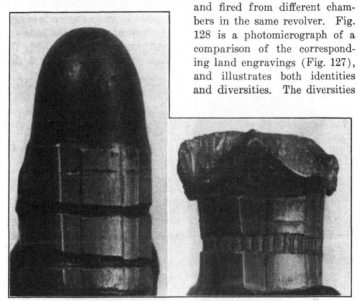

FIG. 124.   FIG. 125.

FIG. 126.

in the signatures are the result of black powder fouling combined with variations in the angle between the axis of the bullet and the axis of the bore at the instant of striking the forcing cone.

Fig. 129 shows the corresponding surfaces of two caliber .38 lead bullets fired from a low-priced revolver and indicates reconcilable diversities in the formation of corresponding land impressions. In a revolver adapted to center-fire ammunition in which the angle of percussion is large, and the firing pin is an integral part of the hammer, if the firing pin strikes the primer off center at four o'clock, the bullet will strike the forcing cone at ten o'clock; particularly is this

FIG. 127.—Corresponding Surfaces of Two Bullets Fired from the Same Revolver.

true when the cartridge fits the chamber loosely. This condition existed in the revolver from which these bullets were fired.

The resultant effect of accidental characteristics on the surface of a land in a Smith and Wesson revolver, caliber .38 S. & W. Sp'l, is shown in Fig. 130, which is a photomicrograph of a comparison of a land impression on a test bullet (left) with a land impression on the evidence bullet (right). In the middle third of the land impressions are a number of identities. These and other identities were found to exist in the signature on the evidence bullet and in the signatures on all test bullets.

In the application of these principles of identification in trials of homicide cases, there are certain matters the importance of which is not generally appreciated. When a bullet which has been removed from the body of a victim is offered in evidence, every link in the chain of possession should be firmly established from the time the bullet was removed from the body until the time of offering it in evidence in court.

In State *v.* Civitan [19] the bullet shown in Fig. 124 was admitted in evidence as being the bullet removed from the body of the victim. After the prosecution had closed its case it developed that the bullet admitted in evidence was not the bullet removed from the body of the victim. The prosecution was therefore placed in the position of having to reopen its case to correct this error and to submit in evidence the bullet which was actually removed from the body of the victim and shown in Fig. 131. It is a difficult matter for a doctor to identify at a later date a bullet which he has removed from a body. He usually scratches his initials on the bullet but this in itself is insufficient to establish the identity of a particular bullet, especially if the doctor per-

Fig. 128.—Photomicrograph of Comparison of corresponding Land Engravings on Bullets shown in Fig. 127.

forms many such operations in a year. If in addition to his initials the doctor would also place a number on the bullet and keep a record of these bullets by number, much of the difficulty might be removed. It is sometimes difficult to find sufficient area on a bullet available for both initials and number. Under no circumstances should marks be placed on any portion of the surface which carries land or groove engravings. The safest procedure perhaps would be to photograph each bullet together with the name of the victim if known at the time.

Fourth type of problem: Given a fired cartridge case and a suspected firearm, to determine whether or not the cartridge case was fired in the suspected firearm.

[19] Hudson County, N. J. 1932.

To solve the fourth type of problem, the procedure is essentially the same in principle as that followed in the solution of the third type of problem, except that the phase relation of the elements does not apply. Instead of a phase relation it is the relative position of the elements of the signature on the cartridge case which plays an important part as set forth in the solution of the second type of problem.

Cartridge cases fired in the suspected firearm are used as the standards of comparison, and ammunition of the same manufacture as

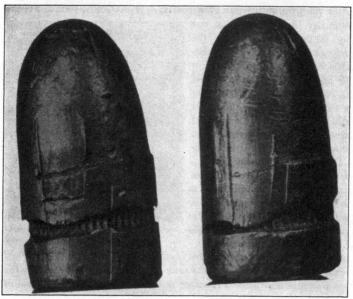

FIG. 129.—Corresponding Surfaces of Two Bullets Fired from the Same Revolver.

that of the given cartridge case should preferably be used. Before any cartridges are used for test purposes they should be carefully examined for scratches and other blemishes which may be found on the cartridge cases in all types of ammunition as well as on the primers in center-fire ammunition.

The signature on the cartridge case is composed of elements any one of which may be found in the signatures of other firearms, but which in any one firearm form a combination the coexistence of which may be highly improbable in the signatures of other firearms.

Each element of the signature is the resultant effect of either a pertinent class or accidental characteristic of the firearm. Some variations in the signature can be attributed to conditions existing in the firearm and play in the moving parts, whereas others are caused by the non-uniformity in ammunition in general.

The first step in the solution of this problem is to compare the signature on the given cartridge case with the signature on a test cartridge case to determine whether or not the given cartridge case was fired in a firearm with the same class characteristics as those of the suspected firearm. The point on the head of the test cartridge case which was at the top of the chamber at the instant of firing furnishes a suitable reference point for determining the relative position of certain elements in the signature. In pin-fire ammunition this point is established by the position of the pin. In an automatic pistol the reference point is established by the resultant effect of the contact made in forcing the cartridge out of the magazine into the chamber.

Fig. 130.

This reference point serves to determine the relative positions of the elements which are the resultant effect of contacts at or near the perimeter of the head of the cartridge case, as well as the relative positions of the elements on the convex surface of the cartridge case which are the resultant effect of contacts with the interior surface of the chamber or with the sharp edges of the magazine.

The impression of the breeching face (or surface of recoil plate) may not always occupy the same relative position with respect to the reference point, but it must always occupy the same relative position with respect to the impression of the circumference or perimeter of the hole through which the firing pin or hammer strikes.

If the requisite agreement in class characteristics is found to exist and the reference point is established on the given cartridge case, the next step is to compare the signatures on two or more test cartridge cases to determine the identities and to reconcile the diversities. This resolves itself into a comparison of the elements occupying the same relative position in two signatures.

FIG. 131.

This comparison having been completed, the next step is to compare the signature on the given cartridge case with the signature on a test cartridge case, determining the congruences and reconciling the incongruences. The relative position of the elements in the two signatures having been previously established by the agreement in class characteristics, it necessarily follows that if all the incongruences can be reconciled, then the signature on the given cartridge case is identified with the signature of the suspected firearm, provided that the congruences establish an individual peculiarity. Obviously, if any one incongruence can not be reconciled, then the signature on the given cartridge case is not identified with the signature of the suspected firearm. For example, a difference in one element of the pattern of the breeching face impression would be an incongruence which could not be reconciled.

The reproduction of a firing-pin impression in the primer is shown in Fig. 132, which is a comparison of the firing-pin impressions on cartridge cases fired in two different Smith and Wesson revolvers, caliber .38 S. &. W. Sp'l.

Fig. 133 is a comparison of the primers in two cartridge cases fired in the same Smith and Wesson revolver, caliber .38 S. &. W. Sp'l. The cartridges were taken from the same package. The base of the primer cup is usually slightly convex. Under normal conditions the base of the primer cup is flattened against the recoil plate as indicated by the primer on the left in Fig. 133. With lower pressure the base of the primer cup may not be flattened to the same extent as indicated by the primer on the right in Fig. 133.

The lip in the firing-pin indentation shown in Fig. 65 is a reconcilable incongruence. The incongruence shown in Fig. 67 can be reconciled by forcing a small piece of lead partly into the firing-pin hole in the breechblock (Fig. 62) and shearing it off. A comparison will show that the striae produced on the lead in the shearing process will be in exact agreement with the striae found on the base of the primer cup in Fig. 67.

Fifth type of problem: Given two or more bullets, to determine whether or not they were fired from the same firearm.

It will be assumed that a comparison of the bullets indicates that they were all fired from

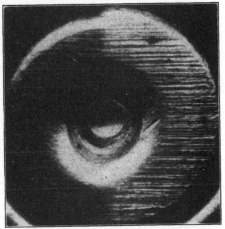

FIG. 132.—Comparison, Firing Pin Impressions, Two Different Revolvers.

firearms with the same class characteristics. To solve the fifth type of problem under these conditions, any one of the bullets is taken as the standard of comparison and the same procedure as in the solution of the third type of problem is followed. If none of the signatures can be identified with the standard of comparison, then the latter is eliminated and the same procedure is repeated with the remaining bullets. This process is continued as many times as may be necessary to solve the problem.

Sixth type of problem: Given two or more fired cartridge cases, to determine whether or not they were fired in the same firearm.

It will be assumed that a comparison of the cartridge cases indicates that they were all fired in firearms with the same class characteristics. To solve the sixth type of problem under these conditions, any one of the cartridge cases is taken as the standard of comparison and the same procedure as in the solution of the fourth type of problem is followed. If none of the signatures can be identified with the

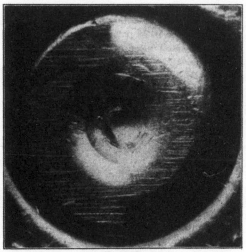

FIG. 133.

standard of comparison, then the latter is eliminated, and the same procedure is repeated with the remaining cartridge cases. This process is continued as many times as may be necessary to solve the problem.

If a man will begin with certainties, he shall end in doubts; but if he will be content to begin in doubts, he may end in certainties.
—FRANCIS BACON.

CHAPTER II

THE SACCO-VANZETTI CASE [1]

PART 1

THE VITAL IMPORTANCE OF THE EVIDENCE ON FIREARMS IDENTIFICATION

IN SOUTH BRAINTREE, MASSACHUSETTS, on the afternoon of April 15, 1920, Parmenter, a paymaster, and Berardelli, his guard, were murdered by two armed men who escaped with two boxes containing the payroll of the shoe factory of Slater and Morrill which the victims were transporting from the main office to the factory. Sacco and Vanzetti were charged with this crime on May 5, 1920, and were indicted on September 14, 1920. The trial began on May 31, 1921, at Dedham, Norfolk County. On July 14, 1921, Sacco and Vanzetti were found guilty of murder in the first degree.

What was the case against Sacco and Vanzetti? Sacco was in the continuous employ of a shoe factory at Stoughton and was absent from work on April 15. Vanzetti was an independent fish peddler at Plymouth. When arrested Sacco and Vanzetti were subjected to a rigorous examination by police officers and by the prosecuting attorney. They told a number of deliberate lies upon material matters, and Sacco asserted that he had never been in South Braintree. However, it is undisputed that they were not informed that they were

[1] It is only after much deliberation that the authors decided to present their analysis of the "unfortunate" Sacco-Vanzetti case. It is not their intention either to rake up the skeletons of the past or further to disseminate the so-called "Sacco-Vanzetti propaganda." The authors believe that, because of the abundance of interesting material relating to the identification of firearms, this case merits a complete analysis.

The record of the case to which the notes refer is "The Sacco-Vanzetti Case, Transcript of the Record of the Trial of Nicola Sacco and Bartolomeo Vanzetti in the Courts of Massachusetts and Subsequent Proceedings, 1920-7." The record was published by Henry Holt and Co., and "approved" by the following: Newton D. Baker; Emory R. Buckner; Charles C. Burlingham; John W. Davis; Bernard Flexner; Raymond B. Fosdick; Charles P. Howland; Victor Morawetz; Charles Nagel; Walter H. Pollak; and Elihu Root.

under suspicion for the murder of Parmenter and Berardelli, and therefore the falsehoods did not evince a consciousness of guilt upon which to predicate the deduction that Sacco and Vanzetti were guilty of the South Braintree murder. It is possible to explain these lies by a desire to conceal either unpatriotic conduct or any other crime committed in Massachusetts within a reasonable period preceding the examination.

One of the bandits who did some of the shooting wore a cap before the shooting but was bareheaded when he rode away. A cap was found near the body of Berardelli which nearly or quite fitted Sacco's head, and it was of the same general outward appearance and color as Sacco's cap. The recovered cap was of common design and color, and both Sacco and Mrs. Sacco denied ownership, the latter explaining that her husband never wore such a cap with earflaps because "he don't look good in them, positively." [2]

The identification testimony of the eye-witnesses was far from convincing. As the learned Canadian Justice Riddle put it, "The evidence of identification was such as could not have been withdrawn from the jury by any Judge; but it cannot be called strong." For example, of those who saw Sacco in Braintree prior to the shooting, W. J. Heron had little or no reason for being impressed with the characteristics of the man supposed to be Sacco, and had nothing at all to correlate the date with the event. The observations of W. S. Tracy were of a casual nature. He thought the men were waiting for a street car and did not notice a single characteristic feature of either man. It is true that the witness may have honestly thought that Sacco looked like one of them, but the record shows that the descriptions given by the various witnesses will fit the thousands of foreigners to be seen every day on the streets of industrial cities in New England. Lola Andrews had the best opportunity to see the man with whom she talked. Mrs. Andrews testified that she was in search of employment at the Slater and Morrill factory. Before entering the factory she noticed an automobile standing outside the factory, and a "dark complexioned" man was bending over the hood of the car. She went into the factory, and on coming out fifteen minutes later the dark complexioned man "was down under the car like he was fixing something." Mrs. Andrews claims to have inquired of this man the way to another factory. "If he would please show me how to get into the factory office, that I did not know how to go." [3]

2 "The Sacco-Vanzetti Case," Vol. II, page 2065.
3 "The Sacco-Vanzetti Case," Vol. I, page 336.

She subsequently identified the dark complexioned man as Sacco. The witness was thoroughly discredited by four witnesses who contradicted her story, which on its face is very weak, i.e., why didn't she seek directions in the Slater and Morrill factory? Why did she inquire of a man lying underneath a car when other persons were more available? Mrs. Campbell, who accompanied Mrs. Andrews, testified that the latter did not inquire of a man working on an automobile.

Among other of the important witnesses relied upon by the state to identify Sacco was Pelser, a young shoe cutter in the factory, who testified that when he heard the shooting he pulled up his window and saw the man who murdered Berardelli. Three fellow workmen testified that they were in the same room and that Pelser did not look out of the window to view the shooting. He admitted on cross-examination a failure to make an identification of Sacco immediately after the latter's arrest. The witnesses who purported to identify Sacco as an occupant of the get-away car may be substantially discounted because they obtained a relatively distant view for but a few seconds of the occupants of the car.

The testimony of the witnesses for the commonwealth who purported to identify Vanzetti as one of the bandits was far from persuasive. For example, Levangie, the gate tender of the New Haven Railroad, was on duty at the South Braintree grade crossing on the day of the murder. He testified that he was forced to raise the gates for the bandits at the point of a gun. Levangie identified Vanzetti as the man who was driving the car, a fact denied by all other identification witnesses on both sides. A locomotive fireman testified that he had a conversation with Levangie three-quarters of an hour after the murder, during which the gateman said he could not identify any of the bandits. Dolbeare testified that he saw a car going past him in South Braintree containing five people, one of whom he identified as Vanzetti. He secured only a glance at the profile of this particular occupant in the back seat under circumstances not calling for detailed attention. Yet he made a positive identification.

The prosecution sought to connect Vanzetti with the crime by tracing the possession of Berardelli's revolver into the hands of the defendant. It would have been an especially difficult feat in the Braintree murder for Vanzetti to secure Berardelli's gun as the guard was in the habit of wearing his revolver underneath a coat, and no one saw Berardelli draw a gun and no one saw a murderer take the gun—in fact no one could even testify that the guard was armed at the time of the murder.

The evidence on the identity of the revolver was surprisingly unsatisfactory. Mrs. Berardelli testified that three weeks before the murder her husband's gun needed a new spring and that she accompanied her husband to the Iver Johnson Sporting Goods Co. in Boston where the gun was left for repairs. The "receipt check" was given to Mr. Parmenter, who actually owned the gun. An employee of the Iver Johnson Co. entered on the record "38 Harrington & Richardson revolver, property of Alex Berardelli, was brought in for repairs, and sent up to the shop on March 20, 1920." [4] There was no entry of delivery on the record.

The foreman of the gun shop who repaired most of the revolvers testified that he had made two entries on his record between March 22 and March 19, 1920. One entry was: "H. & R. revolver, 32 calibre, new hammer, repairing, half an hour." [5] The other entry: "New main spring, new friction spring, repairs, an hour and a half on two." Referring to the latter entry, the witness testified: "There was two of them marked together, two H. & R. revolvers tied together, that is, two in one repair job." [6] The repair number given to the Berardelli revolver referred to the revolver mentioned in the first entry, a caliber .32, H. & R. revolver, whereas the revolver in evidence was of caliber .38. To reconcile this discrepancy, it was stated that these two models of revolvers were mounted on the same frame and were identical in all respects except for two not readily noticeable features, the bore diameters and the number of chambers in the cylinders. The witness also testified that in his opinion the revolver in evidence contained a new hammer because "the firing pin does not show of ever being struck." [7]

The best the state's witness could do at the trial was to testify that the revolver in evidence was of the same general description as the revolver of Berardelli, but no one could give an accurate detailed description including the size of the bore. At the trial no vigorous effort was made to set up the fact that the revolver in evidence contained a new hammer, whereas the defense experts contended that the hammer was no newer than the rest of the weapon. The prosecution failed completely in carrying the burden of proof on this issue of the identity of the revolver in evidence.

There was both direct and circumstantial evidence against both

[4] "The Sacco-Vanzetti Case," Vol. I, page 814.

[5] "The Sacco-Vanzetti Case," Vol. I, page 818.

[6] "The Sacco-Vanzetti Case," Vol. I, page 819.

[7] "The Sacco-Vanzetti Case," Vol. I, page 822.

of the defendants; however, there were glaring deficiencies in the commonwealth's proof. The testimony of the eye-witnesses was not convincing upon the question of the presence of either Sacco or Vanzetti or both in Braintree on April 15. The evidence of their participation in the shooting was exceedingly unreliable and contradictory. The defendants presented numerous alibi witnesses and other witnesses who testified that either the defendants were not visible or that they could not be identified as the murderers. There were inherent weaknesses in the proof upon the issues of Berardelli's gun, the cap, and the conduct of the defendants subsequent to the crime. Only two of the five engaged in the crime were apprehended. The stolen money was never traced to the possession of Sacco or Vanzetti or to any associates thereof. The defendants were not in need of funds and there was nothing in their histories manifesting a disposition to commit murder in cold blood for private profit. The character of the robbery pointed to the work of professionals rather than to that of amateurs. It is plausible, then, that the jury could not have been convinced beyond a reasonable doubt of the guilt of either or both of the defendants on any or all of the evidence offered at the trial.

On what, then, did the jury rely as the basis for their verdict? The answer is inevitable — the jury must have put much reliance on the identification of the Colt automatic pistol found on Sacco at his arrest as one of the weapons used in the crime. If Sacco's pistol was fired in Braintree on April 15, it is persuasive evidence of Sacco's presence in Braintree; and if one of the bullets extracted from the bodies of the victims was discharged from Sacco's Colt, that is convincing evidence that Sacco was a participant in the murder.

(Surrounding the conduct of the trial there was substantial courthouse gossip to the effect that Mr. Moore, of counsel for the defendants, erroneously believed that he would be able, on the ground of self-incrimination, to prevent the state from making use of evidence to identify the Colt pistol found on Sacco as the one used to discharge any of the fatal bullets or the recovered cartridge cases. According to the gossip Mr. Moore asked Sacco whether this Colt pistol had been used at the scene of the crime, and Sacco replied that *that* gun wasn't there. Further gossip was to the effect that after the trial the jurors stated that in arriving at their verdict they relied only upon the expert testimony submitted on firearms identification.)

Once the jury was convinced that the Colt found on Sacco did participate in the murder, the other elements of evidence, such as the cap, the eye-witnesses, and the conduct after the crime, would have corroborated the fact of the defendant's guilt. Once the guilt of Sacco was established the jury would have had some basis for connecting Vanzetti with the murder through Vanzetti's association with Sacco, the evidence of eye-witnesses, and his conduct indicative of a consciousness of guilt.

The identification of Sacco's gun was the natural determinant of the guilt or innocence of Sacco. Rationally and intelligently handled, this would have been the subject of the most reliable form of evidence. The decision of the issue, properly tried, would have rested on physical data which could have been intelligently observed by the jury, upon being properly presented and interpreted by competent experts. Here was an issue based in scientific principles, independent of the unreliability of eye-witnesses with their faulty recollections and doubtful credibilities. The identity of Sacco's pistol could have been determined by an examination of observable physical data: the existence or absence of significant marks on the bullets and cartridge cases used in the crime when compared with the markings on the bullets and cartridge cases fired in Sacco's pistol for test purposes.

PART 2

ANALYSIS OF THE FIREARMS EXHIBITS

THE evidence obtained at the *situs* of the crime and from the possession of Sacco and Vanzetti was marked as follows:

EXHIBITS

18–21	Bullets from Berardelli's body.
18	Bullet No. III, caliber .32, full metal case. Discharged from a weapon with left-hand twist rifling.
19	Bullet No. II, caliber .32, full metal case. Discharged from a weapon with right-hand twist rifling.
20	Bullet No. I, caliber .32, full metal case. Discharged from a weapon with right-hand twist rifling.
21	Bullet No. IIII, caliber .32, full metal case. Discharged from a weapon with right-hand twist rifling.
24–25	Bullets from Parmenter's body.
24	Bullet X (No. 6), caliber .32, full metal case. Discharged from a weapon with right-hand twist rifling.
25	Bullet No. 5, caliber .32, full metal case. Discharged from a weapon with right-hand twist rifling.
27	Caliber .38 revolver found in Vanzetti's possession at time of his arrest.
28	Caliber .32 Colt automatic (auto-loading) pistol found in Sacco's possession at time of his arrest.
30	Four discharged cartridge cases, caliber .32, automatic pistol. Two manufactured by the Peters Cartridge Co., one by the Union Metallic Cartridge Co., and one by the Winchester Repeating Arms Co. Found at the scene of the shooting.
31	Thirty-two cartridges taken from Sacco at the time of his arrest. All caliber .32, automatic pistol, full metal case bullets. Sixteen manufactured by the Peters Cartridge Co., three by the Remington Arms Co.–Union Metallic Cartridge Co., seven by the United States Cartridge Co., and six by the Winchester Repeating Arms Co.
32	Five caliber .38 S. & W. cartridges, lead bullets. Two manufactured by the Remington Arms Co.–Union Metallic Cartridge Co. and three by the United States Cartridge Co. Taken from the revolver found on Vanzetti at the time of his arrest.

The caliber .38 revolver found on Vanzetti (Exhibit 27) and the five caliber .38 S. & W. cartridges (Exhibit 32) are eliminated at the outset because all the bullets extracted from the bodies of the victims were of caliber .32. Furthermore, there was no tangible evidence to show that the revolver found on Vanzetti had been in the possession of Berardelli. Without a knowledge of the serial number on the revolver, or some peculiar mark of identification known to have been on the revolver, it would have been very difficult, if not impossible, under the circumstances, to prove that the revolver found in Vanzetti's possession had been taken from Berardelli; but if there had been available one or more bullets known to have been fired in Berardelli's revolver, it would not have been very difficult to prove whether or not the revolver found on Vanzetti had been taken from Berardelli.

The only caliber .32 weapon in evidence was Sacco's Colt automatic pistol, Exhibit 28. This pistol had a left-hand twist rifling, thus eliminating bullets Nos. II, I, IIII, 6 and 5 (Exhibits 19–25), all of which showed a right-hand twist rifling.

The four discharged cartridge cases, Exhibit 30, were all from caliber .32 automatic pistol cartridges, and Exhibit 31 shows that Sacco had in his possession cartridges of makes corresponding to those of the recovered cartridge cases.

Attention is therefore focused on Exhibits 18, 28, and 30: bullet No. III, Sacco's pistol, and the four discharged cartridge cases recovered at the scene of the crime.

Was bullet No. III fired from Sacco's pistol? Were any of the recovered cartridge cases discharged in Sacco's pistol? These were the only pertinent questions to be answered.

Bullet No. III was stamped with a "W" just above the cannelure indicating that it had been manufactured by the Winchester Repeating Arms Co. To determine whether this bullet was fired from Sacco's pistol, the procedure outlined in the solution of the third type of problem, Chapter I, page 84, should have been followed.

With respect to determining whether or not any of the four cartridge cases, Exhibit 30, had been fired in Sacco's pistol, the procedure outlined in the solution of the fourth type of problem, Chapter I, page 98, should have been followed. It should be noted that any breeching-face impression on the Exhibit 30 cartridge cases revealing concentric circles should have been immediately discounted as not having been fired in a Colt automatic pistol. It appears that such an examination would have revealed that the W.R.A. cartridge case of Exhibit 30 was the only one of the four having the class characteristics of a Colt, caliber .32, automatic pistol.

PART 3

EXPERT TESTIMONY AT THE TRIAL

THE additional exhibits referred to in the testimony were as follows:

Fig. 134 is reproduced from page 3732*r*, and Fig. 135 from page 3732*t*, of the record. Fig. 136 is a composite photograph of the Lowell Test Shell No. 3 and the Fraher Shell W (the W.R.A. cartridge case of Exhibit 30) in Fig. 134. Fig. 137 is a photograph of a bullet of the identical type and make as bullet No. III, fired from a Colt automatic pistol, caliber .32, on April 4, 1932.

The testimony given at the trial by the two experts for the commonwealth was as follows.

[AUTHORS' NOTE: *The testimony is reproduced exactly as it appears in the printed record.*]

MR. WILLIAM H. PROCTOR, SWORN

[884] Q. (By Mr. Williams) What is your name? A. William H. Proctor.

Q. Where do you live, sir? A. Swampscott, Massachusetts.

Q. Are you connected in some way with the State Police? A. I am.

Q. And in what capacity? A. I am Captain in the Department of Public Safety, in charge of the Division of State Police.

Q. How long have you occupied that position? A. I have been a captain, —the name of it changed two years ago. I have been a captain for 16 years, and in the Department for 23.

Q. In the State Police Department for 23 years? A. Yes.

[885] Q. Your office is now at the State House, I understand. A. It is.

Q. At some time, Captain, after the South Braintree shooting, did you go to South Braintree and to Brockton? A. I did.

Q. And did you receive at South Braintree and at Brockton certain articles? A. I did.

Q. Do you know Mr. Fraher of the Slater & Morrill Company? A. I do.

Q. Did you receive anything from Mr. Fraher? A. I did.

Q. What did you receive from him?

MR. JEREMIAH McANARNEY: That I object to, if your Honor please, for the sake of the record.

Q. You may answer, Captain. A. I received four empty shells.

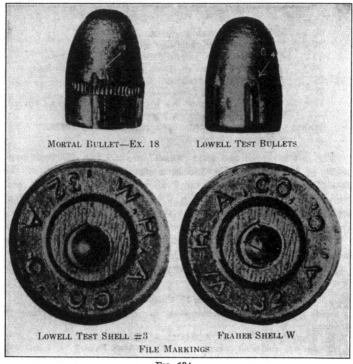

MORTAL BULLET—EX. 18   LOWELL TEST BULLETS

LOWELL TEST SHELL #3   FRAHER SHELL W

FILE MARKINGS

FIG. 134.

THE COURT: That may be stricken out. I sustain that objection.

MR. WILLIAMS: The question, if your Honor please—

THE COURT: I will see counsel at the desk.

MR. WILLIAMS: These are cartridges, four cartridges were found at the scene of the shooting.

THE COURT: Shells,—he means cartridge shells? All right. That may stand.

Q. Will you look at the envelope which I am now showing you and ask you if you can identify what is found inside (handing envelope to the witness)? A. I can.

Q. And what are they? I mean, without describing them, I mean from what source did you receive those articles that I have shown you? A. They were given me by Thomas Fraher at the Slater & Morrill factory at East Braintree,—South Braintree, on April 15, 1920.

Q. Were those the shells which you referred to in my preliminary question? A. Yes, sir.

Q. And have they been in your possession since that time? A. Until I turned them over to the sheriff in this court.

THE COURT: The witness has testified, as I recall it, of picking up some shells.

MR. WILLIAMS: Mr. Bostock testified to picking them up and he turned them over to Mr. Fraher and Mr. Fraher to Mr. Proctor.

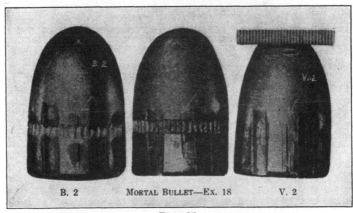

B. 2   MORTAL BULLET—EX. 18   V. 2

FIG. 135.

THE COURT: All right.

(Mr. Williams shows the shells to counsel for the defense.)

MR. WILLIAMS: I offer these, if your Honor please.

MR. JEREMIAH MCANARNEY: They are excepted to, if your Honor please.

THE COURT: Admitted.

(Four empty shells are admitted in evidence and marked "Exhibit 30.")

Q. Captain, what, in general, has been your experience during your term of service with the State Police in reference to firearms and to ammunition for firearms? A. I have examined the rifling of pistols, and re[886]volvers, all revolvers of 38, and 32 calibre that will shoot a S and W cartridge, by pushing a bullet through and examining the rifling on the bullet, measuring the width of the lands made, measuring the width of the groove made by the lands in the pistol or revolver. I have also examined the different makes of automatic pistols that shoot the 380 cartridge, the 32 calibre and 25, and measured the grooves made by the lands in the pistols on the bullets.

Q. Are you familiar with different revolver and pistol cartridges of 32 and 38 calibre type? A. I have examined all the cartridges that would apply to a case I was working on. I qualified as an expert and testified twenty years ago in the Best case at Salem, and I have been examining bullets and cartridges and pistols and revolvers ever since and have testified in over one hundred capital cases.

Q. Have you made personal examinations of these various kinds of revolvers, pistols and ammunition for the same? A. I have. More particularly

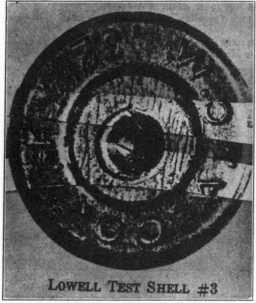

FIG. 136.—Composite Photograph of Lowell Test Shell
#3 and Fraher Shell W in Fig. 134.

I have made an examination of the bullets that I have passed through the barrels, to obtain the difference in the rifling that is shown on the bullets.

Q. And your examination and experiments by way of pushing the bullet through the barrels has extended over what period of time? A. Over twenty years.

Q. Now, Captain, will you look at those four empty shells before you which you say were given you by Mr. Fraher and tell the jury what they are? In other words, describe them to the jury. A. (Witness examines shells.) Well, there are two Peters make. These are shells that the ammunition is ammunition adapted to automatic pistols, 32 calibre. There are two Peters and one U. M. C., and one W. R. A., Winchester.

Q. W. R. A., Winchester? A. Yes.

Q. Two Peters, one U. M. C. What is "U. M. C."? A. That is the Remington, Union Metallic Cartridge Company.

Q. Now, Captain, is there any way that you could mark these, with the assent of the Court, so that we can identify the different marks later, in referring to them? A. It is marked on the—

Q. It is marked on them? A. Right on the bottom of the cartridge, the shell.

Q. I see, "W. R. A., Peters, and U. M. C." I see. Are they all 32 calibre? A. They are.

Q. What is the dent which is noticeable on the end of each cartridge or shell? A. The dent in the centre of the bullet, you mean?

Fig. 137.

Q. Yes. A. That is where the primer from the pistol struck the primer and explodes the primer and fires the cartridge.

Q. What do you call the "primer"? A. Call the "primer" that little round,—that is set into the shell.

Q. Yes. What is the firing pin? A. That is in the pistol.

[887] Q. Showing you Exhibit No. 28, I ask you if you can show the jury in that exhibit where the firing pin is, or, is it easily to be shown? A. It isn't, unless you take it all apart.

Q. Can you show the jury in this revolver, which is Exhibit No. 27, where the firing pin is? A. That is it,—the hammer.

Q. The firing pin is that point in the hammer? A. Yes. It goes through. That is the firing pin there (indicating).

Q. I wish you would tell the jury and show them with that revolver, as long as this pistol cannot be taken apart very well, the various parts of the gun which are instrumental in firing the cartridge? A. You put the cartridge in the cylinder, and this hammer strikes the firing pin that hits the cartridge and explodes the shell.

Q. And again, where is the firing pin that you speak of? A. In here (indicating).

MR. WILLIAMS. (To the jury) The witness points to that (indicating). Beyond there is the firing pin.

Q. And what is that part of the gun through which the firing pin protrudes? A. I do not know as I can tell you all the scientific parts of the gun.

Q. Well, is there a part of the gun called breech-lock? A. I do not know as I could go into that.

MR. JEREMIAH MCANARNEY. I did not hear your answer.

THE WITNESS. I say, I do not know as I could go into that feature.

Q. Now, Captain, did you receive any articles purporting to have been connected with this shooting at the Brockton police station? A. I did.

Q. And where and in what manner did you receive them? A. I received them upstairs in the Brockton police station. John Scott,—State Officer Scott was present, and some Boston,—Brockton police officer handed them to him and he to me.

Q. And what were they? A. The automatic pistol in question, and some cartridges, automatic cartridges, 32-calibre.

Q. Do you remember how many of the automatic cartridges there were? A. 32.

Q. What kind of automatic pistol was it? A. It was this pistol (indicating).

Q. A Colt automatic? A. A Colt automatic.

Q. Has that been in your possession since then? A. Until I turned it over to the sheriff here.

Q. That is what I mean. A. In court.

Q. Will you look at this envelope of cartridges and see if you can identify those (handing envelope to the witness)? A. (Witness examines envelope.) That is the same envelope and it looks like the same amount of cartridges. I can tell by counting them.

Q. Were they in your possession until delivered to the sheriff? A. They were.

[888] Q. Have you examined those cartridges, Captain? A. I have looked them over.

Q. And are those cartridges of different makes? A. They are.

Q. Now, will you tell the jury of what makes those cartridges are, and how many of each make are contained in the envelope? A. There are four different makes. There is a Peters, a Remington,—U. M. C., and there is a Winchester, say W. R. A.

Q. Instead of just giving the initials will you give the full names of the makes of those cartridges? Go a little slower. A. On the 16, Peters Cartridge Company; 3 Remington U. M. C. Cartridge Company; 7 United States Cartridge Company and 6 Winchester Cartridge Company, made by Winchester Arms Company, Repeating Arms Company, the full name.

MR. WILLIAMS: I offer these.

(32 cartridges are admitted in evidence and marked "Exhibit 31.")

MR. JEREMIAH MCANARNEY: Save our exception to the admission.

THE COURT: Admitted. Just one question. When you say, Captain, the police officer handed them over to you, to Captain Scott and Captain Scott handed them over to you, were the shells handed over to Captain Scott in your presence?

THE WITNESS: They were.

THE COURT: And were they handed over,—they were handed over to you at the same time?

THE WITNESS: They were.

THE COURT: All right. Admitted.

Q. What calibre are those, did you tell us? A. 32.

Q. Now, did you receive a revolver at any time? A. I did.

Q. When? A. At the same time.

Q. That is this revolver which is Exhibit 27. That is right? A. Yes.

Q. Did you receive in connection with that revolver any revolver cartridges, lead cartridges? A. I did.

Q. Were they in the revolver or separate? A. Separate.

Q. Were you handed those at the same time those other things were given you? A. I was.

Q. I show you these cartridges and ask you if you can identify them (handing cartridges to the witness)? A. (Witness examines cartridges.) I can identify the envelope, and the cartridges look just about the same.

Q. Were they in your possession until you turned them over to the sheriff? A. They were.

(Mr. Williams shows cartridges to counsel for the defense.)

MR. JEREMIAH MCANARNEY. I assume they are.

MR. WILLIAMS. I offer these five revolver cartridges, if your Honor please.

MR. JEREMIAH MCANARNEY. We object, if your Honor please.

[889] THE COURT. Your objection goes to the competency of the evidence, I take it, rather than to any question about the identification? In other words that the pistol was the pistol found upon the person of one of the defendants at the time of his arrest?

MR. JEREMIAH MCANARNEY. I do not question that, if your Honor please.

THE COURT. That is what I want to get at.

MR. JEREMIAH MCANARNEY. I do not question that.

THE COURT. And there is no question about the—what they call the revolver that was found on the person of Mr. Vanzetti?

MR. JEREMIAH MCANARNEY. Pardon me a moment, your Honor. As to the revolver or the pistol, there is no question there as to the identity, but as to the identity of some of these unmarked shells, I would not waive any rights there.

THE COURT. All right. Then you must bear that in mind with reference to evidence tending to prove identity. I do not know but you have.

MR. WILLIAMS. If I have not, I will try to check it up later, if your Honor please.

THE COURT. All right.

MR. WILLIAMS. This exhibit becomes Exhibit 32, five revolver cartridges.
(Five revolver cartridges admitted in evidence and marked "Exhibit 32.")

Q. Will you look at those cartridges, Captain, and describe them to the jury? Go slow, if you will, in any of these descriptions.

(Mr. Williams hands the witness a magnifying glass, who examines cartridges through same.)

THE WITNESS. There are two. Those are S. & W. cartridges, 38 calibre. Two of them are made by the Union Metallic Cartridge Company,—Remington Union Metallic Cartridge Company. The other three are made by the United States Cartridge Company.

Q. What kinds of bullets are in those cartridges? A. Soft lead bullets.

Q. What calibre are the cartridges? A. 38.

Q. What kind of bullets are in this lot of cartridges marked "Exhibit 31"? A. They are full metal cartridges.

Q. And how are the full metal patch bullets different from the lead bullets? A. Well, there is a patch made by the combination of copper and zinc that is put right over the lead.

Q. And in the lead bullets there is no patch over them. Is that the distinction that you make? A. Yes. There is a metal patch put over it. It is a composition. I do not just know technically what the composition is, but it makes a metal patch over it.

Q. Captain, I now show you Exhibits 19, 20, 21, 24 and 25, which are bullets testified to by physicians as being found in the body or bodies [890] of the victims of this shooting. Will you look at bullet No. 1 first and tell the jury what kind of a bullet that is, the make, if you can tell it, and any further description which occurs to you? A. That is a full—

Q. Let me ask you first, Captain, you have examined at the request of the District Attorney those bullets before going on the stand, have you not? A. I have.

Q. And have examined them, yes. A. That (indicating) is a full metal patch bullet, 32 calibre, adapted to auto cartridges. That is, cartridges fired in an auto pistol.

Q. When you say "auto" what do you mean? A. Automatic. And my opinion is that it is Winchester make.

Q. Winchester make? A. Yes.

Q. That is No. 1, is it? A. Yes.

MR. KATZMAN. Get the exhibit number.

MR. WILLIAMS. Exhibit No. 20.

Q. If you have any memorandum about your examination and investigation of these bullets, you may refer to it, Captain, to save time. Now, can you tell by looking at it through what make and type of gun that bullet has been fired? A. I can.

Q. Will you tell the jury through what make and type of gun, in your opinion that bullet has been fired? A. A Savage.

Q. Well, that is the make. What type? A. That is the make. Automatic pistol, 32 calibre.

Q. Now, what is the basis for your opinion in that respect? A. By measuring the width of the grooves on the bullet caused by the lands in the pistol.

Q. Now, what are the "lands" in the pistol? A. The "lands" are the raised-up places in the pistol that have grooves between them, and when a bullet is pushed in the pistol, when it goes through the pistol, it gives it a twist, holds the bullet and it is given a twist coming out of the barrel.

Q. The lands are the raised-up— A. Places in the barrel.

Q. Can you open up that pistol so that I can allow the jury to look through it and see just what you mean by referring to the raised portions of the barrel? (Witness opens up pistol.) Those raised up portions and the other lowered portions of the barrel are what is commonly referred to as the rifling? A. Yes.

Q. And that rifling, with the twist in it, gives what sort of motion to the bullet as it is fired? A. Well, I can tell better with what result it makes on the bullet.

Q. I am not asking you that, but I mean the bullet going through the air, what happens to it by reason of the rifling and the twist in the rifling on the bullet? A. I think the idea is to make it go straighter. It gives it a twist before coming out of the gun.

[891] Q. Have you that small one, Captain, the glass, the small magnifying glass? A. It is in the room, in the outer office.

Q. Do you know where it is? A. It is in the bag, I guess.

(Court officer leaves the room and gets magnifying glass.)

(Mr. Williams passes pistol barrel to jurors, who examine it with a magnifying glass.)

MR. JEREMIAH MCANARNEY. (To the Foreman) May I assist you (passing paper behind barrel)?

THE FOREMAN. No, it doesn't help.

MR. WILLIAMS. (To the jury) I wish to call your attention, gentlemen, in looking through it, not only to the raised and lowered portions but to the twist of the rifle.

Q. You say those raised portions we see in the barrel are the lands? A. Yes.

Q. And what do those lands do to a bullet when it is pushed or fired through the barrel? A. The lands make a groove on the bullet. It reverses where the groove is in the pistol. It makes,—it is just the reverse in the pistol than it would be on the bullet.

Q. Now, on a Savage what is the width of the lands? A. It makes a groove on the bullet .035 of an inch.

Q. It makes a groove on the bullet .035 of an inch, and what sort of a twist does a Savage automatic pistol give to the bullet? A. Right-hand. In standing the bullet up, it slants to the right.

Q. Now, have you measured the width of the groove on No. 1 bullet? A. I have.

Q. What is the width of the groove made by lands of the barrel? A. .035 of an inch.

Q. And in which direction is the twist on the bullet? A. Right-hand.

Q. That is, as you look at the bullet from the base to the nose, the grooves— A. Slant to the right.

Q. Slant to the right. Is there any other gun which makes a .035 groove on a bullet and a twist to the right? A. There is not.

Q. How certain are you as to your opinion that No. 1 bullet was fired through or from a Savage automatic pistol? A. I have measured the grooves made on bullets of all the makes that I can learn of, and I cannot find two just alike, the width of the grooves made by the lands in the pistol, and I have measured this and made it 35. I haven't got any other such measurement for any other pistol.

Q. How certain can you be then of your opinion that that bullet was fired from a Savage automatic 32? A. I can be as certain of that as I can of anything.

Q. Now, while the jury are looking at that, will you look at No. 2 bullet and tell us if you have an opinion as to the make and type of weapon from which that bullet was fired? A. I have.

[892] Q. And what is your opinion? A. My opinion is it is a U.M.C. make, and it was fired by a Savage pistol.

Q. And in your opinion it was fired from a Savage pistol is based on the same grounds you have just described in regard to bullet No. 1? A. Yes, sir.

MR. KATZMAN. What is the exhibit number?

MR. WILLIAMS. That is Exhibit 19.

Q. Have you measured the grooves made by the lands of the bullet on No. 2? A. I have.

Q. And have you noticed the twist? A. I have.

Q. What is the width of the grooves made on No. 2? A. .035.

Q. Of what nature is the twist on the bullet? A. Right,—right hand.

Q. Will you look at bullet No. 4 and tell us if you have an opinion as to the type and make of weapon from which that bullet was fired? A. I have.

Q. What is your opinion? A. My opinion is it is a Winchester make and was fired from—this No. 4 fired from a Savage.

Q. That is a bullet which is of Winchester make? A. And fired from a Savage.

Q. Savage what? A. Automatic pistol, 32 calibre.

Q. Based on the same reasons you have given us in regard to the other two bullets? A. Yes.

Q. Have you measured the groove caused by the lands on that particular bullet? A. I have.

Q. What is the measurement? A. .035 of an inch.

Q. And have you noticed the direction of the twist as shown on the bullet? A. Right-hand.

MR. KATZMAN. What is the exhibit?

MR. WILLIAMS. Exhibit 21.

MR. KATZMAN. 21.

Q. And now, will you examine bullet No. 5, Captain, and tell us if you have an opinion as to the make and type of weapon from which that weapon was fired? A. I have.

Q. What is it? A. I think a Peters bullet and fired from a Savage automatic pistol, 32 calibre, Peters make.

Q. Your opinion as to the kind of weapon is based on the same reasons you have given us before? A. Yes.

Q. Did you measure the width of the grooves caused by the lands? A. I did.

Q. What are the measurements? A. .035 of an inch.

Q. What is the twist, if any? A. Right-hand.

MR. KATZMAN. The exhibit number?

MR. WILLIAMS. Exhibit 25.

[893] Q. Will you examine bullet No. 6,—which is Exhibit 24, gentlemen, —and tell us if you have an opinion as to the make and type of weapon from which that bullet was fired? A. I have.

Q. What is it? A. U.M.C. make and fired by a Savage automatic pistol, 32 calibre.

Q. Based on the same reasons you have heretofore given? A. Yes.

Q. What is the width of the grooves in that bullet? A. .035 of an inch.

Q. And the twist? A. Right-hand.

MR. WILLIAMS. Gentlemen, in referring to bullet No. 6, so there won't be any confusion, you will notice that is the bullet marked with a cross. It was not formerly numbered on the envelope as 6. It is Exhibit 24.

Q. Now, I call your attention, Captain, to Bullet No. 3, and ask you if you will look at it and tell us if you have an opinion as to the make and type of weapon from which that weapon was fired? A. I have.

Q. And what is your opinion? A. That it is a Winchester make, W.R.A., and that it was fired by a Colt automatic revolver, 32 calibre or pistol, I mean.

Q. Automatic pistol, 32. A. Colt automatic, 32 calibre.

Q. What is the basis for your opinion? A. In the first place this bullet has got a left-hand twist instead of a right. In the second place, the grooves made by the pistol while passing through on this bullet are .060 of an inch.

Q. .060? A. Yes.

Q. And what does that signify to you? A. It signifies to me that it was fired by a Colt automatic pistol, 32 calibre.

Q. Do you know of any other automatic pistol that gives a right-hand twist to the bullet? A. All give a right-hand twist but the one, all I have ever seen.

Q. I mean left-hand twist, pardon me. A. Not any give left-hand.

Q. This is left-hand twist? A. Yes.

Q. It was my slip, then. Do you know of any other automatic pistol that gives a groove of the width of .060 of an inch? A. I do not.

MR. WILLIAMS. (To the jury) I show you bullet No. 3, gentlemen, and ask you to note the left-hand slant to those grooves, and possibly you can do it better with—

Q. What kind of bullets do you say that was? A. W.R.A.

Q. What does that mean? A. That means Winchester Repeating Arms Company.

Q. How do you ascertain that is a W.R.A. bullet? A. I formed my opinion and I got my conclusion from comparing that bullet with all makes of bullets and measuring the length of them and making a comparison, weighing them, but in that particular bullet it is the "W" which is above the cannon-lure, one of the ends of the grooves that distinguishes the Winchester bullet. They put "W" on their bullets.

[894] Q. Point them out to me. I should have asked you before I started it around through the jury. There is a "W" above the what? A. Above the cannon-lure. That is the ring that goes around it.

Q. Show it to me. Perhaps I can— A. Got to have pretty good light to see it.

MR. KATZMAN. Exhibit number?

MR. WILLIAMS. 18.

THE WITNESS. You can touch the groove there just above the cannon-lure.

THE COURT. Perhaps we had better take our morning recess at this time. (Short recess.)

MR. WILLIAMS. I don't know whether you can see this, gentlemen, but you will notice the "W". You see, gentlemen, this bullet here lop-sided, so that there is more or less of a pronounced point, a sharp bend, and that "W" appears right above—

Q. What do you call it? A. The cannonlure.

MR. WILLIAMS. —the cannonlure, right over that sharp point. The light is so bad I don't know whether you can see it. The three things I call to your attention are the grooves, the slant, that is, the slant of the grooves, and the letter "W".

(The bullet is examined by the jury.)

Q. Captain, did you participate in some experiments with the Colt automatic which is in evidence, at Lowell on Saturday where there were present Captain van Amberg, representing the District Attorney, and also Mr. Burns representing the defense? A. I did.

Q. Mr. Burns is the gentleman seated beside Mr. McAnarney? A. Yes.

Q. What was the character of the experiments there carried on by you three gentlemen? A. We fired the automatic pistol in the case into sawdust, and we recovered the bullets themselves, that sprung from the pistol. I got six empty shells that I picked up that came from the pistol.

Q. How many— A. Then van Amberg fired six and Mr. Burns fired eight.

Q. And did you recover the shells from the six cartridges which you and Captain van Amberg fired? A. I did.

Q. Have you them there? A. I have.

Q. Will you produce them?

(The witness produces some shells.)

Q. (Continued) What make of cartridges did you and Captain van Amberg fire? A. There were three W.R.A. and three Peters.

Q. When you say "W.R.A." do you mean Winchester Repeating Arms? A. Winchester.

Q. What kind of a bullet would you say bullet No. 3 was? A. W.R.A., Winchester Repeating Arms Company.

[895] Q. Now, will you examine for a moment the four shells received from Mr. Fraher? Have you an opinion as to how many weapons were used to fire the four shells which you received from Mr. Fraher? A. I think two.

Q. And can you tell us, in your opinion, how many of those shells and which shells were fired from one and how many and which were fired from the other? A. I think there were three fired from one and one from the other.

Q. Will you tell us which, in your opinion, were fired from one? A. From the right.

Q. Well, can you describe it? I want to show it to the jury. A. It is a W.R.A., Winchester.

Q. What are the other three? A. Two are Peters and one U.M.C.

Q. The W.R.A. was fired from one gun, and the other three, in your opinion, from another? A. It is.

Q. And on what do you base your opinion? A. Well, the looks of the hole in the primer that the firing pin struck that exploded the cartridge.

(Mr. Williams shows the shells to the jury.)

MR. WILLIAMS. Gentlemen, the one in my right hand is the one that he says, in his opinion, was fired from one, and the other three in my left fired from the other.

Q. Now, will you examine that W.R.A., that we will call for the minute the Fraher shell, and compare it with the six shells fired by you and Captain van Amberg, of which, I understand, three are W.R.A. and three are Peters, and ask you if, in your opinion, the marks on those seven shells are consistent with being fired from the same weapon? A. I think so, the same make of weapon.

Q. And on what do you base that opinion? A. Well, there is a similarity between the W.R.A. and the other cartridges that were fired.

Q. A similarity where? A. In the looks of the hole in the primer, which does not exist with the other three.

MR. WILLIAMS. Now, gentlemen, I wish when you pass these around you would keep these six in the right hand, and keep this, which is the so-called Fraher shell, in the left, just keep them apart, and compare the holes made in the primer on those six Lowell shells with the one Fraher shell. I will tell you what I will do. I will let you look at the Peters shells first in comparison with that. I am sorry to be so slow, if your Honor please, but I am afraid it cannot be avoided.

THE COURT. All right.

MR. WILLIAMS. If it will help you any to look at it with a microscope you may. If any of you want the microscope, I have it here.

(The jury examines the shells.)

MR. WILLIAMS. I now, gentlemen, wish to show you the so-called Fraher shell, which is a Winchester, and at the same time to examine the three Winchesters fired at Lowell by this Colt automatic. So there will be no confusion, I am going to ask the consent of the defendants if I may [896] have a slight scratch on the Fraher shell, simply for the purpose of identification. I will make one scratch. That is for the purpose of identification. I have made a mark on the Fraher Winchester shell for purposes of identification.

MR. McANARNEY. That is exhibit what?

MR. WILLIAMS. That is one of four in Exhibit 30. Now, gentlemen, will you examine the Fraher Winchester shell, and compare it with the three Lowell Winchester shells which, if you recall, were shot from the Colt Automatic in evidence? Notice the size of the hole in the primer made by the firing pin, and also notice the position of the hole in the primers of all four shells with reference to the middle of the primer.

(Mr. Williams shows the shell to the jury.)

MR. WILLIAMS. If your Honor please, I offer these six shells fired at Lowell.

MR. McANARNEY. No objection, if your Honor please.

MR. WILLIAMS. May I have these marked as two exhibits, if your Honor please, the three Winchester and the three Peters, in separate exhibits.

MR. WILLIAMS. The three Peters will be 33, and the three Winchesters will be 34.

(The three Peters shells are placed in an envelope marked Exhibit 33, and the three Winchesters in an envelope marked Exhibit 34.)

Q. Captain Proctor, have you an opinion as to whether bullets Nos. 1, 2, 5 and 6 were fired from the same weapon? A. I have not.

Q. Have you an opinion as to whether bullet 3 was fired from the Colt automatic which is in evidence?

MR. McANARNEY. Will you please repeat that question?

Q. Have you an opinion as to whether bullet 3 was fired from the Colt Automatic which is in evidence? A. I have.

Q. And what is your opinion? A. My opinion is that it is consistent with being fired by that pistol.

Q. Is there anything different in the appearance of the other five bullets— A. Yes.

Q. Just a minute, I had not completed. —the other five bullets to which I have just referred, which would indicate to you that they were fired from more than one weapon? A. There is not.

Q. Are the appearance of those bullets consistent with being fired with the same weapon? A. As far as I can see.

Q. Captain, did you understand my question when I asked you if you had an opinion as to whether the five bullets which you say were fired from an automatic type of pistol were fired from the same gun? A. I would not say positively.

Q. Well, have you an opinion? A. I have.

Q. Well, that is what I asked you before. I thought possibly you didn't understand. What is your opinion as to the gun from which those [897] four were fired? A. My opinion is, all five were fired from the same pistol.

Q. What is the basis of your opinion? A. By looking the marks over on the bullets that were caused by the rifling of the gun. It didn't seem to cut a clear groove; they seemed to jump the lands, and seemed to make a different mark than the lands would make.

Q. Will you select two bullets, I don't care which, so that I can show the jury what you mean by that mark?

(The witness picks out two bullets.)

Q. You referred to what, Captain? What is the irregularity of the different bullets that you refer to? A. There is marks there other than what it would make if it stayed in the lands all the time. It would make six cuts there. You see there are a whole lot of cuts there.

Q. Do you find that same peculiarity in all those? A. I do. Some don't go so far, but there is a similarity.

Q. How many lands are there to a Savage automatic? A. Six.

Q. And by the "lands" again you mean the raised—the ridges in the barrel? A. Yes.

Q. So, I take it, it must necessarily be that there are six grooves in the bullet? A. Yes.

Q. Caused by those lands? A. Yes.

Q. And you find additional marks other than those six grooves? A. Yes, I do.

MR. WILLIAMS. Gentlemen, I am taking—

Q. What bullets did you take? A. One and two.

MR. WILLIAMS. I ask you to notice the marks on those two bullets.

(Mr. Williams shows the bullets to the jury.)

Q. While the jury are examining those bullets,—Have you got the six bullets which were fired by you and Captain van Amberg at Lowell? A. I have not.

Q. Who has those? A. Van Amberg has three.

Q. All right. I will introduce them when he is called. Have you any of them? A. I have three.

Q. Well, I will take the three that you have.

(The witness produces the bullets.)

Q. (Continued) Are those the three that were fired by you? A. He fired the gun, but I stood right there and picked them up.

Q. These are three what, what kind of bullets? A. Winchester.

MR. WILLIAMS. I offer these three Winchester bullets fired by Captain van Amberg at Lowell. They will be marked Exhibit 35. The stenographer is waiting for a ruling, your Honor. I offer these three bullets shot at Lowell.

THE COURT. Is there any objection?

MR. McANARNEY. No objection.

(The three bullets are placed in an envelope and marked Exhibit 35.)

[898] Q. You said there were six lands on a Savage automatic? A. Yes, sir.

Q. How many lands on a Colt? A. Six.

MR. WILLIAMS. You may inquire.

CROSS-EXAMINATION

Q. (By Mr. McAnarney) Let's see, you have been testifying, you say, for a number of years in these cases, firearms cases? A. I have.

Q. And you have had quite a lot to do with the Colt revolver? A. Not much.

Q. How much? A. Well, what I have had to do is pushing bullets through and measuring the grooves on the bullets.

Q. How long have you been making observations and tests on Colt revolvers? A. Oh, I commenced 20 years ago, pushing bullets through.

Q. And you have been pushing them through ever since, more or less? A. More or less.

Q. And testifying in cases with reference to the results? A. Yes.

Q. Now, I understand your evidence, if I followed you, you said that the Colts had a left twist? A. It has.

Q. And by that, you mean that the twist is from left to right? A. I mean by that—

Q. Well, do you mean from left to right in the barrel? A. I mean by that that the grooves on the bullet has a left slant. You hold the bullet up square, and on the Colt the groove will have a left slant on the bullet.

Q. That is, from— A. From the base to the top.

Q. From the base to the top, looking at a photograph, the slant would be from the left? A. Looking at the bullets. I don't know about a photograph.

Q. All right. Looking at the bullet, as you would look at it, and taking it from the base to the apex of the bullet, the slant would be to the left? A. Yes.

Q. And I understand you say that is the only weapon— A. The only weapon to my knowledge. As far as my examination of bullets goes, that is the only gun that has the left twist.

Q. Have you made effort to find whether that opinion of yours is well founded? A. Well, I have pushed bullets—

Q. Have you? Yes or No? A. I have.

Q. When did you make that investigation? A. Well, I have— The last time was about a week ago.

Q. And you made what you thought was a thorough search to ascertain— A. I did.

Q. —whether there was any other automatic that had a left twist? A. I did.

Q. And you did not find any? A. I did not.

Q. Are you familiar with the Bayard pistol? A. I am.

Q. That is a left twist, isn't it. A. It is not.

[899] Q. The Bayard pistol is not a left twist? A. I don't think so.

Q. You said it was not. Now— A. I am familiar with the Bayard pistol, .25 calibre.

Q. Pardon me just one minute. You said it was not, and your next answer was "I don't think so." A. If you will let me explain just a moment, I will tell you.

Q. Just a minute, sir. I will repeat the question, and I will take your answer. Has the Bayard pistol a left twist or a right twist? Will you answer that Yes or No? A. Any 32 calibre——

Q. One moment——

MR. KATZMANN: One moment, if your Honor please.

Q. Strike that out, and take this question. Has the Bayard pistol a left twist? Can't that be answered by Yes or No?

MR. KATZMANN: One moment, I object to comments by the interrogating attorney.

MR. McANARNEY: I am not commenting at all.

MR. KATZMANN: Whether a question should be categorically answered by Yes or No is purely within the discretion of the Court.

MR. McANARNEY: It seems to me I may ask your Honor's discretion on that question.

THE COURT: Pardon me. If the witness can answer the question Yes or No intelligently, he may so do.

THE WITNESS: I cannot. I would like to have named the calibre.

Q. Thirty-two. What is the answer? A. No.

Q. No what? A. Not any 32 calibre.

Q. What is the twist of a 32 Bayard? A. 32 what?

Q. 32 Bayard. A. I never saw a 32 Bayard. 25 is the only one I am familiar with. You asked me if I am familiar with the Bayard. I am with the 25 calibre.

Q. What is the twist of a 32 Bayard pistol? A. I didn't say there was a 32 Bayard.

Q. Do you say there is not? A. Not to my knowledge.

Q. That there is not a 32 Bayard? A. Not to my knowledge.

Q. Well, you made an examination to find out, didn't you? A. I have.

Q. To the best of your knowledge and belief, there is no such thing as a 32 Bayard? A. Yes.

Q. So, whether it is a right or left twist you don't know? A. Not of a 32. I never found one.

Q. You examined a 35, did you? A. 25.

Q. 25 Bayard? A. Yes.

Q. You say that was what twist? A. A right.

Q. A right twist? A. Yes.

Q. When did you examine that? A. What is that?

[900] Q. When did you make that examination? A. I looked at all the bullets I ever pushed——

Q. Just answer the question. A. About a week ago.

Q. And passing from that just a minute, you say that there are in one of the exhibits, where there was several Savage—several bullets, you say three from a Savage automatic and one from a Colt you have in mind the group of bullets I am talking about? A. Yes.

Q. Now, you say those came from a Savage gun. Why do you say that? A. On account of the width of the grooves made by the lands in the pistol.

Q. What do you say the width of the grooves is? A. 35/1000.

Q. What instrument did you take to make that measurement? A. I have a micrometer and frame to measure with, the same one I have used to measure bullets for twenty years.

Q. So you get 35/1000 as the measurement of a Savage automatic? A. Yes.

Q. Are you familiar with the Steyer gun? A. No.

Q. Then you don't know what the measurement of that is? A. No.

Q. There have been a lot of those in circulation since the war, since the boys came back? A. Never saw it.

Q. You never saw one? A. Never heard of it, and never saw one.

Q. Are you familiar with the Stauer? A. No, sir, never heard of it.

Q. So that the measurements of the lands and grooves you don't know? A. Never heard of it.

Q. Now, taking— Now, does this gun appear to be used to any great extent, from your observation of it?

(Mr. McAnarney hands a pistol to the witness.)

A. I looked through the muzzle up to Lowell, and there were some pits.

Q. Whatever observation you have made, from your observation, what is your opinion, whether this is a gun that has been used much or not? A. I couldn't say. It might have got those pits by being fired only once and then laid away. I couldn't form any opinion on that.

Q. Will you take it apart, please?

(The witness attempts to take the pistol apart.)

THE COURT: Aren't you taking up too much time? Can't you let somebody else do it?

THE WITNESS: I can't see the mark on it.

Q. Are you familiar with the Colt automatic? A. I carry one all the time in my pocket. I don't take them apart very often. I know how.

Q. You don't take them apart very often? A. No, I don't.

Q. When did you last make an examination of a Colt revolver? A. I looked through this one at Lowell. Do you mean, to take it apart?

Q. When did you ever take it apart? A. Oh, a month ago, I guess, —two months ago.

[901] Q. Have you ever taken half a dozen Colt revolvers apart? A. I don't think so, not different ones.

(The pistol having been taken apart, it is returned to the witness.)

Q. Now, you look at that gun, and tell us whether you think it has been used much or not. A. I couldn't tell you whether it has been used much or not. It has been cleaned out, brushed out, and fired. There is a few pits in close to the lands, but there was when I looked through it at Lowell.

Q. From your experience as a gun expert, are you able to give this jury any opinion as to the extent to which that gun has been used? A. I cannot.

Q. Very well. What is the weight of those various bullets you have been handling this morning? Take the U.M.C., have you in mind what the weight is of that? A. I will look and see. I have it officially here. (Witness refers to memorandum.) If you will ask the question again, I will try and answer it.

THE COURT: Read it, Mr. Stenographer.

(The question is read.)

A. This U.M.C.? There were two U.M.C. Smooth.

Q. No. 2? A. Smooth. "2," you mean with reference to the exhibits?

Q. Yes. Have you weighed the various exhibits? A. I have.

Q. Then you may give it. If you are looking now at data with reference to the weight of an exhibit, I will put the question in the form to include what you have before you. What are you looking at now? A. I am looking at the weights you asked me for.

Q. I asked you what was the weight of a U.M.C. generally? A. I think they vary. I have the weights of the U.M.C.'s in the case here.

Q. Well, I would like those. A. Bullet No. 1, 70.54 grams. Bullet No. 4 weighs 72.395.

Q. All right, continue. Give the weight of the others, if you have them. A. No. 1 measures 72.19. No. 2 measures 70.54.

Q. Do you mean "measures" or weighs? A. Weighs. No. 3 weighs 73.59. No. 4, 72.395. No. 5, 72.05. No. 6, 69.94.

Q. Have you in mind what the maximum pressure pounds per square inch on a Colt automatic 32 is? A. I don't know anything about it.

Q. You don't know anything about it? Your answer was you don't know anything about it? A. The pressure on it?

Q. Yes. A. I don't know anything along that line. My experience has all been in finding bullets and finding the right bullet to measure and examine.

Q. You don't know anything about the maximum pressure? A. I don't know anything about that line at all.

Q. What is the penetration power—— A. I don't know anything about that.

[902] Q. What is the composition of the core? A. I don't know anything about the component parts of either ammunition or the revolver or pistol.

Q. Well, as an expert during these 32 years, haven't you got inside of the revolver at some time? A. I am only an expert on what is necessary for me to learn, and that is what I have learned.

Q. Then, it has not been necessary about the insides of the guns much, has it? A. Only by pushing a bullet through and examining what impression was made on the bullets.

Q. That is where you limit yourself? A. I do,—the kind of bullets, and by comparison to tell what gun shot them.

(Mr. McAnarney hands a bullet to the witness.)

Q. Will you glance at that? Have you any idea what that bullet is? I am showing you a bullet. A. I can't tell one bullet from another, unless I have a proper chance to compare them and measure them, and have lots of time on it. I have had a year to compare these five or six, and that is the conclusion I reached.

Q. I don't know as I want all that. I asked you a simple question. Your answer is No? A. Not without making a proper examination, and that would take some time.

Q. Would you look at it, please, just glance at it? Look at it. Are you able to form any opinion as to what kind of a bullet that is? A. I wouldn't give any opinion without having proper time to examine it and compare it.

Q. Pardon me. Your answer is No? A. No.

Q. I want to call your attention to the lands and grooves on that bullet. You can tell us what they are, whether they are left or right, can't you? A. They are left.

Q. It looks like a 32, don't it? You may compare it with some of the 32s you have with you. A. It looks like one, yes.

Q. Now, you have before you a bullet with a left groove, and which appears to be a 32, don't you? A. Yes, sir.

Q. Well, that is the first left groove 32 bullet you ever saw outside of a Colt, isn't it? Strike out that question. That is the first left 32 bullet that you have seen, isn't it, other than such as came out of a Colt revolver?

MR. KATZMANN: One moment.

THE COURT: He hasn't said that yet. You better ask him.

Q. Isn't that the fact, that that is the first left groove 32 bullet you have seen except such as came out of a Colt revolver, Colt pistol?

THE COURT: He hasn't said that that did not come out of a Colt revolver.

MR. McANARNEY: That was to be my next question.

Q. Did that come out of a 32 Colt?

THE COURT: You may answer.

[903] A. I couldn't tell just looking at it. I would have to make an

examination. The same answer stands as to the other bullet. I couldn't tell you off-hand whether it did or not.

Q. Would you say whether or not that came out of a Colt revolver? A. My opinion is, it came out of a Colt, yes.

Q. You know of no other 32 left groove bullet, do you? A. I do not.

Mr. McAnarney: Now, I would like to have that marked. Will you put some little private mark on that, Captain, so that you can remember that later? Now, if you will look on the base, will you notice the letter "L" on the bottom?

The Witness: I would not say that was a 32, unless I had a chance to examine it further.

Q. My question is, do you notice a letter "L" on the bottom? A. I do.

Mr. McAnarney: May that for the purpose of identification, be referred to by the mark "L"?

The Court. Is that the only one in at the present time with an "L" on it?

Mr. McAnarney. Yes, it is, your Honor.

The Court: That may be done.

Mr. McAnarney: Defendants' mark for identification.

Q. You have a Colt bullet there, a bullet that came out of a Colt? Kindly get one of those. A. I haven't, only except that is an exhibit.

Q. I mean those that are in exhibit. Take out the one the dimensions of the lands and grooves which you have given, and place it beside this bullet. A. Have you any preference which one it is?

Q. No, Captain, so long as it is a 32 Colt. A. There isn't but one. (Indicating.)

Q. Now, compare the lands and grooves on those two bullets, and give us your best judgment as to whether they were both fired from a Colt. A. I couldn't tell you off-hand. I have got to take time to make an examination before I can give you an opinion on anything of that kind. I may have to go further, and take a picture to match them up.

Q. One moment. Your answer is No.

Mr. Katzmann: One moment. I submit that is not the answer.

Q. Captain, can you give us a better opinion now from your examination,—whether or not you have any opinion as to whether that bullet came from a 32 colt? Now, will you kindly answer that question? Have you any opinion as to whether or not that bullet came from a 32 Colt? A. I couldn't give you an opinion.

Mr. McAnarney: Put that bullet back, Captain, from where you took it. I am taking the "L" one back to the desk.

Mr. Katzmann: Let me see that, please.

(Witness hands a bullet to Mr. Katzmann.)

[904] Mr. McAnarney: Do you want to see the other one, Mr. Katzmann?

The Court: Have you that other bullet that is marked for identification?

Mr. McAnarney: Do you want that I should leave this here?

The Court: Yes; that is the usual practice.

Mr. McAnarney: Have you been leaving your goods or property marked for exhibits, have you been leaving them with the Clerk?

Mr. KATZMANN: We have left some, and kept others.

Mr. McANARNEY: If the rule be to——

THE COURT: I think all the exhibits ought to be left with the stenographer or with the Clerk.

Mr. KATZMANN: Well, the clothing I thought that was distinctly understood we were not to keep it in the court room.

Mr. McANARNEY: Not the clothing, but I think I made the request of you last night for some other things.

Mr. KATZMANN: I think you made the inquiry of me last night.

Mr. McANARNEY: Now, if you will mark this envelope, Mr. Stenographer.

(The bullet with the mark "L" on the base is placed in an envelope and marked Exhibit 13 for identification.)

THE COURT: We will take a recess until two o'clock.

(Noon recess.)

Dedham, Massachusetts, June 21, 1921.

THE COURT: You may poll the jury, if you please, Mr. Clerk.

(The jury are polled and both defendants answer present.)

WILLIAM H. PROCTOR

CROSS-EXAMINATION RESUMED

Q. (By Mr. Jeremiah McAnarney) Captain, I show you a bullet. Will you kindly look at it (handing bullet to the witness)? A. (Witness examines bullet.)

Q. Are you able from the lands and grooves and general make-up of the bullet, to determine what kind of a revolver or pistol it came through? A. Not with this kind of an examination, I cannot.

Q. Are you able to determine now, looking at it, what calibre bullet that is? A. I am not.

Q. Can't you, looking at that bullet,—haven't you any instrument that in any way you can give us your opinion from your 32 years experience what calibre bullet that is? A. Some of them are very near alike. I could not tell you, no, sir. Not on this examination.

Q. What is that nearest like, in your opinion? A. I wouldn't give you any opinion on it until I had a chance to examine it thoroughly.

[905] Q. Does it look like a 38 shell or bullet? A. No.

Q. Does it look like a 35 bullet? A. Possibly.

Q. Does it look like a 32? A. Possibly.

Q. What is the next under 32? A. The next I know of is 25.

Q. You will agree with me it is not a 25, won't you? A. I won't agree to anything until I make a proper examination.

Q. I see. So whether it is a 25 or 35 or 32 bullet, you don't know? A. I won't give you an opinion until I have the proper chance to examine it.

Q. How do the lands and grooves on that bullet compare with the lands and grooves on a Savage Colt automatic 32? A. (Witness examined bullet.) That has got a right-hand twist.

Q. How do they compare in the bullet diameter? A. I couldn't tell you within a thousandth or perhaps two-thousandths or three-thousandths of an inch under this kind of an examination.

Q. I see. What was the reason you gave when you said that all of these bullets which you enumerated came from the 32 Savage automatic? What was the reason you gave as to how you knew that? A. The reason I gave was the measurement of the groove made by the lands in the pistol, —the measurement.

Q. Whether the lands and grooves on this bullet correspond to those, without a proper examination you could not say? A. I cannot.

MR. JEREMIAH MCANARNEY: I would like to have this marked for identification. It has a well-defined "H" on the base. I do not think there is any "H" gone in yet.

(The bullet is marked "Exhibit 14 for identification.")

MR. JEREMIAH MCANARNEY: I am leaving this with the clerk.

Q. By the lands and grooves you determine it came from a Savage automatic? A. By the lands,—the grooves on the bullet made by the lands in the pistol, the barrel of the pistol.

Q. Wasn't there something else that enabled you to say it came from this gun, or was there? A. Talking about a Savage?

Q. Yes. You said something more than that. Didn't you say something about that there was,—where the bullet entered the barrel that there was some trouble you thought you noticed? Didn't you notice something else about those Savage bullets or bullets you say came from a Savage gun? A. Well, that wasn't any reason for saying it came from a Savage.

Q. What was there? A. The other marks. I said there were other marks besides the marks made by the lands where it jumped the rifling of the pistol.

Q. What causes that? A. I am not up on that line, you know. I am not a gun maker or an ammunition maker. I do not do any test for that purpose, so I plead ignorance to the reason what made it.

Q. Then it may be a natural condition on any bullet going through a Savage, may it not? A. For aught I know.

[906] Q. Then for aught you know, it is the normal condition of any bullet going through any Savage automatic? A. I wouldn't say so, sir.

Q. I thought you said for aught you knew it was. What will you say? A. I never made any tests on firearms.

Q. What is your answer to that question? A. Repeat the question.

MR. JEREMIAH MCANARNEY: Kindly give him the question.

(The question is read.)

MR. JEREMIAH MCANARNEY: I will strike that all off.

Q. The fact is, the significance of any irregularity or any appearance of those bullets which you say came through the Savage gun, they mean nothing to you? A. I do not know the cause of them.

Q. That is your answer, they mean nothing to you, in that you do not know the cause?

MR. KATZMANN: One moment, if your Honor please.

Q. Isn't that right?

MR. KATZMANN: One moment, Captain.

THE COURT: Separate those.

Q. You do not know the cause, do you? A. I know——

Q. Will you answer that question? Haven't you just said you did not know the cause? A. The cause of what?

Q. What you are talking about? A. I know that those marks are made by jumping the rifling, but I do not know what causes them to jump the rifling.

Q. True. Whether the jumping of the rifling is a normal condition in a Savage automatic you do not know, do you? A. That is right.

Q. Very well. Then, I say that for all you know the appearances of those bullets, which you say came from a Savage automatic, have the normal appearances of any bullets that come from any Savage automatic. That is right, isn't it? A. For all I know, yes.

MR. JEREMIAH MCANARNEY: Then your knowledge is our knowledge now.

REDIRECT EXAMINATION

Q. (By Mr. Williams) Captain, in reply to Mr. McAnarney's questions you said you are unable to reach a conclusion in regard to one or both of those bullets he showed you without an examination. Now, what examination did you refer to, or did you mean? A. I don't get your question.

Q. Strike it out, please. You stated, as I understood you, in reply to Mr. McAnarney, that you want some time to examine the bullets which he submitted to you? A. Yes.

Q. Before you would give an opinion as to the kind of bullet or from what gun it had been fired? A. Yes.

Q. Did you not? A. Yes.

Q. Will you tell the jury what kind of an examination you would want or wish to make before giving your opinion as to those bullets. A. [907] I would want a chance to measure them and study them; take some little time. I want to get the exact measurements of the grooves, of the grooves on the bullet to compare them with other bullets that are pushed through automatic pistols.

Q. And are you able to make such examination as you stand here on the witness stand? A. I am not.

Q. What do you mean by pushing a bullet through a gun? A. I mean pushing it through, through the barrel of the gun.

Q. And what method do you take? A. Entering it at the breech and pushing it through to the muzzle, putting a bullet into the breech, put pressure on, enough to take it through and out the muzzle.

Q. Does that give the same marks as would be given if a bullet was fired through a gun? A. It does, to the best of my knowledge.

Q. When you say "pushed through" do you mean pushed through by hand? A. No. Only you have to start it with a little wooden mallet, or something, but after it gets going it goes easily.

Q. No mechanical contrivance is necessary? A. No.

Q. You push it through by hand in some form? A. Yes.

Re-Cross-Examination

Q. (By Mr. Jeremiah McAnarney) Captain, if you had a micrometer here, could you give us some information or some idea? A. I could not make a proper examination of bullets here. I have got to have more time and I have got to have my own instruments.

Q. My attention has been called—will you kindly look at 1, 3 and 4 bullets,—1, 3 and 4? A. (Witness does so.)

Q. Are those three bullets the same bullet or the same make of bullet? A. (Witness produces memorandum from pocket.) That is my opinion.

Q. You feel quiet sure about that, Captain? A. I do.

Q. I show you a photograph. Will you kindly look at it (handing photograph to witness)? A. (Witness examines photograph.)

Mr. Jeremiah McAnarney: Do you wish to see it, Mr. Williams?

Mr. Katzmann: I can't hear the question.

Mr. Jeremiah McAnarney. I say, "I show you a photograph"—perhaps I better show it to the District Attorney, Mr.——

(Mr. McAnarney shows photograph to Mr. Katzmann.)

Q. Now, you may glance at those, Captain; and I call to your attention, take the different series, and may I say to you, unless I am stopped by objections, that these purport to represent two, four, six different photographs of the same bullet turned in six different positions, and you will notice them and you see from your experience, if you kindly look at them so you understand them before I put the next question. A. (Witness does so.)

Q. You have examined them? A. Yes.

[908] Q. Now, I call your attention, Captain, to the appearance of the bullet marked——

Mr. Jeremiah McAnarney. May I have this marked for identification as an exhibit?

(The series of photographs is marked "Exhibit 15 for identification.")

Q. I am calling attention to photograph of bullet No. 1 on defendant's for identification 15, and I call your attention to the one marked 3 and the one marked 4, and ask you if you will examine those three, 1, 3, and 4, and give us your opinion if they are the same in appearance to you, the same kind of bullet.

Mr. Katzmann. I object to that question, if your Honor please.

Q. I would like it, if it would assist you any, that you examine the six different photographs.

Mr. Katzmann. To that question I object, if your Honor please.

Mr. Jeremiah McAnarney. I submit it.

The Court. Read the question.

(The question is read.)

The Court. Are you able to tell?

The Witness. I am not by looking at them there if they are the same kind of bullets.

Q. I call your attention to the knurl on 3 and the knurl on 4——

The Court. The what?

Mr. Jeremiah McAnarney. The knurl.

Q. How do you spell that, Captain? I do not use that word very often. A. Nor I.

Q. You don't, either? A. No.

MR. JEREMIAH MCANARNEY. Well, the stenographer will have to spell it.

Q. Now, look at the knurl of 1, 3 and 4 in this, the second. I call your attention to the knurl in 1, 3 and 4 on this third picture. I call your attention to the knurl on 1, 3 and 4 in this fourth view, noticing particularly the knurl on No. 3. I call your attention to the knurl on 1, 3 and 4, again calling particular attention to the knurl on No. 3. And I call your attention to the knurl on 1, 3 and 4, the last one of this group, again directing your attention to the knurl on 3. Are you able to form an opinion as to whether those three bullets are the same kind of bullets or not? A. From that picture I could not form an opinion.

Q. You have no idea? Is that right? Now kindly—— A. I might have an idea but I wouldn't form an opinion, not to express it.

Q. Perhaps your idea might help out. What is your idea? A. My idea is that bullet No. 3 is fired by a Colt and those others are fired by some other kind of firearm.

Q. That wasn't my question. Did you understand that for one minute as the question I asked you? I was asking, Captain, whether or [909] not they were the same kind of bullet and shell? A. I couldn't tell by looking.

Q. The same kind of cartridge? A. I couldn't tell by that picture.

Q. Well, I will come back to what is the true situation, the bullet, the same kind of bullet. Is there anything on there that instructs you? Will you tell me, please? A. I can't tell anything by that picture, with the exception that one of those bullets was fired by a different gun than the other five.

Q. How do you know that by the picture. A. I know by the width of the groove on the bullet and the slant of it.

Q. You see No. 3 is a left groove, don't you? A. Yes, but it does not look so good on there as it does on the bullet itself.

Q. I did not ask you that, Captain,—kindly answer the question I have put to you. You see on this photograph that No. 3 is a left groove, don't you? A. Slightly.

Q. Well, it is what you call a left groove? A. Some.

Q. Is it what you call a left groove? Will you kindly answer that question? A. (Witness examines photograph.) Well, the way the bullet sets there it has a left slant. It might be cocked some way to make that left slant, so I won't answer any pictures, from the picture what bullets they are.

Q. Have I asked you that question, Captain? A. Without seeing the bullets themselves.

Q. Have I asked you that question, Captain? A. I don't know as you have.

Q. Well, I would like to have you answer the questions I put to you. A. All right, let them come.

Q. They have been coming. Now, isn't that the photograph of what you call a left twist bullet, or are you unable to say? That will let you out;

if you are unable to say, please say so. A. I do not know what that is a photograph of. I did not see them set up, and I do not know anything about them.

Q. Don't you know that is a left twist bullet? A. No, I do not.

Q. What did you mean a minute ago when you said you knew it came out of a Colt revolver because it was a left twist? A. That is the way it looked to me.

Q. Then you know it is a left twist, and from that knowledge, as you see it there, you said it came from a Colt revolver, didn't you?

MR. KATZMAN. One moment, there.

THE COURT. Did he say the picture or did he say the bullet itself? Look that up and see.

(The witness' answer is read as follows: "A. My idea is bullet No. 3 is fired by a Colt and those others are fired by some other kind of firearm.")

[910] Q. That is your answer, isn't it? A. Bullet No. 3, yes.

Q. Yes. And you say that because of the left groove, don't you? A. I don't say because of the left groove on that picture. I say the left groove on the bullet.

Q. Pardon me. You weren't looking at the bullet when you were answering me, were you? A. I had the bullet in mind.

Q. I wasn't showing you the bullet, was I? A. You was not.

Q. Were you answering me, Captain, from the bullet or photograph I was showing you? A. From the photograph.

MR. JEREMIAH McANARNEY. That is all.

THE WITNESS. Having it in mind.

MR. JEREMIAH McANARNEY. That is all, sir.

REDIRECT EXAMINATION

Q. (By Mr. Williams) Captain, do you know how a certain slant to a line would appear in a photograph? A. I do not.

Q. Suppose a line in an object slanted to the right, do you know how that same line would appear in a photograph taken of it? A. All the photographs I have had taken and looked at——

MR. JEREMIAH McANARNEY. Pardon me. That is not responsive.

THE COURT. Just answer.

THE WITNESS. I do.

Q. How would it appear? A. If the slant was to the line onto the bullet, it would be the slant to the left on the photograph, the picture.

Q. Now, if a similar line was shown in a mirror, how would the image appear in the mirror?

MR. JEREMIAH McANARNEY. Wait; I object. THE WITNESS. I don't know about that.

THE COURT. Does he know?

THE WITNESS. I don't know.

MR. JEREMIAH McANARNEY. While I had the examination in mind I meant to ask the jury to view these as a chalk, if your Honor please.

THE COURT. Just see if they got through with the examination of the Captain.

MR. JEREMIAH MCANARNEY. Oh, yes, excuse me.

THE COURT. He is through. Now, you may.

MR. JEREMIAH MCANARNEY. (Showing series of pictures to the jury). You jurymen may look at this series——

MR. KATZMAN. One moment, if your Honor please.

THE COURT. He wants them admitted as a chalk.

MR. KATZMAN. Well, I should like to know if he proposes to introduce them as an exhibit before he shows them to the jury as a chalk or otherwise.

MR. JEREMIAH MCANARNEY. That has been the rule, and I am within the rules.

[911] MR. KATZMAN. I haven't heard you make any such statement.

MR. JEREMIAH MCANARNEY. I haven't heard you make a statement, either.

THE COURT. That is not the question, Mr. McAnarney. The question is now whether you propose to offer evidence tending to prove that these are a direct reproduction of the bullets.

MR. JEREMIAH MCANARNEY. Yes, your Honor.

THE COURT. And that you propose to offer them in evidence?

MR. JEREMIAH MCANARNEY. Yes, certainly.

THE COURT. You may show them then as a chalk.

MR. JEREMIAH MCANARNEY. (To the jury) And call your attention to a bullet in particular I called the attention of the witness to, Nos. 1, 3 and 4 in the series. You may glance at them as 1, 3 and 4.

THE COURT. I think that is all, Captain.

CHARLES VAN AMBURGH, SWORN

Q. [By Mr. Williams.] What is your full name? A. Charles J. Van Amburgh.

Q. How do you spell your last name? A. V-a-n A-m-b-u-r-g-h.

Q. Where do you live, sir? A. Bridgeport, Connecticut.

Q. I have been calling you Captain, and ask you now if you have that title so that I may not miscall you on the stand? A. I might explain I held that rank during the war for one and one half years during the war, and now in the Reserve Corps.

Q. What is your occupation or profession, Captain? A. At present, I am an assistant in the ballistic department, Remington U.M.C. Company.

Q. And where are you located? A. In Bridgeport, Connecticut.

Q. Captain, would you give us briefly your experience in reference to firearms and ammunition for firearms, dealing with them, measuring them, and so forth? A. For nine years I was connected with the Springfield Armory,—that is, eight years of it at Springfield Armory, the manufacture of the United States rifle, or, at least, where the rifle is manufactured; one year at Frankford Arsenal, Philadelphia, where our United States and our government ammunition——

Q. Will you indicate as you enumerate the various places where you have been employed the nature of the work which you there did, beginning with Springfield? A. The Springfield Armory, experimental work. That is, the experimental department, experimental firing with rifles, mainly military rifles and automatic pistols, machine guns. Also inspected ammunition. I might explain that particular point as it is an arms factory. That is, headquarters for inspection and for armament or ordnance district, and inspectors were frequently detailed from there to visit ammunition factories who were under contract for ammunition for the United States Government, and I at one time was detailed on such an errand.

[912] One year at Frankford Arsenal, entirely in the testing of ammunition. Leaving there, I was with the New England Westinghouse Company for about two years, in the manufacture of the Russian rifle, in the test department, by the way. I was assistant proof master. Leaving there during the war, I was about one year in the employ of the Colt Patent Firearms Company in one of their branch plants,—Meriden, Connecticut, to be specific. Then I went,—accepted a commission in the Army. Coming out of the Army——

Q. How long were you in the Army? A. One and one half years. I was instructor in small arms firing or marksmanship with rifle and pistol. Upon discharge, I went with the Remington U. M. C. Company in ballistic work, which I might explain pertains to tests of arms and ammunition, ballistic.

Q. You are still employed? A. Still employed.

Q. —with the Remington people. At the request of Mr. Katzmann, have you examined six bullets which have been given to you by Captain Proctor? A. I have.

Q. And are those bullets which are before you now, Captain? A. They appear to be, sir.

Q. Will you take those bullets, one after another, beginning with the bullet numbered one, and describe those bullets to the jury, with what marks you see upon them? A. My examination of bullet No. 1 indicates that it is a .32 calibre automatic bullet.

Q. When you say "automatic bullet" what do you mean? A. I mean from an automatic cartridge. It is from a .32 automatic cartridge.

Q. Yes. And will you further describe—and this pertains to all the bullets—the number of grooves upon them which are attributable to any lands and tests of any bullets? A. The rifling marks indicate that there are,—in fact, show very clearly that there are six grooves, six groove cuts, and six land cuts, on this bullet.

Q. Yes. And the width of those cuts? A. My measurement of number one, for the groove cut, averaged .123 of an inch. For the land cut, between .035 and .036.

Q. And did you weigh that bullet? A. I did.

Q. What was the weight? A. That bullet weighed slightly over 72 grains, —72.2, approximately.

Q. Now, will you take number two bullet (Exhibit 19), going along the same way? A. Number two, also its rifling marks have six land cuts and six groove cuts. Its weight I found to be 70½ grains.

Q. The width of the lands? A. The width of the land, .125, .125. That is an approximation. I found I had to approximate it slightly, because it was a little——

Q. The width of the grooving? A. The width of the grooving, the widest portion on the bullet, .125.

[913] Q. I thought you did not understand. You said the width of the lands was .125? A. By force of habit, shop practice, my reference to groove pertains to the groove in the rifling.

Q. I see. A. The groove cut, as I refer to it.

Q. If you will give it to us in the terms of the bullet, Captain, perhaps we can keep it a little more clearly in mind. A. As it appears on here?

Q. Yes. The width of the groove in the bullet is what? A. .125.

Q. Well, and the width of it? A. I must say it will confuse me somewhat if I change my name, but, however, the narrower cut——

Q. Let me get it straight in my mind. In the rifling of a barrel, there is a raised portion and a sunken portion, as I understand? A. Yes.

Q. The raised portion or ridge is called what? A. Called the land.

Q. The sunken portion is called what? A. A groove.

Q. And the raised portion or land in the barrel makes a corresponding groove in the bullet, I take it? A. Pardon me. I was answering a question pertaining to the barrel a moment ago.

MR. JEREMIAH MCANARNEY. We do not hear you down here.

Q. Well, perhaps we better start again. A. I wish you would.

Q. Take bullet number one for the moment, so we will be sure we are talking about the same thing. Take bullet number one for the moment. Did you measure the raised parts of the bullet and the sunken parts? A. I did.

Q. Now, what in the bullet do you call the raised part which would correspond to the groove in the barrel? A. According to the system you have suggested, it would be the land on the bullet.

Q. The land on the bullet? A. The land on the bullet would be the raised portion of the bullet.

Q. And the groove would be the sunken portion on the bullet? A. Yes, sir.

Q. Well, will it bother you to consider the bullet in giving your dimensions which you have found? A. I will try to make it clear, sir.

Q. Now, take bullet number one again, sir, and simply give us the width of the land and groove in that bullet, Captain. A. The width of the land on bullet number one, .123. Of the groove, between .035 and .036.

Q. Yes, now take bullet number two. A. Bullet number two. The land on the bullet, .125, and the groove on the bullet, about .036.

Q. Now, skip to bullet number four, if you will. A. Number four?

Q. Just a minute. Mr. Katzmann reminds me you have not given the weight or calibre, the kind of bullet on number two. A. Number two, from an automatic, from a .32 calibre automatic cartridge.

Q. The weight of the bullet? A. The weight of bullet number two, 70.5 grains.

Q. Now, if you will take bullet number four. A. Bullet number four, the width of the land on the bullet, between .122 and .121. Of the groove on the bullet, about .036.

[914] Q. The description of the bullet? A. It is a .32 automatic.

Q. The weight? A. The weight of number four, 72.39 grains.

Q. Now, if you will take number five bullet, Captain. A. From number five, the land on the bullet is between .123 and .125. The groove on the bullet, about .036. Number six——

MR. JEREMIAH MCANARNEY. What is the weight of number five?

THE WITNESS. The weight of number five, 72 grains.

Q. Now, number six, Captain. A. Number six, the width of the land on the bullet, between .121 and .125. The width of the groove on the bullet about .036.

Q. What is the weight? A. Weight, 69.9 grains.

Q. What is the calibre on number five and number six? A. It is from a .32 automatic cartridge.

Q. That pertains to both? A. And the same on the number six.

Q. Will you describe to the jury the mechanism of a Colt automatic, so far as the firing part is concerned? I am referring to the firing pin and the mechanism which causes the bullet or the cartridge to explode. A. It is a .32 calibre automatic pistol, and by "automatic" is meant really self-loading, differing from the revolver which is perhaps better known chiefly because of the fact that the loading is done by action of the recoil. The firer merely presses the trigger to release the firing mechanism. A pistol,—the mechanism performs the loading function, so they are self-loaders. The whole cycle is completed,—the cartridge carried into the mechanism by the forward travel of the slide, the empty shell extracted through recoil or when the slide recoils. Have I given that clear enough?

Q. That is perfectly clear to me, Captain, but what I wish you would describe and point out is the parts of that automatic pistol to cause the cartridge to explode and give us the names of those parts. A. That causes it?

Q. Yes. Perhaps if you step down in front of the jury they can see better. Step right down in the middle, if you can. A. (Witness steps in front of jury rail.) This portion here (indicating) is the firing part. The end of it can be seen here (indicating). Our hammer strikes against that pin against which my pencil is resting. The hammer striking against that. The firing pin is visible on this end. It protrudes. The cartridge is resting against that portion and the primer directly over the firing pin hole receives the blow of the trigger. The primer is the cap, that little round central portion on the cartridge head. The hammer strikes down here (indicating).

Q. Just push it so that they can see the firing. A. Here is the firing pin on this end here (indicating to the jury).

(The witness returns to the witness stand.)

Q. Now, in addition, Captain, to examining those bullets and weighing them, and so forth, did you participate in some experiments at Lowell [915] where were present Captain Proctor and Mr. Burns of Boston, representing the defense, and yourself? A. I did.

Q. Will you tell the jury what acts were formed at that time and place? A. Mr. Burns fired eight cartridges into a large box of oiled sawdust.

Q. Keep your voice up a little. You speak a little bit low. A. Mr. Burns fired eight bullets into a large box of oiled sawdust, for the purpose of catching them unmarred for examination. Captain Proctor and myself fired six. A total of fourteen was fired.

Q. What kind of cartridges were those you and Captain Proctor fired?
A. Fired three Winchester and three Peters.

Q. Did you recover the bullets which you and Captain Proctor fired?
A. I did.

Q. And did you examine those bullets after you had recovered them?
A. I did.

Q. Have you further examined four certain shells, cartridge shells, of .32 calibre, which purported to have been found near the scene of this shooting? A. I have.

Q. We have been calling them the Frahar shells. Have you examined——

THE COURT. Did you ask him from what kind of a gun the shots were fired at Lowell?

MR. WILLIAMS. I will ask him.

Q. What gun did you use in firing the shots at Lowell? A. A .32 automatic, Colt .32 automatic.

Q. Is it the one you have in front of you there? A. It is this particular pistol.

Q. I neglected to ask you in regard to bullet number three, Captain. Will you now take bullet number three and describe that to the jury? A. Bullet number three measures across the land of the bullet between .107 and .108; across the groove of the bullet about .060 of an inch.

Q. Now, Captain, will you describe the rifling in the barrel of a pistol to the jury? A. Describe the purpose of it, did I get your question?

Q. Yes, the purposes; what in general is the nature of rifling in barrels of guns, and what result does it have upon the course and flight of the bullet? A. Rifling is necessary to give rotation to the bullet; otherwise the bullet would fly out and would naturally tumble end over end. The rifling imparts rotation. Our bullet then partakes something of the nature, we are told, of a gyroscope. It travels end on and in short it adds stability to its flight.

Q. How does rifling impart rotation to the bullet? A. The bullet takes the inclination of the rifling.

Q. Well, I wish you would just tell us how it takes the inclination, as you call it, of the rifling. What happens, in other words? Just describe to us what happens, if you will, when a cartridge is exploded. A. [916] When the cartridge is exploded, the bullet moves from rest into the barrel under pressure generated by powder gases. As it travels up the barrel a ways, it is expanded under the pressure of those powder gases until it should and it perhaps does fill or touch the bottom of the grooves in the barrel. It is then taking the rifling, to use the shop term. That is, the rifle marks are then registered on the bullet. It has taken the rifling, and from that point on it must rotate at the rate or number of turns there are in the barrel. It will then spin at a high rate, whatever the rate happens to be.

Q. Something has been said here about a twist to a bullet. A. Twist is merely another name for spiral.

Q. The rifling in a barrel is of a spiral nature or character? A. It is, yes; a true spiral, as a rule.

Q. Which way does the rifling of a gun twist the bullet? A. If the inclination of a rifling is to the right, that is, clockwise, the way the hands of a clock would travel as to the right, if the inclination is to the right, then we speak of it being a right twist.

Q. And what mark would be made on the bullet by that right twist? A. They will be inclined to the right, as we look at the bullet in front of us.

Q. From the base to the nose? A. From the base to the nose. And if the direction is counter-clockwise, that is, the opposite way to which the hands of a clock would turn, to the left, we would say then our bullet would rotate to the left.

Q. What kind of marks will be made upon the bullet? A. In that case, the marks on the bullet would be inclined to the left.

Q. You have examined number three bullet? A. I have.

Q. Can you tell us what make that number three bullet is? A. Number three bullet, I believe, is positively identified as a Winchester Repeating Arms Company make.

Q. And what is your reason for making that positive identification? A. There is a letter W on the bullet, which is a point which is peculiar to bullets of Winchester make,—jacketed bullets.

Q. Is there anything about the bullet which identifies it as a Winchester make? A. Not to any great,—not with any great surety. The letter W is almost,—is very good proof.

Q. Have you an opinion as to the type and make of weapon from which number three was fired? A. I have.

Q. And what is your opinion? A. I believe number three bullet was fired from a Colt barrel. That is to say, Colt automatic pistol.

Q. By the way, Captain, you mentioned among the various places where you had worked, the Colt Rapid Firearms Company, if I heard you right. Is that the company that manufactures the Colt automatic? A. It is.

Q. How long were you with them? A. About one year.

[917] Q. What is the basis, or what is your basis for the opinion you have formed as to the type and make of weapon from which number three was fired? A. Well, the left inclination. That is to say, the rifling marks on our bullet indicates clearly that it was fired from a barrel having a left twist.

Q. Yes. And does the width of the lands have anything to do with the opinion which you have formed? A. It has. From my experience I believe that the width .060 agrees with that of the width,—land width in the barrel of a Colt automatic .32 calibre.

Q. Are you familiar with a Savage automatic? A. I have used them.

Q. Do you know the width of the lands and grooves impressed upon a bullet fired from a .32 Savage automatic? A. I have measured them.

Q. Are the five other bullets,—when I say "other" I mean those other than number three, which I understand you have examined, from their physical appearance consistent with the fact of their having been fired from a Savage or from Savage automatic pistols? A. I would say that they are.

Q. Have you an opinion, Captain, as to whether those other five bullets, and I mean one, two, four, five and six, were fired from the same weapon? A. I believe that they were.

Q. What is the basis for your belief? A. There are peculiar markings on numbers one, two, four, five and six, which seem to occur so uniformly and on all, it inclines me to the belief strongly that they were fired from the same or through the same barrel.

Q. Could you show that on any two of them, or more, I don't care, those markings, I mean, so that the jury could see the markings which have assisted you in coming to an opinion? If you cannot, just say so, but if that can be demonstrated to any extent, I would like to have you do it. A. I will try.

Q. I will hold the light, if you will step down in front of the jury. Take any two, I don't care what two. A. I will take,—the markings are all small and there may be some difficulty in seeing them.

Q. If you will just indicate the kind of markings, the character of the markings you refer to, that is what I mean. You might step down in front of the jury and I will bring the light over. A. The glass would be better.

MR. JEREMIAH McANARNEY. Will you tell us which ones you are now talking about?

Q. Which ones have you now? A. Numbers one and two. Too small to hold with any steadiness to show. I mean, the bullets themselves are too small, but I will try.

(The witness leaves the stand and goes in front of jury rail.)

THE WITNESS. On the left of that land (indicating) there, there is a double cut. The double cut, which occurs on this bullet at the same place, at the commencement of the land cut.

[918] MR. KATZMANN. Captain, I suggest you let the jurymen adjust the lens themselves, and you hold the bullets. They may have different eyes from yours.

THE WITNESS. Yes. (Doing so.)

MR. WILLIAMS. I would be glad to have you all look at it, gentlemen.

A JUROR. Bring it over here without the light.

(The witness does so, and then returns to witness stand.)

THE COURT. It is a warm afternoon and you may have a recess now. These hot afternoons that we sit late, I will try and plan to have two intermissions.

(Short recess.)

Q. (By Mr. Williams) I am reminded to ask you, Captain, that I forgot to ask you for the description of No. 3 bullet.

THE COURT. Do you care to wait for Mr. Moore?

MR. McANARNEY. No, if your Honor please.

Q. Will you describe No. 3 bullet as to calibre and kind? A. No. 3 bullet, I believe, is from the 32 calibre automatic cartridge.

Q. What is the general characteristics of all those six bullets? A. They are all from 32 calibre automatic cartridges.

Q. What about the jacket? A. Oh, they are jacketed bullets, jacketed. I may explain that?

Q. I wish you would, please. A. A jacketed bullet is one on which we find the lead core covered with a harder metal jacket. In this case, I believe that the jackets on these bullets are known as gilding metal, largely copper, and tin plated.

Q. What is the tin plating for? A. More for appearance than anything else.

Q. To make them look shiny? A. To make them look shiny.

Q. Is that why you see the copper under the tin? Is that why you get that copper tinge through the groove? A. It is,—the plating worn off where they are scored.

Q. How do those bullets differ from, we will say, these five bullets which have been called "revolver bullets" in previous examinations? A. These revolver bullets are not jacketed, that is the chief difference. This lead bullet alone is about the same as the core of the 32 automatic bullet.

Q. For what use are these jacketed bullets designed? A. For automatic pistols.

Q. Have you compared, Captain van Amburgh, No. 3 and its markings, with the markings on the six bullets which you and Captain Proctor fired from that Colt automatic at Lowell on Saturday? A. I have.

Q. Have you compared the shells called the "Fraher" shells, those that are purported to have been found in the street in South Braintree—— A. I have.

Q. —with the shells which you have retained from the bullets fired at Lowell? A. Yes, sir.

[919] Q. Do you notice any similarity between any one or more of those Fraher shells with the shells from the bullets fired by you and Captain Proctor at Lowell? A. There is a difference noticeable. May I compare the six?

Q. What have you there? A. I have the four Fraher shells.

MR. McANARNEY. I don't hear you.

A. I have the four Fraher shells; I would like the six others for comparison.

MR. WILLIAMS. I think they are 33 and 34. 34 are the Winchesters. I am going to write "Winchester" on that, so there will be no mistake, if I may. "Winchester" is on 34, and "Peters" on 33.

Q. My question, Captain, was as to any similarity between any one of the so-called Fraher shells with the shells from the bullets fired by you and Captain Proctor at Lowell? A. There is one of these so-called Fraher shells, the one marked "W.R.A. Co.", meaning Winchester Repeating Arms Company, and three that were fired in Lowell, W.R.A. Co., shells, a very strong similarity.

Q. Will you step down before the jury, and show what that similarity consists of? A. Shall I explain it?

Q. Yes. That is what I want you to do as you go along. A. The indentation is off centre slightly.

Q. Is what? A. Is off-centre slightly in all four.

(The witness leaves the stand, and explains to the jury.)

THE WITNESS. They are about the same diameter and depth, that is the distinguishing mark. The principal distinguishing mark is the same diameter, the uniformity of diameter. The uniformity in diameter.

(The witness returns to the stand.)

Q. And is there any other distinguishing mark which you noticed on those four? A. There is.

Q. And what is that? A. In addition to the similarity in diameter, there is a slight set-back, so-called—A shop term, by the way—which means a slowing back of the metal around the point or end of the firing pin. That is present in the so-called Fraher shell and in the three Winchester shells which were fired in Lowell.

Q. Could you illustrate with the revolver better than with the Colt what you mean by the set-back? If you could not, I wish you would take the Colt and illustrate. A. It would be difficult to illustrate with either the pistol or the revolver.

Q. Well, illustrate with a diagram or anything, Captain, because, I venture to say, all of us do not know what a set-back is. A. I could with a diagram, the idea, perhaps.

(Witness makes a sketch.)

Q. You may step down right in front of the jury, if you will, Captain. If you will step a little farther away, they can all see it better.

(The witness leaves the stand.)

[920] THE WITNESS. A set-back on a primer is a little flowing back of the metal beyond the true surface. Here is the true surface of the metal, that portion of it which is bearing against the breech block.

Q. What is the breech block? A. That is the point from which the firing pin protrudes.

Q. The firing pin comes through the breech block? A. It does.

Q. And strikes the cartridge? A. It does.

Q. Similar to the way my finger is now indicating? A. Yes. The set-back takes place around the firing pin sometimes. It did in this case.

Q. And what causes the set-back? A. Largely, a little opening—In my experience, I have found it to be usually a little opening in the mouth of the firing pin hole. Do I make that clear?

Q. Is that something that occurs in all guns? A. It is not an unusual thing, but it does not occur in all guns.

Q. Does it have anything to do with the Colt revolver or Colt pistol? A. It doesn't happen in all Colt pistols.

Q. Now, how does that set-back evidence itself on the shells? A. A little ridge of metal is showing around the rim of the hole made by the end of the firing pin.

MR. WILLIAMS. Now, one I am showing is the Fraher cartridge, and one a shell fired at Lowell, both Winchesters. Will you notice, gentlemen, the little ridge around the hole of the firing pin.

Q. Now, that you say is caused by the heated metal flowing back, or pressed back—— A. Not heated. The primer metal is unsupported at that point, due to the opening or chamfering of the mouth of the firing pin hole. The internal pressure of the shell exerting itself, as it does, on the primer cup finds itself on the primer cup and drives the metal back. That is the conclusion I have come to.

MR. WILLIAMS. If your Honor please, is there any objection to my just passing along the front of the second line of jurors? I don't know as I ever saw it done.

THE COURT. If the jurors don't object, or counsel on the other side.

MR. MCANARNEY. We can stand it if the jurors can.

Q. Now, Captain, having in mind the similarities of the shells of those cartridges, having in mind your examination of No. 3, your examination of the six bullets fired by you and Captain Proctor at Lowell, have you formed an opinion as to whether or not the No. 3 bullet was fired from the Colt automatic gun which you specifically have in front of you? A. Will you just state that question again please? I don't quite get it all.

Q. You may strike that out, please. Have you formed an opinion, Captain, as to whether or not No. 3 bullet was fired from that particular Colt automatic? A. I have an opinion.

Q. And what is your opinion? A. I am inclined to believe that it was fired, No. 3 bullet was fired, from this Colt automatic pistol.

[921] Q. And when you say "this", you mean the one that you have before you? A. The one I have before me.

Q. Now, what is the basis for your opinion, Captain, or bases? A. My measurements of rifling marks on No. 3 bullet as compared with the width of the impressions which I have taken of No. 3 or of this particular barrel, together with the measurements of the width or dimension of rifling marks in bullets recovered from oiled sawdust in Lowell, inclines me to the belief.

Q. Now, what marks have you observed which occasioned you to have that belief? A. You mean, in addition to the dimensions of rifling marks?

Q. Yes, I mean, are there any peculiarities or irregularities of those bullets which you have observed which assist you to form an opinion? A. There are.

Q. And will you describe them to the jury? A. There are irregularities evidently caused by similar scoring or irregular marks in rifling which appear on all bullets which I have examined that I know have been fired from this one automatic pistol which is before me.

Q. Yes. And what about No. 3? A. No. 3 bullets, I find on No. 3 bullets such evidence of scoring in the barrel. It takes on the bullets the form of a, well, a long streak bordering close on the narrow cut, the land cut, on the bullet.

Q. Is there anything in the barrel of that revolver which can be shown to the jury so they can see it which will help them to understand what you mean by irregularities caused by something in the pistol? A. I believe it can be shown to the jury.

Q. Well, will you show it, if you will, and I will manipulate any lighting apparatus which is necessary. You are at liberty to step down from the witness stand and do anything that is necessary.

(The witness leaves the stand.)

THE WITNESS. It is a difficult matter to point it out. I can indicate it.

MR. WILLIAMS. If you can tell them what you see in there, then, possibly, by looking themselves they will be able to see part at least of what you see.

THE WITNESS. Close to the land which is now on the bottom portion of the bore of the barrel, on the right side as you look in, you will see a rough track.

MR. WILLIAMS. Would a microscope assist?

THE WITNESS. It might.

Mr. WILLIAMS. Mr. Katzmann suggests that the jury be allowed one by one to go to the window and look at the bullet.

THE COURT. I would suggest that the witness go to the window and see if it may be explained better at the window.

THE WITNESS. I believe the light is better.

MR. WILLIAMS. Don't you think that will help?

[922] THE WITNESS. Yes.

MR. WILLIAMS. You can hold it in proper position. The jury may step up one by one.

THE COURT. Explain to the jury what you have in mind, but explain it in such a way that all can hear.

MR. WILLIAMS. Did you hear, Captain, the Judge's suggestion?

(The jurors go one by one to the window to examine the barrel as shown by the witness.)

THE WITNESS. On the bottom portion of the barrel as you look into it, beside that land on the right side of it is a rough track, at the bottom of the barrel.

THE COURT. Would you like some fans, any of you? Oh, you have them, all right.

THE WITNESS. At the bottom of the barrel is a rough track.

THE COURT. Mr. Foreman, you might come right around and be ready, and then follow right around so you will be there.

THE WITNESS. On the botton a rough track. At the bottom of the barrel is a rough track.

(The witness returns to the witness stand.)

Q. Captain, what, in your opinion, has caused that rough track which you have just been showing? A. This appears to be what is generally known as a pit. I really believe that it is caused by allowing powder fowling to stand in the barrel, and the matter of rust allowed to stand and eat its way in, and finally pits occur.

Q. What effect does it have on bullets? A. The bullet has got to drag— If it touches, expands to barrel size, it has got to drag along. It will be scored in travelling over a rough track.

Q. And have you found on those bullets you have been speaking of marks corresponding with the rough track which you have shown to the jury? A. I have.

Q. I notice that you showed to the jury what you designated as "a rough track." Is there more than one rough track in the barrel? A. Yes. There is a general—Yes, I would say there are quite a few streaks of roughness and quite a collection of pits in there, but it seems to be more pronounced in the corner of the groove in the barrel the corner of the groove.

Q. Are there any marks upon the bullets consistent with the tracks which you have seen in the barrel? A. I know of no others, I can think of no others, that I have noticed, for the moment, other than those caused by those rough tracks in the barrel.

Q. And how many of those are there on the bullets? A. There seems to be one which has impressed me very much, one streak along each bullet fired through this exhibit gun—very pronounced. Others are not quite so prominent.

[923] Q. What is the weight of No. 3 bullet? A. No. 3 bullet I found weighed 73.59 grains.

Q. Captain van Amburgh, when did you first consider, at our request, this matter of the firing of No. 3 bullet? A. I don't quite get that.

A. I say, when did you first take up and consider, at the request of the District Attorney, this matter of No. 3 bullet? A. Last December.

Q. At that time I understand you were not allowed to fire the gun because Captain Proctor had no such instructions? A. Exactly, we were not allowed to fire at that time.

Q. You have been up here for a few days. A. Since last Thursday.

Q. And have you been giving it your careful study and attention during this time? A. I have given it my best study and attention.

Q. And is what you are now telling the jury your best and mature judgment on the matter? A. It is.

MR. McANARNEY. Pardon me just one minute.

CROSS-EXAMINATION

Q. (By Mr. McAnarney) You have now been telling us about the pitting of the barrel of a pistol, haven't you? A. I have some knowledge——

Q. That is what you have been telling this jury about. By "pitting" you mean an accumulation in there of rust, fowling and rust in the barrel? In other words, the barrel is permitted to remain uncleaned? A. No, I don't mean that exactly by "pitting".

Q. What do you mean by "pitting"? A. Pitting is the effect of rust. The rust may have been removed, but the pits, or little cavities, will remain in the metal. That is my best explanation.

Q. It has the effect of rust? A. Yes.

Q. And you have found a condition inside of the barrel of that automatic caused by allowing the gun to rust, haven't you? A. It could be caused.

Q. In your opinion, is that not the cause? A. I, for the moment, without knowing the history of the gun, would ascribe the cause of that pitting—would ascribe rust as the cause of the pitting.

Q. You will say it was? A. That is generally what causes pitting in most cases, therefore, I am inclined to believe it was caused by rust.

Q. You believe and you are inclined to believe. Will you tell us just what the situation is? Isn't it that the pitting in that barrel, in your opinion, was caused by rust? Isn't that the situation you are talking about? A. To the best of my knowledge, I would say so.

Q. That is what I want. That is the situation, isn't it? A. To my best knowledge.

[924] Q. You have got a gun right there, and you have been working on it several months, the barrel of which has been permitted to become somewhat rusty. A. It perhaps did.

Q. Isn't it more than "perhaps"? isn't it the fact? A. I believe that it was caused——

Q. You know it was, don't you, by your months of study, that has been rusty? A. I believe it has been caused by rust.

Q. When you say "I believe", have you anything back of that that you don't quite feel sure of? A. Yes, I have a slight reservation. I have known pitting to be in metal when it has come from the mill, and it was due to a flaw or some little imperfection in the metal, and, therefore, I would not say every time I see a pit in such a piece of metal it was caused by rust.

Q. Pardon me, I have not called for all that.

MR. KATZMANN. One moment.

MR. MCANARNEY. If you want to go on, continue.

MR. KATZMANN. Will the stenographer please read the answer?

MR. MCANARNEY. I said he may continue.

MR. KATZMANN. I ask your Honor, please, that the stenographer be allowed to read the answer so that the interruption may not affect the witness.

THE COURT. Can you go right on, Mr. Witness, where you left off.

THE WITNESS. My last phrase, perhaps, is not clear to me now.

THE COURT. Give him the last phrase, Mr. Stenographer.

(The answer is read as follows: "And, therefore, I would not say every time I see a pit in such a piece of metal it was caused by rust.")

THE WITNESS. That, I believe, explains my reservation, mental reservation, on the matter of rust.

Q. Have you formed an opinion as to what caused the condition in the barrel of that gun? A. To the best of my knowledge and belief, it was caused by rust.

Q. And it is rare, indeed, that you get a revolver a year or two old that has not got a condition in the barrel caused by rust? That is a rare thing to find, isn't it? A. I would not say rare.

Q. Well, you usually do find that there is more or less signs of the barrell's having become rusty, don't you? A. I would say it is quite common.

Q. It is a common thing, isn't it? A. It is common, yes.

Q. Now, in this particular gun, you found one particular place where there seems to be a little more the effect of rust than in any other place, haven't you? A. I have.

Q. So, you have described the land and groove. And speaking of the gun, it is in the groove of the gun, is it not? A. In the groove of the gun, yes, sir.

[925] Q. In where the groove makes up against the shoulder of the land? A. That is the place, the junction of the groove—we may say, the corner of the groove, the junction of the groove and the land.

Q. Now, the centre of the groove might be more apt to be wiped off by the firing of the gun? A. It perhaps might be.

Q. That being the fact, the centre place would be wiped off by the discharge of the bullet, and you are more apt to get the accumulation at the shoulder where the groove and the land make together? A. In my experience I have found such invariably the case.

Q. And that is the case in this case, isn't it? A. That appears to be it.

Q. So now, we have a condition brought about by rust in the usual place where rust accumulates in a gun under ordinary conditions, that is right, isn't it? Is that right? A. I believe that is right.

Q. I show you, marked for identification, "15", a series of pictures.

(Mr. McAnarney shows pictures to the witness.)

Q. (Continued) Would you glance at those, if looking the several ones over might assist you to see what they are. A. Is there a question?

Q. I said, you might look at the six so as to get familiar with what I am going to ask you. I might say, if it will assist you, Mr. Witness, those purport to be six different views of the same six bullets. You probably know that from your examination. Now, I call your attention——

MR. KATZMANN. Mr. McAnarney, do you mean 36 pictures?

MR. McANARNEY. I didn't hear you.

MR. KATZMANN. I didn't quite understand your explanation. Do you mean 36 pictures of the same bullet?

MR. McANARNEY. No, six pictures.

MR. KATZMANN. Each picture represents six views of the same bullet?

MR. McANARNEY. Yes. Each picture is a different bullet. Six bullets in six different positions. Six different bullets and six different pictures.

Q. Now, calling your attention to those numbered 1, 3 and 4 on the pictures—You may look at the whole set, but examine 1, 3 and 4 carefully. A. I have examined them.

Q. Now, are you able to form an opinion as to what kind of bullets those are, from what shell? A. Not what kind of bullets they are, no, sir.

Q. Or from what shell they came from? A. No, sir, I wouldn't be able to form an opinion from the pictures, any more than that they are—No, I can't form an opinion as to classification; I can't classify them as to calibre or anything of that sort.

Q. Well, whether or not they appear to be from the same, never mind what it is, but do they appear as though they were from the same cartridge, whether they would be a Winchester or some other? Could you form an opinion on that? A. I would say that there are——

[926] Q. One, 3 and 4, your interrogation is? A. 1, 3 and 4 hold many points of similarity as to construction—there is nothing distinctive or outstanding to indicate that 1, 3 and 4 here, say, are of different makes.

Q. Or whether the same make. Your position is, you can't tell from that photograph? A. I am not certain. A photograph is not——

Q. Your position at this minute is, you are not able to determine from these? A. Not from the photographs.

Q. Now, what did I understand you to say with reference to the Savage pistol or the Savage—did you say anything whatever about the Savage gun? A. The Savage gun?

Q. Yes. A. I don't recall what I said.

Q. Did you do any work on that? A. On the Savage?

Q. Yes. A. I have done some firing of the Savage.

Q. Then, it is understood in your remarks here this afternoon you have said nothing about the Savage gun? A. So far, I believe—I don't recall I mentioned the word "Savage."

Q. Now, coming back, you spoke about the Fraher bullets, or shells from the tests at Lowell last Saturday. When you started your answer, I didn't quite get the meaning of it. "There is a difference noticeable". That is, between the Fraher shells and the Lowell shells? Did you use such words as that, or did you have anything like that in your mind? A. The difference in shells which I must have had in mind—I have not got the entire answer, but the only difference I have noted in shells is a difference in the size of the firing pin indentation.

Q. So that you do notice the difference in the pin indentation in the Fraher shells and the— A. Pardon me. What is the name of those exhibit shells?

Q. Fraher, F-r-a-h-e-r, I think, or "a-i-r".

A. Fraher. Well, among the Fraher shells—May I explain it?

Q. Yes, please. A. There were four all told. Three of them are similar in point of indentation in primer, and one is alone of its kind. Of those fired in Lowell, they agree very closely with that one.

Q. What is the diameter of the Colt automatic 38 firing point? A. I have no factory figures.

Q. You have no what? A. I have no factory figures, that is, standard figures as would be used in the factory, but I have figures of my own. As I recall, the last measurement of it I did not record. I rather not trust to my memory to give you the figures.

Q. Have you any idea what is the diameter? A. About .075.

Q. About .075? A. That is my impression.

Q. How much do you allow on that— A. I am trusting to memory. I haven't any notes with such a figure.

Q. I don't care what you haven't got, but what have you in your mind now, if anything, as to your best judgment as to the diameter of a [927] firing pin of a 32 Colt automatic? A. I would say it would be in the neighborhood of a—The outside diameter, I think, ought to be around .075.

Q. Ought to be? A. Ought to be.

Q. How about the Savage, the size of that? A. The Savage is somewhat near it,—perhaps about 70/1000.

Q. Is that your best judgment on that? A. That is my best judgment on that. I have nothing accurate on that point.

Q. Would it assist you any to look at these shells? A. No, I think not.

Q. Did your official duties require you at any time to deal with a 32 automatic, and if so, when? A. My official duties have required me to handle a 32 automatic.

Q. When? A. Well, on many occasions.

Q. Well, just going back to the last time that you can recall in furtherance of your duty in your work that you handled a 32 automatic? A. Well, probably six weeks before coming to this case, before coming here.

Q. In what connection was that? A. Observing tests of amunition of that calibre.

Q. Where? A. In Bridgeport.

Q. In Bridgeport? A. Yes.

Q. Have you in mind in the last year doing anything other than that with a 32? A. I have made several tests in the past year with the 32 automatic.

Q. With reference to the firing pin? A. That is always observed in any of our experimental firing.

Q. Now, what is the rate of velocity of the twist in a 32 Colt? A. The velocity of twist?

Q. Yes? A. I don't get the meaning.

Q. Why, your twist imparts— The rifling of your barrel, the lands and grooves, affect a twist of the bullet, don't they. A. Yes.

Q. Now, what is, if you know, the rate of velocity of twist imparted to the bullet in the 32 automatic? A. Oh, the velocity of rotation?

Q. Yes. A. That is a matter of computation.

Q. Have you ever computed it? A. I have.

Q. What is your knowledge of the 32 automatic? A. I have no notes.

Q. Can you tell us off-hand anything approximately? A. No, I wouldn't venture an opinion.

Q. None whatever? None whatever. Well, now, you spoke something about the firing point, a little opening, that it is not an unusual thing. You spoke about something around where the pin strikes on the cap, and you have spoken about that rust. Have you any of the bullets that were taken as a result of the Lowell experiment? A. Have I?

Q. With you? A. I have.

Q. May I have one, please?

[928] (The witness produces two bullets.)

A. These are some.

Q. Take one, any one. A. This is an odd shell I just pulled out of my pocket just now.

Q. Take the bullet. Now, the bullet you have in your hand, that is one of the bullets that was fired from the Colt automatic before you in the test last Saturday, am I right? A. It is.

Q. Now, looking at that bullet in your hand, any way that you want to answer this question, is that not a pretty clean bullet, and is not the lands and grooves remarkably well defined? A. They are not perfect.

Q. Did I ask you that question? You are an instructor, are you not, some place?

MR. KATZMANN. Wait a minute. One question at a time. Which one are you asking?

MR. McANARNEY. I am putting the last one.

MR. KATZMANN. You withdraw the other one.

Q. You are an instructor——

MR. KATZMANN. I object.

THE COURT. Do you withdraw the other one.

MR. McANARNEY. Yes, if your Honor please.

Q. You are an instructor, are you? A. I rate as an instructor.

Q. You do instruct someone? A. I do at times.

MR. McANARNEY. Kindly read the last question.

(The question is read as follows: "Now, looking at that bullet in your hand, any way you want to answer this question, is that not a pretty clean bullet, and is not the lands and grooves remarkably well defined?"

MR. KATZMANN. I object to the question, if your Honor please.

THE COURT. Read it again.

(The question is again read.)

THE COURT. Leave out, "in any way you want to answer it."

(The question is read as follows: "Now, looking at that bullet in your hand, is that not a pretty clean bullet, and is not the lands and grooves remarkably well defined?")

THE COURT. You may answer.

MR. KATZMANN: I object.

THE COURT: You may answer.

MR. KATZMANN. If your Honor please, may I be heard?

THE COURT. You may.

MR. KATZMANN. There are two questions, "is it not a pretty clean bullet," and "are not the lands and grooves remarkably well defined," is another question.

MR. MCANARNEY. It seems to me they go pretty near together.

Q. You understand that question, don't you?

[929] THE COURT. If there are two questions there, Mr. Witness, answer them, if they are all included in one question, you may answer them as such.

A. It is a clean bullet.

Q. Now, the last part of it? A. The lands and grooves show very well.

Q. Now, I have not a chance of looking into that barrel that you showed the jurymen. May I see it, please?

(The witness hands the barrel to Mr. McAnarney.)

Q. There are quite a few pit marks up through on the grooves there, are there not? A. Large and small, I would say that there are some.

Q. Quite a few, aren't there? A. Well, quite a number. I wouldn't say how many, of course.

Q. Are you able to say—I don't think I have asked you—from your observation of this gun, whether it is a gun that had been used much? A. A difficult question to answer.

Q. Give us your best judgment, Mister. What is the answer, please? A. Judged by my own standards of the use of a gun, I would say it had not been used a great deal.

(Mr. McAnarney hands a cast to the witness.)

Q. I show you for examination that cast. Are you familiar with the use of those casts, and how they are made? A. I am.

Q. Examining that, what does it appear to be?

MR. KATZMANN. Mr. McAnarney, we can't hear you over here.

Q. Examining that exhibit, what does it appear to be? A. A cast of what is probably sulphur.

Q. It appears to be of what calibre gun? A. Well, a casual glance of course, would not be enough to make up one's mind thoroughly.

Q. What would be your best judgment? A. May I ask which is the muzzle portion of it? Q. Are you able to determine from the appearance of it yourself? If not, I will have to ask somebody else. A. Yes.

Q. Can you tell from that which it is? A. Yes, I have it.

Q. Now, looking at that cast and comparing that cast with the inside of the revolver that you have—I don't mean the revolver, the pistol—how do they compare? A. For appearance?

Q. Yes.

THE COURT. You may have a recess, gentlemen, now, for five minutes.

(Short recess.)

Q. Have you a cast there that I showed you? A. I have.

Q. From your examination of that cast and your examination of the rifling on the Colt automatic before you, whether or not you are able to affirm that this is a cast of the inside of that gun? A. I believe that it is.

Q. And whether or not it fairly represents the inside of that gun. A. In appearance it brings out some of the defects of that gun.

[930] Q. Well, does it not appear to be a fair representation of the inside of the barrel of that gun? A. To some extent, I think, it is a fair representation.

Q. Well, to some extent that it is not a fair representation? A. It is— May I qualify that with an explanation?

Q. Yes, you may. A. It differs from being a good or a fair representation due to being chipped slightly.

Q. Is that chipping caused by imperfections in the barrel? A. I think not.

Q. Are you quite sure of that? A. It is my opinion that little flakes have been knocked off from there that were, well, just incident to the delicacy of the cast and some little handling through which it went, possibly.

Q. Well, other than that as you describe it, isn't it a fair representation of the inside of the barrel? A. Other than that, I would say that it is a fair representation, other than that.

Q. And the taking of casts as of the manner of that one is one way of getting the condition of the inside of the gun, isn't it? A. It is one way of bringing out things that cannot be seen by glancing through.

Q. And that is about the best way, isn't it? A. I know of no better way than to take a cast.

Q. Did you measure that automatic gun? A. Did I measure the automatic?

Q. Yes. A. In what respect?

Q. What are lands, the diameter? What are the diameters of the lands in that Colt automatic? A. Diameter of the lands, do I get you clearly?

Q. Well, the width of the lands? A. The width of the lands? From measurements made, I believe the width of the lands to be about 60/1000.

Q. About 60/1000. Well, now, you made those measurements yourself, didn't you? A. I did.

Q. And that is no approximation, is it? That is an accurate measurement made by what? What instruments? A. Accurate by my standards.

Q. Well, your standard is an accurate one? Your standard, I say, is an accurate one? A. My standard is an accurate one?

Q. Yes, isn't it? A. It is a matter of opinion.

Q. When I said, "Is it accurate" you said "by your standard." Now, I say to you, your standard is an accurate one, isn't it? A. As accurate as it is possible for me to bring it to.

Q. Now, what instrument did you measure that with? A. I used a scale, dividers, micrometer and depth gauges.

Q. And you say you got the width—what is the width of the lands on that 32? A. 60/1000.

Q. What is the width of the grooves? A. What is the width of the grooves?

[931] Q. Yes.

MR. KATZMANN. Do you mean of the grooves in the barrel or the bullet?

MR. MCANARNEY. I haven't said anything about a bullet yet.

MR. KATZMANN. The witness has been speaking of two separate grooves. Do you mean the groove of the barrel?

Q. You don't understand I mean the width of the groove of the bullet?

MR. KATZMANN. I am objecting.

THE COURT. You may answer the question.

Q. You understand I am talking about the gun? A. My understanding is, you mean the barrel.

Q. Now, what is the width of the groove? A. 107/1000 of an inch.

Q. What is the diameter of the groove? A. The diameter of the groove, .3115.

Q. What is the diameter of the lands? A. I didn't get the diameter of the lands or bore measure.

Q. Well, did you overlook that? A. I overlooked it. I didn't consider it important, because——

Q. The lands in your gun make the grooves upon the bullet, don't they? A. It makes the little groove on the bullet.

Q. Your land in your gun makes the groove on the bullet? A. It does make the groove on the bullet.

Q. So that the bullet you have been talking about and giving dimensions of, you have not ascertained the width of the land of that Colt? A. The width?

Q. The diameter of the land. Have you or have you not? I understood you to say you did not. You say you overlooked it, is that right? A. I didn't measure the diameter between lands.

Q. You did not? A. The diameter of my circle, do I get you clearly?

Q. Yes. A. I didn't take that measurement.

Q. Give us the diameter of the lands of the bullets fired at Lowell? A. Do I understand now you mean the lands of the bullets representing the grooves in the barrel?

Q. Yes, that is right. A. I made them to be across their greatest diameter about .311.

Q. What was the width of the land? A. I made the width to be about .107.

Q. .107 the width of the land? A. The width of the land on the bullet,— the groove of the barrel.

Q. Are you quite right about that? A. .107.

Q. Now, take the groove— A. May I ask a question?

Q. Yes. A. Are you referring to the firing gun in Lowell and this exhibit gun? Q. The test was made with the exhibit gun? A. Yes.

[932] Q. You are talking about the bullets fired from that gun? A. Yes, that is it.

Q. Now, come to the diameter of the groove. A. The diameter of the grooves,—do you mean the diameter of the circle?

Q. Yes,—the diameter of the groove. A. I measured the width of the grooves only.

Q. You didn't make two measurements? A. No, I didn't make two measurements.

Q. What did you make the width of the grooves? A. The width of the grooves approximately 60/1000.

Q. Have you that measure in your notes? A. No, sir, I haven't the accurate figures in my notes.

Q. You have been rolling the pages over. Is there anything in your notes that shows the width of the grooves of the bullets? A. There is no special notes.

Q. The question is, is there anything in your notebook that you have been turning over while I have been asking my questions that shows the width of the groove? A. There are figures that gives me a key to those dimensions.

Q. There are figures that give you a key to those dimensions? A. Yes, sir.

Q. What figures are they? A. They are headed "Pistol."

Q. Headed "Pistol". What else is there that gives you a key to the width of that groove? A. My remembrance of what that measured.

Q. Well then, you said there were figures there that gave you a key, and on there you say there is what? What is there on that book, I say? A. Under the heading of "Colt automatic pistol".

Q. What is there that gives you a key? A. There are the figures.

Q. What figures are there that give you the key to the width of the groove? A. The exact figure is there.

Q. Then, you haven't understood my previous question. Do you say now that you have in your book there the exact width of the groove? A. I have a figure here which represents what I found of those bullets.

Q. Have you a figure there of what you found the width of that groove to be? A. I have that figure.

Q. What is it? A. .107.

Q. Will you kindly show me that, show me the .107? A. Right here (indicating book).

Q. Now, this is under the heading of "Pistol"? A. Under the heading "Pistol," yes, sir.

Q. Now, will you read what you have there? A. "Pistol, Colt automatic, calibre 32. Groove diameter .3115, and six grooves. Left twist. Eight-shot magazine. Grooves .107 in width. Lands 60/1000."

Q. Now, did you measure that with an instrument when you got that width of that groove? Are you prepared to say that the width of [933] that groove is not .112? A. From my own measurements I would say it is not .112.

Q. Well, now, you measured it, didn't you? A. I did.

Q. And you are now testifying in regard to that measurement, and you say that the width of that groove is not .112? A. According to my measurement, it is not .112.

Q. Your measurement is made for the purpose of ascertaining that fact, isn't it? A. It is.

Q. I haven't asked you.—Do you know of any left-hand twist .32 automatic other than a Colt? A. I haven't met up with any in my experience.

Q. Do you know of any? Well, you are familiar with the 765 mm. shell, aren't you? A. I have handled quite a few of them.

Q. That is the same dimension as the 32 calibre Colt? A. The 765 mm.——

Q. Pardon me, is that the same diameter as the 32 Colt? A. Are you referring to pistol or cartridge?

Q. I believe you used the word "Colt". A. We speak of them in the vernacular as the same, 32 revolver, 32 shell, 32 cartridge. Those are the same, are they not? A. The 765 mm. as made at our own factory is the same as the 32 "A.C.P." By "A.C.P." I mean, "Automatic Colt Pistol." That is our abbreviation.

Q. Now, as to the Savage gun, do you know of any other gun the same width of land and groove as the Savage? A. I haven't observed any.

Q. Are you acquainted with the Steyr gun? A. I am not familiar with it.

Q. Are you familiar with the Bayard automatic? A. I have heard of it, but have not found any. Those are foreign guns you have mentioned.

Q. Don't you know there were a lot brought back from overseas? A. I haven't seen any.

Q. Did your work in connection with the Army take you across? A. No, sir.

Q. Are you acquainted with the Stauer? A. I have not handled a 32 Stauer.

Q. You have handled others that were not 32? A. Not the 32.

MR. MCANARNEY. I think that is all, Mr. van Amburgh.

THE COURT. Anything in re-direct?

RE-DIRECT-EXAMINATION

Q. (By Mr. Williams) Why could you not tell the kinds and make of bullets from the photographs which were shown you, Mr. van Amburgh? A. The photograph does not enable me to visualize an article that I may be familiar with, that is to say, in any detail as fine as here on those bullets. May I qualify that? It is necessary to recognize certain small details on [934] bullets to form a proper judgment, and are sometimes not present in photographs.

Q. Do you know of any gun which combines a left twist with lands of .061 in width? A. I don't know of any.

Q. For the purposes of the record, can you describe, with reference to any mark on that revolver, where this particular pitting is located which makes or has made the irregularities on the bullets you have described? A. I think it is brought out on this cast.

Q. I was coming to that. Can you show on that cast which has been submitted to you by Mr. McAnarney, the irregularity of the pitting which you think has caused another irregularity on the bullets? A. I can.

Q. Well, if you can, will you step down in front of the jury. You can step right up to the window, if that will help you. I know His Honor will accommodate you.

(The witness goes to the window, where he shows the cast to the jury-men.)

MR. MCANARNEY. While the jury are looking at that I would like to have it marked for identification. We desire to offer it later.

THE COURT. That may be done when the witness is through with it. Of course, I assume you will follow the usual rule by showing that is, in fact, a cast of the pistol in question.

MR. MCANARNEY. Yes, if your Honor please, I will offer that.

(The cast is placed in an envelope marked Exhibit 17 for identification.)

MR. WILLIAMS. Mr. Katzmann would like to see that.

(The witness shows the cast to Mr. Katzmann, and then returns to the stand.)

Q. Mr. McAnarney asked you what you had said in regard to a Savage, and you said you didn't recall. Do you remember testifying in reply to my inquiry about the five bullets from the Savage? A. What was the question? I would answer it again.

Q. I don't know as it is of any importance. Do you recall testifying in regard to bullets 1, 2, 4, 5? A. I recall some testimony about them.

Q. And that was in respect to Savage, as I remember? A. It was mentioned, I believe.

Q. Will you look at the six shells which you have from the Lowell firing, three of which, if I recall, are three Peters and three Winchesters,—am I right? A. Three Peters and three Winchesters.

Q. All three having been fired from the same gun. Is there a difference, in your opinion, in the appearance of the dents in the three Peters and the dents in the three Winchesters? A. There is a difference in appearance.

Q. And, as we assume they are made by the same firing pin, will you tell the jury what causes that difference in appearance in the dents? A. [935] In my opinion the appearance in the indentation could be caused, one, by a difference in the chamber pressure generated by the powder charge, bearing in mind they are two different makes, and, secondly by a difference in thickness of the metal in the primer cap, or the metal of which the primer is made. A heavier would not respond to the blow of the striker.

Q. You find evidence of what you call a set-back on the three Peters? A. On the three Winchesters especially.

Q. No. My question is this——

MR. MCANARNEY. Pardon me. You don't mean to contradict your own witness.

MR. WILLIAMS. No. He didn't understand my question, that is all.

Q. My question is, do you find evidence of a set-back on the three Peters shells? A. No. There is no evidence of set-back on the three Peters.

Q. And why does the evidence of the set-back occur on the Winchesters and not on the Peters? A. In my opinion, there is perhaps more pressure in those Winchesters which caused the primer metal to set back with greater force.

Q. Do you recall my friend Mr. McAnarney asking you about the velocity of rotation? A. I recall his question.

Q. Did you ever hear the term before? A. Oh, yes.

Q. What affects the velocity of rotation? A. Well, the most direct influence would be the muzzle velocity, that is, the velocity at which the bullet is travelling.

Q. And what does that depend upon? A. That depends upon the pressure, and it depends upon the velocity of the cartridge.

Q. Then, can you answer a question as to what the velocity of rotation of any particular gun is without knowing what the powder charge of the bullet is? A. That is a matter of computation, a figure which is not published in any handy form, and, therefore it is one which I would not, at any rate, have at hand.

Q. Now, I find I have omitted to put in the three bullets which you shot and which I understand Captain Proctor and you have in your possession. Are these the bullets shot by you? A. They are the three bullets fired at Lowell.

Q. Those are what make? A. Those are the Peters.

Q. Captain Proctor fired the Winchesters? A. Captain Proctor fired the Winchesters.

MR. WILLIAMS. I offer these three bullets.

THE COURT. They may be admitted.

MR. McANARNEY. Are they Peters?

MR. WILLIAMS. He says they are.

[936] Q. Any private mark on them, Captain? A. There are no private marks on them.

(Mr. Williams shows the shells to counsel for the defendants.)

THE COURT. Your next question, Mr. Williams. I mean, that you should go on with the question and allow the examination to go on while— Kindly allow the gentlemen to have them.

MR. WILLIAMS. To have them back?

THE COURT. Yes. He gave it to you because of my suggestion. You may make an examination, and you go on with some additional questions, if you have any.

MR. WILLIAMS. I don't believe I have any more, if your Honor please.

THE COURT. Well, if that is the last, we will wait.

Q. What do you understand by "diameter of the lands of a revolver?" A. It is an indefinite term, although I believe I know what is meant.

Q. What do you think was meant by Mr. McAnarney? A. The diameter of the bore, being the measurements from one land to its opposite land.

Q. Now, was it of any importance to you in making your observations here to get the diameter of the lands? A. I felt it was of no importance.

Q. It would not have helped you in any way? A. No.

Q. Will you mark this——

THE COURT. Put them in an envelope, and have them marked.

(The three Peters bullets are placed in an envelope and marked Exhibit 36.)

THE COURT. Have the other three been put in?

MR. WILLIAMS. The other three have been put in, your Honor.

THE COURT. Does that finish the re-direct? Any cross-examination, Mr. McAnarney?

MR. McANARNEY. Just one moment.

Q. (By Mr. McAnarney) In speaking of those automatic bullets, you spoke of their being plated. You didn't mean that, did you? They are simply washed, isn't that all? A. I referred to that as plating.

Q. Plating is something different? A. It has a meaning of electroplating, or something.

Q. These bullets were simply washed? A. These bullets were simply washed, those I noticed.

Q. You used the word "plated"; I didn't know whether you used it advisedly. A. I meant a thin covering.

MR. MCANARNEY. That is all.

The testimony given by the experts [8] for the commonwealth was incompetent; it was not based upon a reliable method of proof. It fell far short of proving that either bullet No. III or the W.R.A. Fraher cartridge case had been fired in Sacco's pistol. The individual peculiarities of the Sacco pistol, with respect to the markings on bullets and cartridge cases fired therein, were neither established nor contemplated by the testimony. The experts did not use any proper instruments and, on the contrary, a low-powered magnifying glass was represented to the jury as being an adequate means of magnification. It was thought that a reliable method of making comparisons was first to examine the surface of one object under the glass, and then to examine the surface of another object under the glass, and thereby ascertain whether the same markings appeared upon the surfaces of both objects. The experts for the commonwealth made crude, disagreeing measurements, and neither of them knew of an automatic, other than a Colt, having a left-hand twist. They were "gun-shy" of photographs (yet photographs were exhibited to the more inexperienced jurors).

The lack of qualification of the first witness for the commonwealth was made glaringly evident by his failure to distinguish between class and accidental characteristics; his assumption that pushing a bullet through a barrel produced the same effect as firing the bullet; his unfamiliarity with firearms and ammunition; and his failure to recognize any distinguishing marks upon the surfaces of either bullet No. III or the Fraher cartridge cases. Unquestionably his opinions were useless. His redeeming quality was an apparent frankness.

He testified that the No. III bullet was fired from a Colt automatic pistol, the basis of the opinion being the direction of the twist of the rifling and the width of the land mark; the pitch of the rifling

[8] The proper function of an expert witness and the theory upon which his opinions are admissible in evidence are discussed in Chapter III, *infra*, page 247.

was not mentioned. He did not know whether there were any other automatic pistols which would produce the same width of land mark with a different pitch of rifling.

The basis of his opinion, that bullet No. III was fired from the Colt automatic pistol in evidence, was not developed. It did not appear that he was reasonably sure of any accidental characteristics which would tend to affirm or refute his opinion of "consistency." As a basis for the opinion that the other five fatal bullets were fired from the same Savage automatic pistol, there was no adequate foundation. He said: "There is marks there other than what it would make if it stayed in the lands all the time. It would make six cuts there. You see there are a whole lot of cuts there." He was unable to explain the causes of the "irregularity," and he referred to it in the unscientific language of "jumped the rifling."

The question of the Fraher cartridge cases was handled in the same unsatisfactory manner by this witness. The similarity between the W.R.A. Fraher cartridge case and the six test cartridge cases fired in Sacco's Colt was based on "the looks of the hole in the primer." Research has demonstrated the weakness of the position and size of the indentation in the primer cup as differentiating features in themselves. No mention was made of the impression of the breeching face or the extractor and ejector marks. The first expert for the prosecution endeavored to establish a similarity between the W.R.A. Fraher cartridge case, the three Peters test cartridge cases, and the three W.R.A. test cartridge cases. The other experts were influenced by the "flow-back," which did not occur on the three Peters test cartridge cases, so they only identified the W.R.A. test cartridge cases with the W.R.A. Fraher cartridge cases.

The second expert presented by the state was experienced in the field of testing arms and ammunition. Consequently he exhibited a knowledge of the component parts of arms and ammunition but he missed making a demonstration of expertness in firearms identification by a magnificent margin. His testimony was subject to the same limitations as that of the previous expert with the exception that he pointed out additional elements in the signatures on the bullets and cartridge cases, but these additional elements did not throw any substantial light on the individual peculiarities of the Sacco pistol.

He called attention to evidence of scoring in the barrel: "It takes on the bullets the form of a, well, a long streak bordering close on the narrow cut, the land cut, in the bullet." Such a mark was not an individual peculiarity, and there was no evidence to show that this element appeared in approximately the same relative position on the test and fatal bullet. In this connection a sulphur cast of the inside of the barrel of the Sacco Colt was introduced in evidence to illustrate the contour of the surface. A sulphur cast has a definite limitation because such cast might be made to show the condition of the bore at the breech or muzzle, but it must be made in such a short length that it will drop out of the barrel upon cooling, on account of shrinkage. If the cast sticks and has to be pushed in being removed, it is useless, as the surface of the cast will be scarified. The conclusion is forced that the court erred in allowing the experts to use the cast for any purpose but that of showing the condition of the barrel in the proximity of the breech or muzzle. Since the experts presumably knew more about the practical utility of the cast than did the court, the blame was upon any and all experts making use thereof. They ought to have known that their use of the cast was an inaccurate, misleading representation.

The second expert for the state also mentioned the ''flow-back'' on the primers of the W.R.A. Fraher cartridge case and the W.R.A. test cartridge cases. This ''set-back . . . which means a slowing back of the metal around the point or end of the firing pin'' depended upon the relative hardness of the primer metal or excessive pressure, and consequently was not even the effect of an accidental characteristic of the Sacco pistol.

It should be noted again that bullet No. III was stamped with a ''W'' indicating Winchester manufacture, and that it was the only bullet with a left-hand twist of rifling and therefore the only bullet that could possibly have been fired from Sacco's pistol. Only one of the four Fraher cartridge cases was of Winchester manufacture, so it was but natural that the experts for the commonwealth should try to prove that bullet No. III of Winchester manufacture and the Fraher cartridge case of Winchester manufacture had both been fired from Sacco's pistol.

The testimony given at the trial by the two experts for the defense was as follows:[9]

[1405]   JAMES E. BURNS, SWORN

MR. JEREMIAH MCANARNEY. This witness is a little hard of hearing. I may have to talk loud. If you do not hear any question I ask or the District Attorney asks, tell us and we will repeat it.

THE WITNESS. I will.

[9] The entire testimony of the witness Burns is not reprinted here. His testimony was voluminous, and parts thereof lack substance.

Q. (By Mr. Jeremiah McAnarney) What is your full name? A. James
E. Burns.

Q. What is your present occupation? A. Ballastic engineer, United
States Cartridge Company.

Q. How long have you been working for the United States Cartridge Company in that capacity? A. Thirty years.

Q. What has been your experience with reference to firearms? Perhaps
you had better go back and pick it up, that is. Give us your life history,
in so far as you are connected with firearms. A. Why, with different—all
makes of ammunition.

Q. When did you begin,—Thirty years ago, you said. A. Thirty years
ago I began working for the Cartridge Company. I have been in this business for 30 years.

Q. For whom did you go to work first? A. In the munitions business?

Q. Yes. A. The United States Cartridge Company.

Q. Have you been with them all this time? A. Yes.

Q. So that for thirty years you have been with this same company?
A. Yes.

Q. Now, have you had outside experience other than working for the
Company? A. Why, naturally I am, I like shooting and I have followed
the rifle game in the militia. I have been in the militia eighteen years, followed the rifle game, shot on the Massachusetts team and won distinguished
marksmanship in the United States with the rifle; with the shotgun, an expert,
and shot on the Eastern team against the West in 1893 and beat them.
Pistol,—I won the championship of Massachusetts.

[1406] Q. Well, I realize you are talking about yourself, but we want
your experience. A. I do this and go into this so I can feel the pulse of the
shooter and, you know what he wants. A man advertising ammunition has
got to know what the shooter wants in order to develop the goods to be
satisfactory. That is the idea of going into the shooting game.

Q. So that during your experience during these thirty years in the factory you have kept in touch with the actual use of the various weapons?
A. I have.

Q. In competition? A. I have. Because that is the only way to know.
You can make ammunition to suit yourself, but the other thing is to make
it suit the other fellow, the fellow who is going to buy it, and the only way
you can do that is to feel his pulse, get out and shoot with them.

Q. Now, Mr. Burns, you have been called upon to make an investigation
with reference to the bullets and ammunition that have been produced in this
case? A. I did not get you.

Q. You have made a study of the bullets that were found in connection
with this case, have you not? A. Yes, sir.

Q. Now, I wish you would turn to any data that you have. I will refer
you to this if you need it, as you go along (indicating). Have you that slip
I had the other day, showing the different kinds of right and left twist? A.
You have it there (indicating.)

Q. Before I go into your examination I will ask first,——

MR. JEREMIAH MCANARNEY. Let me have the revolvers please, the revolver and the pistol.

THE COURT. The sheriff has gone into town.

MR. JEREMIAH MCANARNEY. I will pass that by.

THE COURT. What time do you expect him back?

A COURT OFFICER. Expect him back at three o'clock.

Q. Now, I will take up,—you have examined the bullets, have you? A. I have.

Q. That have been introduced as exhibits? A. Yes, sir.

Q. What bullets did you examine? A. I examined six bullets.

Q. I call the witness' attention to "Identification 15". What is that? A. That is a photograph of the six bullets that were called the murder bullets.

Q. The bullets that were taken from the body of the men? A. From the body of the men, as I understood.

Q. Now, I will connect that later. Now, will you turn to your notes and describe what those bullets are? A. Bullet scratched number "I" on base, made by the Peters Cartridge Company; weight was 72.3 grains; lands .040 wide. The diameter of the lands 297 to 302½. That is, thousandths of an inch. The grooves——

MR. KATZMANN. Wait a minute.

THE WITNESS. That is, the imprint that the lands made on the bullet. You might call it the groove on the bullet. We don't. We call it the [1407] land mark on the bullet. The groove, the width was 125. The diameter was .3075. The length of the bullet was .465. There were six grooves and lands on the bullet. Fifty-seven knurl marks in the cannon lure.

MR. KATZMANN. I did not get it.

THE WITNESS. Fifty-seven knurl marks in the cannon lure. The cannon lure was .035 of an inch wide.

Q. That is No. 1? A. That is No. 1 here (indicating), yes, sir.

Q. Now, you have there,—that we may understand these photographs what are those different pictures representing? A. This here (indicating) represents the direct view of one land. The bullet was turned around so that this land on the right showed to the front in the next picture and the same all the way through so you get a complete full photograph, front view of every land that there is on the bullet.

Q. Will you show that to the jurors so they can see it? A. (Witness does so.) Perhaps I better illustrate with the bullet.

Q. Using the two together. A. This bullet,—all photographed as they are. This photograph No. 2, this bullet was turned to the right, so that this lands here (indicating) shows here (indicating), and this one here (indicating) the same way all along the line. Now, the next one, here is this lands here (indicating.) I will use the second bullet so that they can see it better. No. 2 here in this picture here (indicating). This lands here (indicating) shows here (indicating). Do you understand it? No. 3, this lands here (indicating) shows here (indicating). No. 4, this lands around here to the left (indicating) turns to the right.

THE COURT. Can't hear.

Q. Keep your voice up; can't hear you. A. This lands here (indicating) shows here (indicating). Turns to the right. This lands here (indicating) on No. 5 is turned to here (indicating), and that makes a complete circle of the bullet, showing the lands on the front of each and every bullet as you go along.

Q. If you will step around and show it to the gentlemen so they may get the idea. You may walk up here (indicating). A. (Witness goes to rear of jury). One of these is coming apart. Hadn't I better mark them so we will get each view?

Q. You better mark it correctly. A. Yes. (Witness does so.) This photograph was taken with the bullets with one lands to the front all along, No. 1, 2, 3, 4, 5 and X. Just as they were scratched on the base. The next picture here (indicating) you hold that, please. The next picture shows this land here (indicating) turned to the right. In other words, we turned the bullets just 1/6 of a turn to the right so that the next land showed to the front. That was done on each and every bullet of the six, and the same way all through. This here is where a deformity is, just a sixth of a turn for each and every bullet. Do you understand?

THE COURT. You must not ask them any questions.

MR. JEREMIAH MCANARNEY. That will be all right for the present.

[1408] (The witness returns to the witness stand).

Q. So that the photographs you have there show the photographs of the six bullets that were taken from the body of the deceased? A. Yes, sir.

Q. And shows them so that you get an exposure of every land and every groove of every bullet? A. Yes, sir.

Q. Now, I direct your attention—you have given us one. Now, will you give us No. II. A. No. II bullet, U.M.C. bullet, manufactured by the Remington Union Metallic Cartridge Company. The weight is 70.6 grains. The width of the lands, .040. Diameter of the lands .305, .302 and .302; taken in three measurements. The width of the grooves is .125. The diameter was 309, 310 and 310, measuring at 3 points. The length of the bullet was .449 of an inch, six lands. There was no cannon lure or no knurling,—smooth bullet.

Q. You have given them all now? A. No, that is the second.

Q. All right. Keep right on. I want them complete. A. No. III? No. III was a Winchester bullet. The top of the lands,—the top of one of the lands was a part of the "W" shot. That is the trademark that the Winchester Company uses on their bullets.

Q. That identifies that as a Winchester bullet? A. That identifies that as a Winchester bullet, and also the weight identifies it. If you notice the Winchester bullets, that is the heaviest one of the lot. The Winchester bullet is 74 grains. They won't go far from 74 grains. The lands at the base, .050. At the top, that is up here (indicating) where it starts in to the rifling, .060. Deformed,—you could not get a diameter, but the two diameters given were .302 and .330 of an inch.

Q. No. III, calling your attention to the twist of the bullet shown on these photographs, and I ask you how the twist on No. III compares with the others shown? A. No. III bullet has got a left-hand twist. All the others are right.

Q. From what gun may No. III be fired, or gun or guns? A. It could be fired from a Colt; fired from a Bayard.

Q. Is there any difference in the measurements of the lands and grooves of a Bayard 32 automatic and a Colt 32 automatic? A. There is.

Q. Give us the difference. How near are they to each other? A. The Bayard is .040 wide; the Colt is 50. That is the width of the lands. Do you want me to read off full measurement, full dope on the two guns?

Q. Yes. I want the comparison. A. Hammer blows and all?

Q. Yes. A. The Colt I used in these experiments,—I did not take the blow of the Sacco——

MR. KATZMANN. Can't hear you.

Q. Then I don't want the blow unless it was of the Sacco gun. Give me the measurements you got from the Sacco gun and then give me your Bayard measurements. Strike that question out. Give me the measure- [1409] ments of No. III bullet, the one with the left twist there. You have given that, haven't you? A. I have given you that.

Q. Very well, now. A. Just a minute. I haven't given you all. I haven't given you the grooves.

Q. Well, I want all. A. I stopped.

Q. I interrupted you. I beg your pardon. You complete bullet No. III. A. The width of the grooves, .105 of an inch. It is deformed. I could not get the diameter. The length is .462. Six lands and grooves. Fifty-four knurl marks. Cannon lures, .047 wide.

Q. By "knurl marks" you mean the indentations? A. The indentations of the knurl, tool marks on the bullet.

Q. The knurling marks, meaning those little indentations that appear there where the shell is pressed onto the lead? Perhaps I better let you complete it. Give us No. 4 now. A. No. 4 bullet, made by the Peters Cartridge Company. The weight is 72½ grains. The lands is .040 wide. The diameter is 299, 300 and 302, measured in three points.

Q. Repeat IIII slowly. A. The diameter of the lands is 299, 300 and 302. Grooves, the width, 120. Diameter, 308, 308, 305, measured at three points.

Q. 308, 308, 305. Does that finish it? A. No. The length is .467. Six lands and grooves. Fifty-seven knurl marks.

Q. Now, take up the next bullet, please. A. And the cannon lures, .035 wide. The next bullet, No. 5. Made by the Peters Cartridge Company. Weight, 72.1 grains. Lands, .040. Diameter of the lands, 302, 300, 3ɔ3. Grooves, widths, 125, 308, 308, 309. I didn't take the——

MR. KATZMANN. That is the diameter, isn't it?

THE WITNESS. Diameter, yes, sir.

MR. KATZMANN. You did not so state.

THE WITNESS. Cannon lures, .035 wide, six lands and six grooves. Bullet marked "X" on the base.

Q. What is that, No. 6? A. No. 6, do you wish to call it? It is marked "X" on the base. Weighs 70 grains. Lands, .040 wide. Diameter, 3023, 3035. Grooves, widths, 125 by 308, 308, measured in two places. The length of the bullet was .443 of an inch. All six lands and six grooves,—in fact, all these pistols are. No cannon lure.

MR. KATZMANN. What is the make?

THE WITNESS. What, sir?

Q. What is the make? A. Union Metallic Cartridge Company, Remington UMC.

Q. That completes this list of bullets? A. That completes that list, yes, sir.

Q. And those which you have already said shows a complete view of all the marks, lands and grooves? A. Yes. I forgot to ask,—to add, those two bullets are taken three diameters, so they are three times the length of the original bullet. The length of the original bullet will aver- [1410] age 450. If you measure that bullet (indicating) or that bullet, or that (indicating) you will find it is three times the original bullet.

Q. It is three times the size of the original bullet? A. Yes.

Q. For a minute I call your attention to bullet No. III, which has a left twist. That is a 32 Winchester bullet, is it? A. Yes, sir.

Q. Is it not? A. It is.

Q. And I was interrogating you when I switched away from it as to what gun that bullet could be fired from? A. Well, it has the characteristics of a Colt, but I wouldn't swear that it was a Colt because,—it is deformed.

Q. Keep your voice up. A. It is deformed.

Q. Tell us about how much deformity there is in that bullet. A. The lead to the rifling is corroded, showed it was corroded and fouled.

Q. Have you anything to show what the leads represent, Mr. Burns? A. I think I let you take it.

Q. Well, it is lost, then. Perhaps I will get it later. You go on and explain. Explain as well as you can without it. A. Why, it is impossible to make a true print of the lands any wider than it is, so there is a strip there. I withdraw that. There is no indications of it, but there was a fouling, corroded fouling there that made an imperfect mark at the point of the lands; and as the bullet travelled up, that gave a false impression of the true lands,—as the bullet travelled up the bore of the rifle, it expanded— I won't put that. Well, the facts of the cases—I will give it to you whether it goes over your head or not.

Q. All right. We will try and follow you. Try and not be too technical. Tell it so every man on the panel will understand it, if you can. A. Yes. As the bullet enters the rifling, the average pressure, gas pressure is, breech pressure is 12,000 pounds per square inch. That will vary from twelve to fifteen thousand pounds. That maximum pressure takes place just as the bullet is getting into the gun, into the rifling, just after it gets out of the lead, which I will explain,—I will explain that later.

Q. (Mr. McAnarney hands paper to witness) What have you here?

MR. JEREMIAH MCANARNEY. I will show it to Mr. Katzmann.

THE WITNESS. As the bullet gets into the lead——

MR. KATZMANN. Wait just a minute.

MR. JEREMIAH MCANARNEY. That your Honor may follow us,

(Mr. McAnarney hands paper to the Court).

Q. Now, you take that (indicating) and turn toward the jury and try to explain what you mean by the "leads". You may show it to them so they will follow. A. (Witness leaves witness stand and goes to the jury) Here is a so-called blueprint of the chamber and bore of a Colt rifle.

Q. Turn it up so they may see it. You were looking at it, but they couldn't see it. A. This (indicating) is the throat. Here (indicating) is the chamber, comes to here (indicating). The chamber ends right there [1411] (indicating). This little taper that comes up in here (indicating) is called the throat. This taper here (indicating) is the bore or lands, and that is what we call the "lead." That is the little taper that comes from here (indicating) and in this case here (indicating). Here you are. Here (indicating) are the figures of the true gun, what it should be. "D" is the lead, .013 of an inch taper. "C" is the throat, .050 of an inch, from here to here (indicating), so it is .063 of an inch from the mouth of the shell into the true rifle.

Q. Show the men on the other end. A. (Witness shows rest of jurymen) Here (indicating) is a blueprint of the gun. This (indicating) is the chamber. This (indicating) is the throat. This (indicating) is the lead, from here to here. In where "D" is, the lead is, I should say, "D" is used. The throat is "C," and this (indicating) is the bore. The figures here give "C," the length of "C," from here to here (indicating) .050 of an inch. The lead is .013 of an inch, therefore .063 of an inch from the mouth of the chamber, or the end of the shell, to where the bullet enters the true rifling. In other words, that is the guide for it. That is what starts the lands. The groove on the bullet. (The witness returns to the witness stand).

Q. Now, you spoke of an imperfect land on bullet No. III. How does that imperfect land manifest itself on bullet No. III so that it may be seen? A. Yes, particularly in No. II.

MR. KATZMANN. Well, wait a minute.

THE WITNESS. No. III, No. IIII is not clear.

MR. JEREMIAH MCANARNEY. Do you wish to object to something?

MR. KATZMANN. I thought he changed to bullet No. II,—picture No. II.

THE WITNESS. Picture,—six views of all the lands.

Q. Now, tell us how it manifests itself? A. Visible to the eye right there (indicating).

Q. Now, in what way is it visible to the eye? A. Wider at the top by .010 of an inch than it is at the bottom.

Q. What does the fact that that groove in the bullet is wider at the top than it is at the bottom indicate? A. It indicates that the lead was corroded and fouled. What I mean by "fouling" perhaps might be metal fouling. Invariably it is metal fouling. If a gun is corroded there, this jacket—which is soft, it is only .013 of an inch thick, thirteen to fourteen thousandths of an inch thick at the base or throughout the whole bearing— of course would collect metal fouling. That is copper and zinc. That jacket is 95 and 5, copper and zinc,—95 copper and 5 zinc,—and that builds up and that would form a false land, a false measurement, and as the bullet travels along the bore, common reason will tell you if it was 15,000 pounds pressure behind it, the bullet is going to be upset at the base. The sluggage is greater at the base of the bullet with that 15,000 pounds behind it, say 12,000 pounds normally. It runs from 12 to 15. [1412] That is why you get a variation in pressure; will cause a variation in sluggage.

Q. Tell us why the groove mark in the bullet is narrower at the base than it is up towards the nose of the bullet, apex of the bullet? A. Because the pressure there upsets the base of the bullet and conforms it to the true lands as it goes along the muzzle, showing that bullet was fired,—come right down to brass tacks, showing that bullet was fired in a gun with a fouled up lead.

Q. If I understand you right the— A. With a perfect muzzle——

Q. —the bullet enters into the lead and where it enters into the lead the lead is irregular? A. Irregular.

Q. And the bullet may take on an irregular conformation? A. Sure.

Q. But as she goes through the nozzle, having a perfect rifling, the bullet then at its base assumes the true rifling of that gun? A. Yes, sir, always measure the base of the true lands.

MR. JEREMIAH MCANARNEY. (To the jury) I want to call that to the jury's attention. Having in mind what Mr. Burns says I call your attention to No. III. With the naked eye you will see that the groove is narrower at the bottom than it is at the top. See it? Notice the bullet III has a narrower groove at the bottom than it has at the top. Bullet No. III, you take it, so you get your own view of it better. Compare the width of the groove at the bottom with the width at the top. This (indicating) is No. III, and set it so the light will be at right angle to your eyes. The width of the groove at the bottom, compared with the width as it goes toward the top. Did you gentlemen get it? Get it so each one of you can get it in the right angle of light. The width of the groove at the bottom and as it compares at the top.

Q. Assuming that the lead of the gun that fired bullet No. III had been clean and normal would you have found such irregularity as exists in bullet III on this photograph? A. I have never found it.

Q. Have you some bullets that have been fired from the—we will call it the so-called Sacco gun? A. Yes, sir.

Q. Will you produce those bullets? A. (Witness does so.)

MR. JEREMIAH MCANARNEY. I am calling the Colt automatic, for the purpose of designation on the record, the Sacco gun.

Q. How many bullets have you that were fired from the so-called Sacco gun? A. I had eight to start.

Q. How many have you lost? A. Two.

Q. Well, you have six with you. All right. Now, do those bullets that you have there show the irregularity that is shown on bullet No. III in these photographs? A. No, sir.

Q. How are the lands and grooves on those bullets, whether regular or not? A. Regular, clean-cut.

MR. JEREMIAH MCANARNEY. (To the jury) Now, gentlemen, you may pass those among you. Wouldn't the glass help out better with this?

[1413] THE WITNESS. Yes.

THE COURT. Take a recess of five minutes now.

(Short recess.)

MR. JEREMIAH MCANARNEY. Now, let me see some of these bullets. Mr. Katzmann and if the Court please, I am now showing to the jury some bullets that came from the alleged Sacco gun, with a glass.

MR. KATZMANN. Are those the ones he fired at Lowell?

MR. JEREMIAH MCANARNEY. These are some of the bullets you fired?

THE WITNESS. I will make sure.

Mr. Jeremiah McAnarney. Well, make sure, and as you look at one put it in the——

The Witness. (Examining with glass). Yes, sir.

Mr. Jeremiah McAnarney. (To the jury) Having in mind, gentlemen, what you saw with reference to the marks on the bullet No. III, it being narrower at the base than at the top, I now ask you to examine four bullets up here (indicating), the top, the markings being narrower at the base than towards the top.

Mr. Katzmann. He testified to that.

Mr. Jeremiah McAnarney. Yes. I call to your attention, gentlemen, and I ask you through this glass to look at the lands and grooves on the bullets that the Foreman has with reference to whether or not the grooves are regular and of uniform width or not.

Q. You have two more of these, have you? See if those are the same. A. (Witness examines) Yes.

Q. Both are? A. Yes.

Mr. Jeremiah McAnarney. (To the jury) Well, gentlemen, here are two more he says are from the Sacco gun. You may use those and pass them along. While the others are going along, I call your particular attention to this photo of No. III, which seems to be a remarkably good one. So that the jury may understand, I would like to state No. III is supposed to be the fatal bullet that was found in the body of Berardelli. III is the same bullet in each.

Q. In the measurement given of bullet No. III, you gave two measurements, base and another, did you not? A. What is that?

Q. In the measurements given of bullet No. III, what was the measurement at the base? A. .050.

Mr. Jeremiah McAnarney. (To the jury) Have you all seen the bullets, gentlemen?

Q. What is the measurement up— A. At the top?

Q. —at the top? A. 60.

Q. So you have .010 greater width at the top of that groove than at the base? A. Yes.

Q. Which is the true measurement of the groove? A. 50.

Q. Well, having in mind the appearance of No. III on the photograph there, bullet No. III, and having in mind the grooves made on the [1414] bullets fired from the Colt automatic and designated as the Sacco gun, have you an opinion as to whether the so-called fatal bullet No. III was fired from the Sacco gun? A. I have.

Q. Was it fired from the Sacco gun? A. Not in my opinion, no.

Q. What is the ground and base of your opinion, and on what do you base that opinion? A. On the eleven bullets that I examined that were fired from the Sacco gun. It doesn't compare with it at all.

Q. In what way does it not compare with it? A. It shows a clean-cut lead all the way through, the same diameter, the same width at the top as it does at the bottom, practically no difference.

Q. What does the fact that the groove is clean-cut all the way through coming from the Sacco gun indicate with reference to the rifling of that gun, the condition of the rifling? A. Clean lead.

Q. That it has a clean lead? A. Yes.

Q. In your opinion was bullet No. III fired from a gun that had a clean lead? A. You mean this gun?

Q. That (indicating) is the Bayard. The other. A. Oh, the other?

Q. I am speaking of No. III; was that fired from a gun that had a clean lead? A. It was not.

Q. Is there any doubt about that? A. No doubt in my mind, no, sir.

Q. Any other reason that you have from the appearance of the bullet that would indicate that it was not fired from the Sacco gun? A. That is all. The main point. The bullet is deformed so you could not get any connecting links as to diameter. Those were wiped out.

Q. The bullet was somewhat deformed? A. The photograph does not show the deformation.

Q. Does it show it in part? A. In part it does.

Q. Kindly indicate where it does. A. To the jury, sir?

Q. Yes. A. If you follow these turn around, 1, 2, 5, 4, 6, every one stands up right. If you will follow the contour of the No. III bullet you will see it begins to tip, see it (indicating); it leans to the left, the bullet does. Again, No. IIII leans more to the left. There is where the great deformity was. On that side it does not show a clear land. That comes back through.

Q. Step up this way, Mr. Burns, and show to these men.

THE COURT. Is there any way of reaching the sheriff?

THE WITNESS. Starting with No. I, turn the bullet around. Six——

THE COURT. Any way of reaching the sheriff?

A VOICE. Telephone his home.

THE COURT. See if you can get him. It seems to me probably both sides will want the bullets and pistols that he has now locked up here, so it is quite advisable that he be reached some way. Now, you may proceed, please.

[1415] THE WITNESS. See, No. III bullet stands up right and if we turn it around it begins to tip. See it? It begins to straighten up again.

Q. What does that indicate? A. Indicates a deformed bullet.

Q. A deformed bullet? A. Yes. Oh, there was another thing about this bullet which I did not quite catch. It has a hollow base.

Q. No. III has a hollow base? A. Yes.

Q. What significance has that? A. The hollow base gives it a greater chance for sluggage or upset with the pressure in going through, and so that you get the true lands at the base.

Q. Now, I show you a set of photographs which have not been marked, and I ask that they go in for identification.

(Photographs of bullets marked "Exhibit 22 for Identification.")

MR. JEREMIAH MCANARNEY. These become for identification, Exhibit 22.

Q. What do those represent? A. No. 1 bullet, No. 2 bullet,—No. 1 bullet was fired in a Steyer rifle or pistol.

Q. What calibre or bore? A. 765 millimeters, which is meant to be a 32.

Q. Is that right or left twist? A. It has a right twist. That is the way the foreign makers designate their guns, in metric system.

Q. Whether or not that corresponds to our measurement of 32? A. It is intended.

Q. May it be fired in a 32 gun? A. Yes.

Q. What is No. 2? A. No. 2 is U.M.C. bullet, fired in the same gun, a Steyer. No. 3 is a U. S. bullet fired in a Bayard.

Q. What twist is that? A. Left.

Q. And then there is a gun, the Bayard gun, that has a left twist? Q. Yes, sir.

Q. The same twist as the Colt automatic? A. Yes, sir.

Q. Well, continue right along there. A. No. 4 was fired with a Steyer. No. 5 was fired with a Steyer. And "X" was fired in a Steyer and that is a U.M.C. bullet. The others are the same.

Q. Of those photographs you have there, how many have the left twist? A. One.

Q. That is fired from a 32 Bayard? A. 32 Bayard.

Q. That is bullet number what? A. No. 3.

MR. JEREMIAH MCANARNEY. (To the jury) I show you, gentlemen, photograph No. 3 as being fired from a Bayard 32, and if you will note the pitch of the groove, it has a left, as compared with the others that have a right twist.

Q. While the jury are looking at that, I want to ask you one question. It is in evidence here last week, I want to just look at it for a second, but,— have you ever seen a Bayard 22 revolver or pistol? A. 22?

Q. A Bayard,—I beg your pardon,—Bayard 25? A. Yes, sir.

Q. What twist has a Bayard 25? A. Left.

[1416] Q. Have you such a gun. A. I believe Mr. Fitzgerald has it.

Q. Is that (indicating) a Bayard 25? A. Yes, sir. It is a 635 millimeters, which corresponds to the 25 Colt automatic.

Q. And the 25 Colt automatic shell may be fired in that gun? A. Yes, sir, they are interchangeable.

Q. Whether or not all the Bayards are left twist? A. All that I have ever seen were left twist.

Q. How many different calibres of Bayard have you seen? A. Three.

Q. What? A. 25, 32 and 380.

THE COURT. May I ask where they are made?

THE WITNESS. Right on there (indicating).

THE COURT. That does not tell me.

THE WITNESS. Belgium.

Q. Whether or not any of those guns are around, or to what extent are these guns around now? A. Why, lots of them. I had one brought to me last week by a police officer.

Q. Well, you say there are lots of them. How long since they have become so you say there are lots of them around? A. They are becoming common since the War.

Q. I see. A. Here is one here (indicating).

Q. Have you a 32? A. Bayard?

Q. Yes. A. The magazine is out.

Q. That (indicating) is a 32 Bayard? A. Yes, 765; the same thing.

Q. I would like the 32 pulled down so they can see the rifling. A. (Witness hands same to Mr. McAnarney).

Q. What have you there? A. Those are casts of this Bayard pistol, taken from the muzzle.

MR. KATZMANN. Which Bayard pistol?

Q. Which calibre? A. 765, or 32, what I refer,—I call it 32.

Q. Yes. A. And the top here where the wire is, is the muzzle.

(Mr. Jeremiah McAnarney shows casts to the Court.)

Q. What does the twist,—what do those casts show the twist of the Bayard to be? A. Left twist.

Q. How does the twist on this exhibit show with reference to the twist in the barrel,—is it the reverse or the same? A. It is the same. You hold it this way (indicating). It would be just exactly as though you were holding a bullet point up. The bullet goes through and there you have it. Hold them by the wire.

Q. Hold them by the wire, and this is the base of the bullet (indicating), and this (indicating) is the top of the bullet, showing the bullet with left twist?

THE COURT. May I see one of the Colts and the Bayards?

THE WITNESS. Bullets?

THE COURT. No, pistol.

THE WITNESS. Pistol, yes (handing it to the Court.)

[1417] Q. What have you here (indicating)? A. Three 25 Colt automatic bullets fired from a Bayard pistol.

Q. What would you say is the groove on that? What is the twist, right or left? A. Left.

MR. JEREMIAH MCANARNEY. (To the jury) See, gentlemen, these here (indicating) from the 25 are a left twist. You take one and look at it. I just want to show you that they are a left twist (showing to the jury).

Q. Now, if you will place your matter aside for the time being, place those out of the way for just a minute. A. Those bullets you have got——

Q. What? A. You have them bullets.

Q. Will you kindly return them.

MR. KATZMANN. Do you mind,—our expert is examining them. I will see that they go back in that box for you.

MR. JEREMIAH MCANARNEY. Here are your Bayard 25s, right in here they belong (indicating).

Q. Now, I am going to divert for a few minutes, and I call your attention to the Exhibit 27, that is what? A. Harrington & Richardson.

Q. Before I go on that— A. .38 Smith & Wesson revolver.

MR. JEREMIAH MCANARNEY. Before I go on that, I would like to call, in these shells, if you gentlemen please,—that your Honor may follow us, I want to call to your attention, and I have shown the jury, that No. III, being one of the series of photographs shown in six different views.

THE COURT. Different views taken of the same bullet?

MR. JEREMIAH MCANARNEY. Yes, and here a bullet fired from the Sacco gun on Saturday at Lowell. Both gun experts went up together. Those are what the jury have seen up to the present minute. They have all of them. I will leave them down there with them. That (indicating) is one of the Sacco gun bullets.

THE WITNESS. One I offered.

MR. JEREMIAH MCANARNEY. Yes.

. . . .

[1419] MR. JEREMIAH MCANARNEY. I am returning this gun to the sheriff.

Q. Now, Mr. Burns, I call your attention to the evidence which you heard while you sat beside me at the desk last week with reference to some marking appearing on the shell. You heard Mr. Van Amburg, as he testified to some back flowing of the metal on the shells. What causes that? A. That is a natural cause, subject to any gun,—common.

Q. Kindly tell us in what way? A. It is the pressure exerted on the fired shell,—forces against the face of the breech block.

Q. All I care about,—is it a usual thing that occurs in the ordinary Colt or any other gun? A. Sure it is, yes, sir. The usual thing.

[1420] Q. Anything you care to say or add to that or in any way explain it? A. You can get those marks and again you can't get them with the same ammunition. It will show plain with one cartridge and it won't with another.

Q. Kindly repeat that answer. A. It will show with one cartridge and again it will not show. It all depends upon the pressure exerted on the face of the breech block, what we call "ball thrust".

Q. I am more concerned with the situation,—is it an ordinary thing to occur usually in guns that have been used? A. Yes, sir.

Q. Something was said also, when the shells were shown to the jury, that where the head of the point struck the primer that one,—some of them appeared to be off center. What causes that to be off-center? A. The flowback of the primer. It is practically impossible to get a plumb center blow, owing to the flow-back of the primer across the firing pin and the anvil which,—the contour of the anvil will not allow it. It flows off one side.

Q. What do you mean by "contour of the anvil"? A. The anvil which is inside of the primer.

Q. Now, I would like that made a little clearer, if you would. Will you just draw a crude,—here is some paper. Roughly draw what you mean by the anvil. A. (Witness draws sketch).

Q. Now, what have you? A. There (indicating) is a rough sketch. There (indicating) is the primer. Here (indicating) is the anvil. When the firing pin strikes here (indicating) there is a flow-back against—it flows back one side to the other, has a tendency to, owing to the looseness of the shell, which has five or six thousandths of an inch play in the chamber, and again get two or three thousandths inch play, and sometimes five or six of the striker,—in the striker channel.

Q. Would you go before the jury and tell it to them as you— A. They are all tipped up now. When the blow is struck on there——

Q. You say "on there." What is "there"? A. On the center of the primer, assuming it was a center blow. Here (indicating) is the primer. Here (indicating) is the primer cup. This (indicating) represents the primer inside and this (indicating) the anvil.

Q. You mean the little brass and copper thing there? A. Yes. This is the cross-section, represents a cross-section. That (indicating) is the anvil, Winchester anvil. They call it the "W" anvil. The support is double up near the firing pin. This (indicating) is the anvil here. This (indicating) is the priming composition. When the firing pin strikes here (indicating) there is a set-back to the shell. The primer explodes, and the pressure is raised from ten to fifteen thousand pounds. Bang! Bang! This comes back and this firing pin is projected out 45 to 50 thousandths of an inch. I would have to verify that. I have it in my notes. This whole thing comes and flows back. That is what causes the side blow. Sometimes you will get it when it lengthens on here and does not show much of a side blow. Again it will flow away down in here, with the same gun.

[1421] Q. Kindly show the men up here (indicating) on this side, that they may see. A. When the blow is struck here,—this is the cup here (indicating) represents the cup.

Q. Is that the primer? A. The primer of the cup. The components of a primer are the cup, priming and the anvil.

Q. Now that we may understand, what is the cup? A. The cup is this outer part, the cup which holds the primer in here (indicating).

Q. The cup is the powder or whatever the composition is? A. In this case it is chloride of potash, fulminate of mercury, sulphide of antimony. And that is closed and ignites the powder and develops 15,000 pounds pressure. The firing pin is projected there and held there with 45 to 50. The striker hits the blow, the primer explodes, the priming explodes, ignites the powder and results into 15,000 pounds being developed. This shell comes back against the firing pin. The projection of the firing pin is from 45 to 50 thousandths of an inch. The ball thrust here of this whole shell comes back with over a thousand pounds force, actually a thousand pounds force; over a thousand pounds force. It all depends upon the pressure that is exerted in the cartridge. The area of that cartridge is .091 of an inch, and, of course, that is again in direct proportion to the pressure to the square inch in the chamber.

Q. You were telling us about the anvil. A. The point of the striker strikes this cup, explodes the primer, ignites the powder, and as this shell,—primer comes back against the firing pin, the projection is 45 to 50 thousandths. The point of the anvil acts as an obstruction to the point of the striker and flows by it. The indentation, of course, goes into the copper cup. I forgot to add that that copper cup will vary from 15 to 17 thousandths of an inch in thickness, so you see it is a thin metal. That anvil forms a bridge, and there is no give to it.

. . . .

[1423] Q. And I omitted to call your attention to the alleged pitting in the Colt gun introduced as Exhibit 28. Whether or not there is sufficient pitting on that gun so that you are able to identify a bullet that goes through that gun? A. Absolutely not.

Q. Why do you say that? A. Because I have shot about 100 bullets; got them right there in the grip.

MR. KATZMANN. What is that answer?

THE WITNESS. About 100 bullets with scored and pitted guns, and it is impossible to identify them from one another.

. . . .

[1424] MR. JEREMIAH MCANARNEY. I am offering the barrel of a 32 Colt automatic, the full complete cast taken showing the rifling and the leads, and a bullet fired through that barrel which I understand to be this bullet marked "A" on the base (showing to the jury).

Q. This (indicating) is the nozzle of the gun? A. No, sir. This (indicating) in this case here is the breech of the gun, but we could not pull the chamber through.

MR. JEREMIAH MCANARNEY. This (indicating) represents the full cast of the inside of this barrel, a Colt 32 automatic, showing the enlarged part there at the breech, and showing, if you will notice, that it is a good deal thinner, the grooving is, back next to where the shell enters to what it is as it goes out the nozzle; and you will also take and may have for your future consideration the shell, the bullet that went through as bearing on how much importance can be attached to some scoring or firing. They will be in evidence as wanted, showing the complete rifling.

. . . .

[1425] Q. Now, you stated you had about 100 bullets that were fired through various guns more or less scored? A. Two different kinds of bullets. There are 1, 2, 3, 4, 5, 6, 7, and 8, 9 or 10 barrels there, counting what is in the other guns. There is a bunch there all pretty well scored. They are all scored right here (indicating).

Q. Now, you made your experiments, and after firing those guns what do you say with reference to whether any man can determine a bullet came from the Sacco gun by reason of any alleged scoring on that gun? A. No, sir.

Q. Why not, if you please? A. Because you can't get a bullet through a gun,—a bullet will come through a gun. You might fire ten bullets and they would all vary. There would be some slight variation in the bullet, unless it was a perfect gun.

Q. Well, take six bullets, say, three bullets fired from the Sacco gun, how would they appear different from those fired from any other gun, in so far as scoring would be concerned? A. The bullet that comes from the Sacco gun that I fired were practically perfect and clean-cut lands.

Q. In that connection I will call your attention to Government Exhibit 36, purporting to be three bullets fired by Capt. Van Amburg from the Sacco gun. Will you examine those and say whether there are any marks on those bullets by which they can be identified as coming from the Sacco gun by reason of any imperfection in the barrel caused by pitting or fouling of the rifling of the barrel?

A. (Witness examines bullets) No, sir, not positive of it.

MR. KATZMANN. What is that answer.

THE WITNESS. It could not be positively done.

Q. Why not? A. In my experience in shooting of all these different bullets through guns the bullets vary so, the marks on the bullets vary so, it is impossible, and particularly with the Sacco gun, the pitting there is so slight that it does not mar the bullet very much and you have a perfect muzzle. My cast showed that.

Q. I do not know that I have introduced your cast. Let me see the cast of the inside of the Sacco gun. A. It is in exhibition. It is an exhibit, I believe.

MR. JEREMIAH MCANARNEY. It is?

THE COURT. Yes, that is it.

MR. JEREMIAH MCANARNEY. That was put in while the Government's case was going in. I can inquire for it later.

. . . .

[1426] Q. And you give as your judgment that bullet No. III, alleged as the fatal bullet, was fired from a Colt or could have been fired from a Bayard possibly? A. Yes, sir.

Q. I believe you have expressed the opinion it was fired from a Colt? MR. KATZMANN. One moment, please.

Q. Well, what is your belief on that? A. It shows indications of [1427] a Colt. Still, I believe it could be fired from a Bayard. Not having the experience with Bayard pistols that I have with the Colt leads me to believe it is possible.

Q. So that you really are in doubt as to which gun it was fired from? A. Yes, sir.

Q. Now, you mentioned a 25 gun. Have you that gun here, a Bayard 25? A. Yes, sir.

MR. JEREMIAH MCANARNEY. I want, if your Honor please, to introduce this Bayard 25 in evidence. Capt. Proctor said there was no such thing as a Bayard 25 with right twist.

THE COURT. Didn't he say right the opposite?

MR. KATZMANN. He said the exact opposite.

MR. JEREMIAH MCANARNEY. I have reversed it.

THE WITNESS. This is a Bayard 635 millimeter, which is the same.

Q. What is the twist? A. Left.

. . . .

CROSS-EXAMINATION

Q. (By Mr. Katzmann) You did not have any trouble putting it together did you? A. No.

Q. That is why you did it so quickly, wasn't it? A. I was a little rattled.

Q. Would you credit another man with being rattled a bit in taking it apart?

MR. JEREMIAH MCANARNEY. Just a minute.

MR. KATZMANN. All right. I withdraw the question if you object.

Q. Mr. Burns, how many bullets at Lowell did you fire through the Sacco gun? A. Eight.

[1428] Q. How many have you produced in evidence? A. Seven.

Q. Where are the seven? A. There is the other one (indicating).

Q. Where are the other six? The six are the six produced originally? A. Yes, sir.

Q. Where is the eighth one? A. I kind of think Mr. Fitzgerald lost it or else he has it in his pocket looking it over.

Mr. Jeremiah McAnarney. I would like to have that seventh one go in with the others, if you don't mind.

Mr. Katzmann. I would be glad to have it go in with the others.

The Witness. I will probably find it before morning.

Mr. Katzmann. Give me the seventh one, will you, please?

Q. What manufacture of cartridge did you fire through the Sacco gun at Lowell? A. "U.S."

Q. What manufacture of cartridge is it your opinion No. III bullet was fired from? A. Winchester.

Q. Was there any reason why you should not have fired eight Winchesters through that Sacco gun? A. Yes, sir.

Q. What was the reason? A. Because our bullet represented that bullet nearer than the present Winchester bullet that we could buy.

Q. Did you make any effort to procure a Winchester bullet? A. Yes, sir.

Q. Was that the only reason? A. Yes, sir.

Q. Is there any difference in size between a Winchester 32 and a United States 32? A. Which Winchester do you mean?

Q. The Winchester of the type that was taken from the body of Alessandro Berardelli? A. It is practically the same as ours, with the exception of a grain, 1-1/10 grains in weight.

Q. Is there any other exception? A. That is all.

Q. Any difference in the length? A. That would make it slightly longer.

Q. Then there is that difference, isn't there? A. Yes, slightly.

Q. Is there any difference in the contour of the nose of the bullet? A. Very slightly.

Q. Is there any difference? A. Not so as you would notice it.

Q. Could you measure it? A. Yes, sir.

Q. What is the difference between the contour of the No. III bullet that came out of Berardelli's body and the United States bullet that you fired through the Sacco gun? A. There (indicating) is the whole measurement of it.

Q. What is the difference? I asked you as to the contour, the difference? A. No difference,—no perceptible difference with the micrometer.

Q. Is there any difference? A. No difference.

Q. What measurements are you now examining on which you pred- [1429] icate the answer, "There is no difference between a United States and a Winchester"? A. Not more than one-thousandth.

Q. You are measuring to thousandths of an inch, aren't you? A. Not on the contour. You can't measure any closer than two thousandths of an inch. I will stretch that to two thousandths of an inch.

Q. You are assuming a difference of two-thousandths of an inch makes no difference in measurement of a bullet? A. It is impossible for me, with my experience, to measure any closer than two-thousandths of an inch.

Q. Is there any instrument known to you by which you can measure exactly? A. No, not on contour.

Q. Is there any measurement known to you by which you can measure exactly the width of the groove or land on the bullet? A. Yes.

Q. What is the name of that instrument? A. Plug gauge.

Q. What? A. Plug gauge.

Q. Plug gauge? A. Yes.

. . . .

[1431] Q. Then you are not saying bullet No. III was not fired from a Colt, are you? A. No, sir.

Q. What is the pitch of the rifling in a Colt 32? A. One to sixteen.

Q. That means, does it not, one turn of the bullet in its flight to 16 inches of distance travelled? A. Yes, sir.

Q. What is the pitch of a Bayard 765? A. I could not tell you directly.

Q. Isn't the pitch of the rifling of the inside of the barrel as measured by the marks therefrom on the bullet fired from that barrel a very important determining factor as between a Bayard and a Colt? A. Somewhat, yes.

Q. And you don't know what the pitch of the rifling is in a Bayard 32, do you? A. I know approximately.

Q. Have you ever measured it in a Bayard 765? A. Yes, sir.

Q. When? A. (Witness examines revolver) I haven't got the record.

Q. What gun? A. Will you allow me to explain?

Q. No. You just answer my questions now, please. A. All right.

Q. What Bayard 765 did you measure it in? A. This one.

Q. When? A. The day that the gun was bought, or very shortly afterwards.

Q. When was that? A. I haven't that record. I did not think it important enough.

MR. KATZMANN. I ask that be stricken out, if your Honor please.

THE COURT. It may be.

THE WITNESS. What is that?

MR. KATZMANN. I ask that remark of yours be stricken out.

Q. Was the purchase of the Bayard 765 millimeter made subsequent to the testimony of Capt. Van Amburgh and Capt. Proctor? A. I feel safe in saying we have had that gun over a year.

Q. Then your answer is, it was purchased before the testimony of those two gentlemen? A. Yes, sir.

Q. Did you have it, was it accessible to you when you measured bullets I, II, III, IIII, 5 and X? Yes, sir.

Q. Did you make any measurements of the rifling pitch in that Bayard 765 when you had those bullets in your possession? A. No, sir.

Q. Is the pitch of the rifling of a Colt 32, from which bullet III was fired—if it was fired from a Colt 32—one in sixteen? A. Again, please.

Q. Is the pitch of the rifling of the Colt 32—if it was a Colt 32 from which bullet III was fired—consistent with the pitch of one in sixteen inches? A. I wouldn't say that, I wouldn't want to say. You can't tell.

Q. Did you not measure that in determining the weapon from which bullet III was fired? A. The bullet? The pitch of the bullet?

[1432] Q. Yes, from the bullet. A. It can't be done accurately.

Q. I did not ask you whether it could be done. I asked you if you did it? A. No, sir.

Q. Why didn't you do it. A. Because it could not be done accurately on the bullet.

Q. Can't you measure the pitch of a bullet from the lands and grooves on the bullet after discharge? A. No, sir, not enough of it to do it accurately.

. . . .

[1434] Q. Does it indicate anything as to the manner in which the bullet took the lead? A. Not plumb.

Q. Not plumb? A. Yes.

Q. That is, it did not jump the rifling perfectly? That is what it means, doesn't it? A. It did not go straight into the lead, perfect center.

Q. Isn't that a shop term, "It did not jump the rifling perfectly"? A. Not straight.

Q. Yes. And that is why you get the double marking at the end, isn't it? At the upper end? A. Yes, you are liable to get that in any gun.

Q. And isn't this splaying out that you have indicated as to bullet No. III, particularly on sheet 2 of the series of six photographs the same result only not so pronounced? A. No, sir.

Q. Do you distinguish between a slippage there and a fouling at that point? A. There is no slippage there. It doesn't show indications of it.

. . . .

[1437] Q. Answer it, if you please, your own way. A. The only—this No. III bullet was a hollow base bullet, with knurls and a cannon lure. As near as I can make out, it had been made some little time. It is not of recent manufacture. The most recent manufacture of Winchester cartridges has got—Winchester ammunition of this caliber has got a smooth bullet. The latest manufacture of this cartridge has got a smooth bullet. Undoubtedly Mr. Van Amburg had the same trouble in getting it, because I noticed he had the smooth bullet and I had the same. That is all I could find, and I thought it would be the last place to find them, at the Winchester Company, because we do not keep old samples of ammunition, and if we do, we keep them just simply for record to go by. Now, this bullet of ours is the nearest thing I could get to this Winchester bullet, because it has got the same jacket and analysis and the contour, as near as I can make out, is the same. That is why we used it. We have a Winchester, and all kinds. Is that satisfactory?

. . . .

[1439] THE WITNESS. Why——

Q. Did you hear anything in that question about the Colt automatic? A. Get that again. I don't quite get you.

Q. Shall I raise my voice? A. Yes. Come over here a little closer.

Q. If I do, you will talk so low they won't hear you. Will you promise to keep your voice up if I do? A. I will try to.

Q. That is a bargain. Do you not predicate your opinion that bullet No. III was not fired by the Sacco gun wholly upon the matter of sluggage? A. Not entirely on the matter of sluggage.

Q. Upon what other basis? A. On the perfect muzzle. That is, for an inch, half or three-quarters of an inch. I can't tell you just now. I should say,—have you got the cast?

Q. Yes. A. I can explain it a little better.

Q. (Mr. Katzmann hands cast to the witness.) A. I should say for half an inch in the muzzle it was near perfect. We call it "upset", to be a bit easier, perhaps. It is a term we use.

Q. Mr. Burns, the upper end of that cast is the muzzle end, isn't it? A. Yes, sir.

Q. The one in which the wire is inserted? A. Yes.

Q. Will you look at the groove made by the land on the inside of that barrel about the center, please? A. The groove?

Q. Yes. A. Yes, sir.

[1440] Q. It that a perfect land mark from the land of the inside of the barrel? A. No, sir.

Q. No. Now, will you take Exhibit No. 28, please. Take out the barrel. A. (Witness does so.)

Q. You know, of course, what I mean by clock-wise, don't you? A. Yes, sir.

Q. Holding— A. Can I brush it out?

Q. Yes, please. You can't brush any pits out with that brush, can you? A. No.

Q. All right. Holding the groove, looking in the muzzle end of that barrel, holding the groove at nine o'clock, do you see a land at substantially six o'clock, holding those grooves at 9 o'clock by the clock? A. That (indicating) is nine o'clock.

Q. That (indicating) is nine o'clock, yes. A. This (indicating) is twelve.

Q. That (indicating) is twelve. Now, I ask you if at six o'clock— A. (Witness examines).

Q. —do you see pitting at six o'clock along the right side of that groove? A. Yes, sir.

Q. Pitting sufficient to make a mark corresponding to it on the right side of the groove when this groove is held in the position I hold it in? A. No (examining barrel). It is six o'clock?

. . . . . .

[1446] Q. I will withdraw that. What is the most important measurement, in your opinion, Mr. Burns, in the matter of identification of a bullet and a gun as to whether or not a bullet has come from the particular gun or not? A. The width and diameter of the lands and the same of the grooves.

. . . .

[1447] Q. I show you, Mr. Burns, one of the three bullets from Exhibit 35, which bullet is one of the three Winchester bullets fired by Capt. Van Amburg at Lowell on the day that you fired eight United States bullets. I call your attention to the groove that is now in that bullet in the center between my two fingers, and ask you if you find a lead at the top of that bullet widening out as in bullet No. III? I don't dare let go of it. A. [1448] You hang right onto it. You fooled me yesterday, and I don't want you to do it again.

MR. KATZMANN. I ask that be stricken out.

THE COURT. That may be stricken out from the record. Don't make those comments to counsel.

THE WITNESS. He showed me the cast from the wrong end.

THE COURT. You should not make comments of that kind. Leave that for counsel later.

Q. Can you take the bullet in your own hand and hold it that way (indicating). I don't want to hold it for you. Take it in your own hand and keep it so that groove is up so I may show it to the jury. A. The metal is turned over.

MR. KATZMANN. I ask that be stricken out, if your Honor please, and the witness required to answer the question.

THE COURT. It may be stricken out. Can you answer the question? If you can, please do so.

THE WITNESS. What was the question?

Q. Do you see a widening at the top end of that groove I pointed out to you on the Winchester fired at Lowell out of the Sacco gun that corresponds to the widening on bullet No. III which you showed photographically to the jury? A. No, sir.

Q. Does the groove that I have pointed out to you widen at the top end of the groove, the nose of it? A. Not perceptibly.

Q. You can't see it? A. I can see it. I have seen it.

Q. Then it is perceptible, isn't it? A. It is a turn-over of the metal.

Q. Do you see the widening? A. No, sir.

Q. You do not see any widening? A. Not any perceptible widening to the eye.

MR. JEREMIAH MCANARNEY. I do not believe these gentlemen here can see.

THE COURT. All right. They may go to the window and see. They may go to the window and see.

(The various jurors go and examine bullet at window.)

MR. KATZMANN. I call your attention, Mr. Juryman, to that groove (indicating) and ask you to observe the width of the upper end as compared with the lower end. Will you adjust that to suit your own eyes? I call your attention to that groove and the width of the upper end as compared with the lower end. I call your attention to the width of the upper end of that groove as compared with the width of the lower end.

Do you want to see it, Mr. McAnarney?

MR. JEREMIAH MCANARNEY. Not now.

MR. KATZMANN. (To the jurymen) I call your attention to the width of the upper end of that groove as compared with the width of the lower end. You can adjust that to suit your own eyes. I call your at- [1449] tention to the width of the upper end as compared with the width of the lower end. Can you adjust that to suit your own eyes? I call your attention to the width of the upper end of the groove there as compared with the lower end. Will you adjust that to suit your own eyes? I call your attention to the width of the upper end of that groove as compared with the width of the lower end. Will you adjust that to suit your own eyes? I call your attention to the width of the upper end of that groove as compared with the width of the lower end. I call your attention to the width of the upper end of that groove as compared with the width of the lower end. I call your attention to the width of the upper end of that groove as compared with the width of the lower end. You can adjust that to your own eyes. Then I call your attention to the width of the upper end of that groove as compared with the width of the lower end.

A JUROR.   Which end?

MR. KATZMANN.   The one where the pen is.

A JUROR.   Away down or through that little——

MR. KATZMANN.   I do not suppose I can answer your question. I call your attention to the width of the upper or nose end of the bullet as compared with the width of the base end of it.

. . . .

REDIRECT EXAMINATION

[1459] Q. (By Mr. Jeremiah McAnarney) Speaking of the bullets that you used in your test with the other experts at Lowell, why did you use the bullet that you used? A. Because it had more of the same contour, same effect as the Winchester bullet that was used, or No. III bullet.

Q. I call your attention to Exhibit 36——

MR. JEREMIAH McANARNEY.   Or have I here, Mr. Katzmann, the three used by Capt. Van Amburg?

MR. KATZMANN.   Not in 36.

THE COURT.   I think 35.

MR. KATZMANN.   They were all fired by Capt Van Amburg, but those are the Peters, not the Winchester. 35 is the Winchester.

Q. I call your attention to these bullets (35). Those are the three bullets fired by Capt. Van Amburg. Now, how do they compare and how do they differ, if they do, from bullet No. III, which we will call the fatal bullet? A. Bullet No. III has a cannon lure in it and it has a hollow base. [1460] This bullet (indicating) has not a hollow base and it hasn't the cannon lure.

MR. JEREMIAH McANARNEY.   (To the jury) I call your attention to bullet No. III, and I call your attention to the bullet used by Capt. Van Amburg, that it is absolutely smooth and has not got that marking there which you see on that bullet, the cannon lure. The bullet used by Capt. Van Amburg, an absolutely smooth bullet, and the bullet No. III fired in that gun had the cannon lure. Bullet No. III, the fatal bullet, so-called, the heavy cannon lure as you see it on the picture. The bullet fired by Capt. Van Amburg was an absolutely smooth bullet, and the bullet fired by Capt. Van Amburg has a flat base, smooth on the bottom. Also that the Van Amburg bullet is smooth on the bottom; the Van Amburg bullet smooth on the bottom.

Q. The bullet No. III that was fired that was found in the body of the deceased and is shown here, what kind of a base did that have? A. Hollow base.

Q. Now, what kind of a base was the bullet,—did the bullet have that you used in your test? A. A hollow base.

Q. Did the bullet you used have a cannon lure? A. Yes, sir.

Q. And could you have used and have procured and used the same kind of bullet that Capt. Amburg used, if you wanted to? A. Yes, sir.

Q. Why did you not use the bullet he used, but used the one you used? A. Because that bullet that I used was more,—corresponded more to the bullet than the other one, than the Winchester.

Q. What would be the difference in the effect of discharging a bullet, two bullets through the same gun, one that was a hollow base and one that was a flat base? What would be the effect? Explain the different effects, if there would be any? A. You get more upset, more sluggage, owing to the hollow base. At the base of the bullet you get a better gas check. The hollow base gives a better gas check at that pressure.

Q. In other words, as the bottom of this magnifying glass, you get a better gas check, a gas grip, on the hollow than you would on a perfectly flat base? A. Yes, that is my experience.

Q. How would that manifest itself on the bullet as the bullet went through the rifling of the barrel with reference to the marks of the lands and grooves on the bullet? A. It would show a perfect print at the muzzle in the base of the bullet. You get more upset with a lower pressure.

Q. Which would you get the truer course of the bullet, the one that had the hollow base or the flat base? A. I would say hollow base.

. . . .

[1464]   J. HENRY FITZGERALD, SWORN

Q. (By Mr. Jeremiah McAnarney) What is your full name? A. J. Henry Fitzgerald.

Q. You reside where? A. Hartford, Connecticut.

Q. What is your present employment? A. I have charge of the testing room at Colt Patent Firearms Company of that city.

Q. Is that where this automatic pistol is made? A. At Colt's factory, yes, sir.

Q. How long have you been connected with the gun business? A. About 28 years.

Q. Prior to going with the Colt people, whom were you employed with? A. By Iver Johnson Company, in Boston.

Q. The Iver Johnson Company? A. In Boston.

Q. What was your work there? A. I had charge of the revolver department.

Q. How long were you in charge of the revolver department at Iver Johnson's? A. I don't remember exactly. Somewhere five to six years. I don't exactly remember.

Q. Prior to that five or six year's work where were you employed? A. At Manchester, New Hampshire.

Q. In the gun business? A. In the gun business, sporting goods, general merchandise.

Q. Well, before we get into the thing too deep, I call your attention to the revolver, Exhibit 27. Directing your attention to the hammer, what have you to say, Mr. Fitzgerald, whether that is a new hammer, new in March, 1920,—what is your opinion? A. May I look at this at the window?

MR. JEREMIAH McANARNEY. Yes.

THE COURT. Haven't you looked at it before?

[1465] THE WITNESS. Yes.

THE COURT. Do you need to look at it again?

THE WITNESS. I want to see that it is the same.

THE COURT. All right. Will this light do here?

THE WITNESS. I guess possibly it will.

(Witness examines revolver under light at desk.) The hammer in this revolver has every indication of being as old and used as much as any other part of the pistol.

Q. Does it bear evidence of having been used? A. You mean, has it been fired?

Q. Had the hammer been fired in use? A. In my judgment it has.

Q. What do you base that on? A. On the condition of the hammer nose.

Q. Hammer nose? A. Yes, sir.

Q. That is where it makes contact with the shell? A. With the primer.

Q. How about where the hammer comes in contact below, with the trigger? A. You mean in here (indicating)?

Q. Yes. A. It shows wear. Shows the amount of wear I would expect, I would expect taking into consideration the condition of the gun.

Q. That is all. Now, passing to other parts of the case, you have made an examination of bullet No. III? A. I have.

Q. And the Colt automatic, Exhibit 28, the so-called Sacco gun? A. I have.

Q. Now, are you familiar with the photographs that have been offered in evidence? Have you studied those? A. I have seen them.

Q. As the result of your study and comparison of the Colt automatic and bullet No. III,—you have seen the bullet itself. A. I have.

Q. And have examined it? A. I have.

Q. And have examined these six photographs of it? A. Yes, sir.

Q. Are you able to form an opinion,—were you able to form an opinion as to whether the bullet No. III was fired from the Colt automatic now before you, Exhibit 28?

MR. KATZMANN. One moment, if your Honor please.

THE COURT. Read the question.

(The question is read.)

MR. KATZMANN. There is nothing, if your Honor please, to indicate he ever looked at the Colt automatic, either in the preliminary question or in that question.

MR. JEREMIAH MCANARNEY. Previous to that he said he had examined it.

MR. KATZMANN. Examined the bullet.

MR. JEREMIAH MCANARNEY. The revolver, I thought.

(A portion of the witness' previous testimony is read.)

THE COURT. Suppose you ask him, to be quicker.

[1466] MR. JEREMIAH MCANARNEY. I think I asked him.

Q. However, have you examined this revolver, this pistol, they all call it? A. I have.

THE COURT. "This pistol." You mean by that the Colt automatic pistol?

MR. JEREMIAH MCANARNEY. Yes, if your Honor please. They have called it pistol. Revolver, Exhibit 28, Colt automatic pistol.

Q. You have examined that, have you? A. I have.

Q. In substance, repeating my former question, are you able to form an opinion—were you able to form an opinion as to whether bullet No. III was fired from Exhibit 28? A. I was.

Q. What is your opinion? A. My opinion is that No. III bullet was not fired from the pistol given me as Exhibit 28.

Q. Kindly now explain to the jury the reasons that you have for that opinion. A. The land marks on the No. III bullet do not correspond, in my best judgment, to bullets I have seen fired from this pistol.

Q. In what way do they differ? A. The land marks on the No. III bullet show a slippage at the top or front of the land mark on the bullet, and it isn't pronounced in the same way on the bullets I have seen.

Q. How many bullets have you seen that were fired through— A. Three that I examined.

Q. And what three were they, please? A. The three I understand fired by Mr. Van Amburg at Lowell.

Q. And you say you have compared those three with the photos here? A. Yes.

Q. Now, will you disassemble that gun and show us the breech. I call it "breech," but— A. (Witness does so.)

Q. Now, let us see, we had a,—where the bullet takes the rifling that appears here (indicating), I assume. It has been mentioned. A. That (indicating) is the lead. This (indicating) is the throat.

Q. Now, in your examination of this gun, did you examine the throat? A. I did.

Q. And whether,—what is the condition of that throat with reference to whether its condition is such as would cause a bullet to jump the lands? A. I find no condition in this barrel that would cause a bullet to jump the lands as shown in bullet No. III.

Q. On the photograph. Now, what causes the bullet to jump the lands? A. A rusty or corroded lead would cause that.

Q. Assuming that you have a condition of a bullet where the metal is turned, on which side of the groove will that manifest itself? As you look at the bullet, will it be on the left-hand side of the bullet—strike that out. As you look at the bullet, will it be on the left-hand side of the groove or the right-hand side of the groove? A. The turn-over of the metal?

Q. Yes. A. Would be on the left-hand side of the groove.

Q. For information, I would like to ask out loud where it is caused [1467] by slippage, does it manifest itself particularly on either side? A. On the right side.

Q. You would get the manifestation of the slippage on the right side? A. On the right side.

Q. Would that be wholly or in part? A. It would be wholly on the right side.

Q. I want to call your attention to the matter of—you were in court when Mr. Van Amburg testified? A. The last part of it, I believe.

Q. Well, take the,—he testified in substance that where the firing pin struck on the head of the cap, that the hole was off center. Will you state whether or not that is a condition,—to what extent that exists in Colt automatics? A. It does not exist in the Colt automatic. It is the anvil in the primer that helps to cause that condition.

Q. Now, will you explain to the jury how that condition comes about? A. The center of the anvil of the primer rests within—I think it is 60 thousandths from the top of the primer. When the firing pin strikes this blow, bringing the two metals together, that blow would cause the slippage of the shell one way or the other enough to make that mark look off center.

Q. I have here the primer,—the anvil, I think. A. This (indicating) would show the anvil. When those two are brought together—

Q. Would you step down? I do not believe they can follow you. Step about here and then the gentlemen— A. (Witness steps down from stand). This (indicating) is the primer, and here (indicating) is the anvil. When the blow is struck here (indicating) if it is struck, we will say, nearly central, as these two metals come together, the anvil is braced and it is liable to throw the firing pin one way or the other on the head of that primer.

Q. Now, down here (indicating). A. This (indicating) is the primer. Inside is the anvil. When the blow is struck here (indicating) bringing those two metals together, igniting the fulminate, the anvil, being braced in the center, would cause the firing pin to be thrown a trifle one side or the other. (Witness returns to the witness stand.)

Q. Mention has been made here that there appears in shells fired to be what has been called a flow-back of the metal of the primer. State to what extent that exists. A. Why, that was nothing strange. That exists in all guns to a certain extent. It depends a great deal on the hardness of the primer.

Q. Does it in any way,—is it brought about in any way by the amount of percussion? A. By the pressure, yes. Ten to fifteen thousand pounds in the gun before me.

Q. Whether that is a common or uncommon thing to find in shells, such as are used in these automatics? A. It is a common thing to find.

Q. From your examination of bullet No. III, are there any indications on that bullet which you can say now from your examination that are caused by any pitting of the gun, Exhibit 28? A. You mean in this gun?

[1468] Q. Yes. A. I can see no pitting or marks on bullet No. III that would correspond with a bullet coming from this gun.

MR. JEREMIAH MCANARNEY. I think that is all for the present. I won't go into the matter,—you are familiar with the Bayard, are you, and other like guns?

THE WITNESS. Yes.

MR. JEREMIAH MCANARNEY. Has the Bayard 25 a right or left twist?

THE WITNESS. A left.

MR. JEREMIAH MCANARNEY. Well, I won't go into that.

CROSS-EXAMINATION

Q. (By Mr. Katzmann) What is the twist of a Bayard 9 millimeter? A. 9 millimeter?

Q. Yes. A. I haven't got a blueprint of that pistol. It is made in Belgium. I haven't got a blueprint of any specifications of that.

Q. Have you a Bayard 9 millimeter pistol, aside from the blueprint? A. Not here, no, sir.

Q. You haven't one at all? A. I haven't one here, no, sir.

Q. Not one here. A. I have handled a great many of them, but I haven't got one here.

Q. Is your answer that the twist of a Bayard 25 is left founded upon blue print or founded upon the Bayard 25 gun? A. Founded upon the gun and bullets fired from the gun.

Q. Now, then, have you seen a Bayard 9 millimeter? A. You mean a 380 Bayard, 380 or 9 millimeter? Yes, I have.

Q. Have you examined the rifling in it to see whether the twist is right or left? A. It is left.

Q. Are you sure of that? A. Yes.

Q. Can you conceive— A. The one I had was left. I do not know about any others. As far as I know they are all left.

Q. Can you conceive of a Bayard 380, or 9 millimeter, manufactured by the Belgium Bayard concern in one gun being left twist and in another being right twist? A. I don't know why they should change it, why they should not make them all left.

Q. It would be very poor shop practice, wouldn't it? A. It may be.

Q. To have some of their 380 left twist and some right twist? A. They might have tried them both ways. What I saw was a left.

Q. To change a Bayard 9 millimeter from a left twist to a right twist manufacture, would that involve changing shop tools? A. Changing your machine, yes, your rifling machine.

Q. And it would be a matter of quite considerable expense, wouldn't it? A. I could not say what their expense would be on the other side.

Q. Well, on this side? A. It would be expensive over here.

Q. Thousands of dollars? A. Possibly.

Q. Is it the fact that for the purposes of directing a bullet on its [1469] true course with the nose on all the time, that a right twist is just as effective for that purpose as a left twist? A. Either one.

Q. One has no benefit over the other? A. No.

Q. And the question of whether any particular make of gun should have a right or left twist is the matter of the choice of the manufacturer, isn't it? A. A great deal, yes.

Q. How long have you been in charge of the testing room at the Colt place? A. Between three and four years.

Q. And before that was your business wholly that of a salesman? A. I had charge of the department of the Iver Johnson.

Q. When you were at Iver Johnson's, would you call yourself a salesman? A. Part of the time. I had charge of the revolver department, and of course I sold pistols.

Q. Did you make any tests as to the effect of the rifling on bullets prior to your going to the Colt place? A. Yes, a great deal, a great many.

Q. In connection with your salesman's work or being in charge of the department? A. In connection with, a great deal with trials for court business and for my own satisfaction.

Q. And do you say, Mr. Fitzgerald, that the hammer in Exhibit 27 is not a new hammer? A. I believe my answer was it showed wear the same as the rest of the revolver.

Q. Do you say that in March, 1920, a new hammer was not put in Exhibit 27? A. I don't know anything about the gun in March, 1920. I only saw it a day or two ago.

Q. From your examination of the gun now before you, Exhibit 27, will you say a new hammer was not put in that gun in March, 1920? A. The hammer shows the same amount of wear as the rest of the pistol.

Q. That is not the question. Will you answer my question. A. May I hear the question again, please.

Q. From your examination of Exhibit 27 will you say that in March, 1920, a new hammer was not put in that revolver? A. (Witness hesitates. I can only answer that as I did before. In my best judgment it shows the same amount of wear as the rest of the pistol.

MR. KATZMANN. I ask that be stricken from the record.

THE COURT. It may be. Now, answer that, if you please. If you cannot, say so, and he will go on with the next question.

THE WITNESS. I can't answer it only as I did before.

THE COURT. You can't answer it? Put the question again and see if you cannot.

Q. From your examination of Exhibit 27, will you say that in March, 1920, a new hammer was not put in that gun? A. I can't say what was done to the pistol in 1920.

Q. From your examination you can't say, can you? A. From my examination it shows the wear of the rest of the pistol.

[1470] Q. But from your examination you can't say that it was not a new hammer in March, 1920? A. I can't say what was done to it in March, 1920.

Q. You can't say a new hammer was not put in from any examination you can make? A. I don't know what was done with the pistol in 1920.

Q. But from your examination— A. That is what my examination shows.

Q. And your examination does not tell you a new hammer was not put in in March, 1920, does it? A. I don't know.

Q. You don't know? A. (Continued) —what was done to it in 1920.

Q. Excuse me. A. I don't know what was done to it in 1920.

Q. You don't know even after examining the pistol, the revolver itself, do you? A. I can only give the same answer I did before.

Q. Examine it as you will, can you say that a new hammer was not put in in 1920? A. Without knowing how much that gun has been used since March, 1920, I couldn't say.

Q. That is the whole nub of the question, isn't it? I will withdraw that. Can any man, Mr. Fitzgerald, other than the man who put the new hammer in, if one were put in, without knowledge of the number of times the gun has been used in the sense it has been cocked and fired, can any man say that a new hammer was not put in that in March, 1920? A. What do you mean by "cocked and fired"? Do you mean with a shell in it, fired with a shell in it?

Q. No, fired without the shell, this way (snapping gun). A. That revolver has been fired with a shell in.

Q. Assume that. But I am talking,—has that revolver been fired with a shell in it since March, 1920? A. I did not say that. Since the hammer was put in.

Q. I did not ask you that, did I?

MR. JEREMIAH McANARNEY. Have the question re-read.

(The question is read as follows: "Q. Can any man, Mr. Fitzgerald, other than the man who put the new hammer in, if one were put in, without knowledge of the number of times the gun has been used in the sense it has been cocked and fired, can any man say that a new hammer was not put in that gun in March, 1920?")

Q. Can any man, other than the man who put a new hammer in, if one were put in, in March, 1920, having no knowledge of the number of times that Exhibit 27 has been cocked and fired, tell whether there was or was not a new hammer in it in March, whether or not a new hammer was put in in March, 1920? A. You mean fired with a shell in it?

Q. I mean fired, as I said before, without a shell in it? A. It can't be fired without a shell in the gun, sir.

Q. You are playing on words, aren't you? A. No, sir.

[1471] MR. JEREMIAH McANARNEY. Pardon me. I object to that.

THE COURT. You may ask him if he is playing on words.

Q. Are you playing on words? A. No, sir. I can't fire a gun without a shell in it.

Q. What do you call that (indicating)? A. Snapping the hammer.

Q. Then you did not understand my question when you said, "snapped"? A. You said "fired." I understood you meant with the shell in.

Q. I will repeat the question. Can any man, other than the man who put in a new hammer in March, 1920, if a new hammer were put in in March, 1920, having no knowledge of the number of times the gun has been snapped since March, 1920, tell whether a new hammer was or was not put in in March, 1920? A. I do not think the man who put it in could tell.

MR. KATZMANN. I ask that be stricken out and the witness answer my question.

THE COURT. It may be stricken out.

Q. Let him answer my question. A. I cannot tell what was done about that gun since March, 1920.

MR. KATZMANN. I ask that be stricken out.

THE COURT. It may be.

Q. Can any man tell? A. I don't know what any other man can do.

Q. Then you mean you can't tell? A. I can't tell how many times it has been snapped.

Q. I did not ask you that, sir. Do you think that is the question? I am asking you if any man, with those assumptions I have made to you three times——

MR. JEREMIAH McANARNEY. I think, if your Honor please, that question ought to be confined to this man. He is not responsible for any other man. It seems as though all he can do is to take care of his own observations.

THE COURT. Well, that is true. I suppose the question includes the man of the ordinary skill in this kind of work, whether he could tell.

MR. JEREMIAH McANARNEY. It seems as though that question is asked of this man and that this man should be asked that question.

THE COURT. Well, if you think so perhaps Mr. Katzmann will confine it to the witness.

MR. KATZMANN. Yes.

Q. Mr. Fitzgerald, this is the fourth time. Can you tell, without any knowledge of the number of times that Exhibit 27 has been snapped since March, 1920, whether or not a new hammer was put in Exhibit 27 in March, 1920? A. I can't tell what was done to the gun in 1920.

Q. Does that mean you can't tell whether a new hammer was put in in March, 1920? A. May I answer that by mentioning the things that cause me to think this was not a new hammer?

[1472] THE COURT. That is not the question now. If Mr. McAnarney desires any explanation, he will undoubtedly call for it when the proper time comes, but now it should be answered by yes or no. It is your duty so to do, leaving the explanation to be brought out by Mr. McAnarney.

THE WITNESS. I can't fairly answer yes or no.

Q. The question is, does your last answer, to wit, "I don't know what was done in March, 1920," does that answer mean that you can't tell whether or not a new hammer was put in in March, 1920? I ask you to answer that question yes or no. A. I can't tell when the new hammer was put in. It doesn't look as though one was ever put in.

Q. I am showing you, Mr. Fitzgerald—have you a glass? I am showing you Exhibit 30, being four empty shells which we have called the "Frahar" shells. Will you tell me if the one of those four I show you is a Winchester shell? A. This (indicating) is a Winchester shell.

Q. Yes. Now, will you help me to keep these so we won't get confused again? A. I will try to.

Q. Now, I call your attention to Exhibit 34, being three empty shells, and ask you if they are three Winchesters? A. These are three Winchester shells.

Q. I now call your attention to Exhibit 33, three empty shells, and ask you if those are three Peters shells? A. Three Peters shells.

Q. I ask you to examine with your glass, if you please, the mark made by the firing pin in the base of the one Frahar, a Winchester shell, and of any one of the three Winchester shells fired at Lowell through the Sacco gun, noting particularly the characteristic of the indentation made in one or all of those three. A. (Witness examines.)

Q. Now, if you will please examine the three Peters shells, which are from the cartridges fired at Lowell by Capt. Van Amburg, as to the indentations on the base made by the firing pin. Have you so examined the seven shells? A. I have examined them.

Q. Do you notice any difference in the flow-back at the base of the three Peters shells fired at Lowell and the three Winchester shells fired at Lowell? A. (Witness examines). A difference in all of them.

Q. Is there any class difference between the three Peters at Lowell and the three Winchesters at Lowell? A. By "class," what do you mean?

Q. I mean a difference that is common to all three of the Peters as distinct from a common characteristic of the three Winchesters fired at Lowell? A. There is a difference in the three Peters and there is also a difference in the three Winchesters.

Q. How does the flow-back on the Frahar shell, a Winchester, compare with the flowback on the three Lowell Winchesters? A. I do not think it does compare.

Q. Is it not the same? A. No, sir.

Q. Do you say, Mr. Fitzgerald, that in the Frahar shell there is not a clear-cut, well-defined, narrow flow-back that is characteristic of the [1473] three Winchester shells fired at Lowell that is not present in the three Peters fired at Lowell, a marked lip at each indentation that is absent from the three Peters fired at Lowell? A. There is no mark in this so-called Frahar shell that I could identify with the marks on these three shells.

MR. JEREMIAH MCANARNEY. What are those three shells?

THE WITNESS. There is a flow-back on all of them.

Q. Yes, there is a flow-back in all of them, but is there any difference in the characteristic outline of the flow-back? A. There is a difference in these three.

Q. Is there any common difference between the Peters at Lowell and the Winchesters at Lowell? A. There is a difference in the three Peters. They are not the same. They don't show the same mark.

Q. Don't the Frahar shell and the three Winchesters fired at Lowell show two concentric rings at the indentations, the indentation ring itself and a narrow lip around it outside? A. (Witness examines) Those three don't show the same mark.

Q. Mr. Fitzgerald, at least three or four times I have asked you specifically about a difference in the ring. Have I understood,—I asked you if the three at Lowell of the Winchester variety show the same. Have I asked you that? A. I believe not.

Q. You have answered it three or four times? A. I have.

Q. You haven't answered the question I put. A. May I hear the question again?

Q. Yes, sir. I show you a rough chalk I have just made, a segment of the Frahar Winchester and the three Lowell Winchesters, showing a little lip, making what appear to be two concentric rings in the Winchesters? A. Not the same shape lip. It is a different shape.

Q. Does it show a little lip or ridge around the outside of the hole made by the firing pin? A. That shows in most all shells, a great many.

Q. Does that show in the three Lowell Peters? A. Shows in one of them, yes, sir.

Q. Does it show in the other two? A. I can't see it in the other two.

Q. And is that the one you say it shows in, handing you one of Exhibit 33? A. That is the one that throws up a ridge.

Q. Around the indentation made by the— A. Firing pin.

Q. —firing pin. You don't see it in the other two, do you? A. I do not see it in these two, no.

Q. You do see that ridge, do you not, in the three Lowell Winchesters and the Frahar Winchester? A. I see that ridge in those three, and I see a ridge with that one shell.

Q. I will take the Frahar Winchester in my left hand and the Lowell, one of the three. It is the same in the three, isn't it? A. Not the same.

Q. Well, take any one of them that shows the same. A. I can't say it shows the same. It looks different to me.

[1474] Q. All right.

MR. KATZMANN. (To the jury) One of you gentlemen take that. You remember that is the Lowell Winchester. That (indicating) is the Frahar Winchester. Please look at the indentations and see if you see around the edge of that hole or depression made by the firing pin a very thin ring of metal, and compare that, if you please, then with the—that (indicating) is the Frahar. Now, that (indicating) is the Lowell. You take the Lowell, if you please. Did you all hear what I said, so I won't have to repeat it again? That (indicating) is the Frahar shell. That (indicating) is the Lowell shell. Now, will you please keep the glass and look at the Lowell one. That (indicating) is the Frahar one. Now, will you wait just a minute so I won't have to travel this course twice. I now take, while you two gentlemen have the glass, one of the Peters, one of the two of the Peters fired at Lowell, and ask you to please observe and see if there is any such ridge around the hole. Now, will you please pass that on. That is the Lowell,—Frahar, Lowell Winchester, Lowell Peters. Will you pass that on, please? That (indicating) is the Lowell Winchester, the Frahar Winchester. That (indicating) is the Lowell Peters. I ask you if on that indentation you see any clean-cut ridge or ring. That (indicating) is the Frahar Winchester. That (indicating) is the Lowell Peters. Now, perhaps I am going to get mixed up here. That (indicating) is the Winchester at Lowell. What is that you have got? And that (indicating) is the Frahar.

Q. Let us see a minute. That (indicating) is the Peters or Winchester, which? A. Winchester.

MR. KATZMANN. That (indicating) is a Winchester. That (indicating) is a Lowell Winchester. Mr. McAnarney, I can't, without—get them mixed up. (Mr. Katzmann shows to Mr. McAnarney).

MR. KATZMANN. If your Honor please, I did not ask your Honor's permission, but with Mr. McAnarney's consent I marked that Frahar one with a mark on the side, so as to distinguish them.

THE COURT. All right.

Q. Is that (indicating) a Winchester? A. Winchester shell.

MR. KATZMANN. (To the jury) I have marked, gentlemen, on the Frahar Winchester a cut, a straight line,—see that shadow? Which one is that, a Winchester? All right. There is the Frahar shell, a Winchester, and please examine that hole. This (indicating) is the Frahar shell. That (indicating) is the Peters at Lowell. What is that, Winchester or Peters? Here is the Lowell Peters. That (indicating) is the Frahar Winchester.

Q. What is that? A. Winchester.

MR. KATZMANN. That (indicating) is the Lowell Winchester. I call your attention to the indentation. See if you see a little ridge around the hole made by the firing pin. That (indicating) is the Lowell Peters. I ask you to examine that hole and see if you see a little concentric ring [1475] around that hole. See if you notice any little ridge around the hole made by the firing pin. That (indicating) is the Lowell Winchester. See if you see similarly a ring around that hole. That (indicating) is the Lowell Peters. See if there is absent such concentric ring on that one. That (indicating) is the Frahar shell that you have. Please notice if you can see any little ridge around the hole made by the firing pin, concentric—that (indicating) is the Lowell Winchester. See if you see a similar ring there. That (indicating) is the Lowell Peters. See if you see an absence of such ring in that shell.

Will you hold that and pass the glass now. You, Mr. Juryman, now have the Frahar shell. Here (indicating) is a Winchester at Lowell. Now, take that, and there you have the Peters at Lowell. See if you notice the absence of such ring. You now have the Frahar shell. You now have the Winchester fired at Lowell, and will you take from your associate the Peters fired at Lowell. You hold that and keep it straight. You have the Frahar (indicating). There (indicating) is the Winchester at Lowell. You have the Frahar. You now have the Peters fired at Lowell. You have the Frahar shell. You now have the Winchester fired at Lowell.

Q. Which are the Peters? A. (Witness shows.)

Q. That (indicating) is a Peters, isn't it? A. Peters?

Q. And those are the two Winchesters. Is that difference, Mr. Fitzgerald, of any significance to you? A. What difference do you mean?

Q. In the concentric rings in the primer portion of the shell. A. Do you mean in regard to these shells?

Q. Yes. A. That some throw out more of a ridge?

Q. Yes. A. That is,—that happens in all guns, to a certain extent.

Q. Is it of any significance to you, is the question. A. Significance?

Q. Yes, in the matter of identification. A. I could not identify those shells.

Q. Have you examined, Mr. Fitzgerald, the sulphur cast that was testified Mr. Burns made of the Sacco barrel? A. I think there were two here yesterday, full length and a short one.

Q. I refer to the one that was put in originally. A. I haven't examined it.

Q. (Handing to the witness) Well, have you examined,—looking into the muzzle of Exhibit 28, holding those gear marks—if I may so speak of them—in clock terms, at nine o'clock, have you examined the right hand side of the lands at what would appear to be at six o'clock? A. May I take this to the window?

Q. Yes, by all means. A. (The witness examined at window).

Q. And will you,—well, come back here. All right. A. At nine o'clock?

Q. Yes. Holding those little gear marks at nine, looking at six o'clock in the barrel. Perhaps you better step back here because of the stenographer.

[1476] (Witness returns to stand.)

Q. Do you notice any pitting at six o'clock when the barrel is held that way? A. It is pitted in all directions.

Q. At six o'clock? A. It is pitted at six o'clock on both sides of the lands, pitted, in fact, all around the barrel.

Q. Have you found more pronounced pitting along the mark of the land when the barrel is held in that position more pronounced at that land than at any other? A. How far from the muzzle?

Q. About one inch? A. About the same number of pits on the,—at that side that there would be on the other side.

Q. You mean that there are on the other side? A. There are about the same number of pits on all sides of that barrel.

Q. You say it isn't any more pronounced when the gun is held in that position? A. You mean the number of pits?

Q. Yes. A. I can't see there is any more on that side than any of the other sides.

Q. Then I ask you to examine Exhibit 17, which is the sulphur cast made by Mr. Burns, and ask you if you find, about one inch down from the muzzle end of that cast, one one of the grooves, being the mark made on that cast by the lands, pitting that corresponds identically to the pitting you see at the nine o'clock land, at the six o'clock land held in that position (indicating) substantially in the middle of the cast? A. I can find no mark on this that would correspond with any pit in the barrel.

Q. Any line of pitting, not any one pit? A. With any line of pitting.

Q. Do you find any evidence of pitting on that cast? A. Yes, sir, I find evidence of pitting.

Q. Do you find it more pronounced in any one groove on the cast? That is, the land mark, than on any other similar groove? A. No, as it is handed to me here, I can't say that I do.

Q. I am asking you about them all. A. It shows evidence of pitting.

Q. No, wait a minute. You answer my question. A. I will try to.

Q. Is there evidence of greater pitting on any one groove in the cast than on the rest of them? A. I find evidence of it on three. I can't fairly answer, can't on one. I find evidence of greater pitting on three than I do on the others.

Q. Take the three on which you find greater evidence than on the other three, and see if any one of those three is greater than either of the other two? A. May I take this to the window, please?

MR. KATZMANN. Yes; yes, sir.

(Mr. Fitzgerald examines at window.)

Q. What is the answer? A. Very little difference in those three. Two are about the same. There is very little difference in the three. They both show——

Q. How is the third one? A. Very nearly the same as the others.

[1477] Q. But a little more than the other two? A. No, about the same.

Q. Well, is it different, or isn't it? A. A trifle different. They are all different.

Q. All right. Now, those three, is there any one that shows more pitting than either of the other two, is the question? A. I can't say that there is, no, sir.

Q. But you say two are very nearly the same, and three, that is different? A. Not the same. They are all different.

Q. Did you say, when you came back from the window, you saw two about the same? A. About the same.

Q. That is, you meant by that, did you not, the third was different from those two? A. A trifle different on one side.

Q. Is it different in greater degree of pitting or lesser degree? A. On one side it is about the same.

Q. And on the other? A. Not quite so much.

Q. Not quite so much. Do you find in those two that have the greater pitting on one side which you say are about the same, do you find a greater degree of pitting on one side or the other of either of those two, greater than its fellow? A. No, sir.

Q. They are identically the same? A. About the same, as near as I can judge.

Q. Will you say that one is not greater than the other? A. My last answer, about the same, as near as I can give to that.

Q. Will you not say one is greater than the other? A. About the same.

Q. Will you say it is not greater? A. I can't say.

Q. Can't say? A. I will answer, about the same.

Q. The condition you speak of, as to the pitting on the sulphur cast, must necessarily, must it not, be the same in the barrel? A. Not necessarily.

Q. Why, the cast is a cast of the interior of that barrel, isn't it? A. When a pit occurs——

Q. Wait a minute. The cast is a cast of the interior of that barrel, isn't it? A. As nearly as can be taken out.

Q. And is it its purpose to show, as Mr. Burns said to a ten-thousandth of an inch the condition there? A. You can't show a pit.

Q. You can't show a pit? A. Not always, no. If I can explain that?

Q. No, I haven't asked you for an explanation. I have asked you for a direct answer. Then that cast that is made, in your opinion, does it or does it not show the interior of that barrel? A. It doesn't measure accurately.

Q. Accurately to a ten-thousandth of an inch? A. It doesn't show all the pits.

[1478] Q. Of course it doesn't show all the pits, because it is only a part of the barrel. A. It doesn't show as far as it goes.

Q. Doesn't show all the pits to the length to which the cast goes? A. It doesn't.

Q. Do you mean you can see pits in that barrel as you look at it with that glass that are less than one ten-thousandths of an inch. A. I couldn't say that I could.

Q. You know that you couldn't, don't you? A. I could not with what I have here, with the glass I have here to do with, I could not.

Q. Do you agree,—have you ever examined it with any better glass, the cast? A. No.

Q. Do you agree with Mr. Burns in his statement of yesterday the cast shows things to ten-thousandths of an inch,—to one ten-thousandth of an inch? A. I thought he said ten-thousandths of an inch.

Q. Was he talking about that difference here when he said ten thousandths of an inch,—wasn't he talking about the difference in size of the groove on bullet No. III at the upper end? Wasn't that what he was talking about, ten thousandths of an inch or one-hundredth of an inch? A. I don't remember.

Q. You don't remember. Does that cast show to one ten-thousandth of an inch the condition of the interior of the barrel? A. It shows very nearly. I won't say it is one ten-thousandth, exactly.

Q. Without being able to see to one ten-thousandth, do you say to this jury the pits displayed by markings on the cast do not contain everything that the barrel shows, as to pits? A. They could not contain a deep pit. That would be scraped off when the cast is taken out.

Q. It would contain the imprint of the upper surface of the print, wouldn't it? A. It may, possibly.

Q. You mean, it would not show the extreme depth of it? A. That would

depend on how quick the cast was taken out. If it was taken before the
sulphur was a little cool, it wouldn't.

Q. Are there any pits, from your examination of the interior of that
barrel, that are so deep that they do not show on the cast? A. I can't pick
out any particular pit as showing on that sulphur cast. I can't look through
the barrel and pick out any particular pit and pick the same thing on your
cast.

Q. Yet you were willing to answer to this jury, weren't you, "The cast
does not show all the pits in that barrel" in the extent to which the cast
penetrated the barrel? A. I can't say it does.

Q. You did say it, didn't you? You did say it a moment ago, didn't you?
A. That it doesn't show?

Q. Yes, that it doesn't show all the pits? A. It may not show all the
pits.

Q. Did you say a moment ago it did not show all the pits. A. I think I
said it may not show.

[1479] Q. Did you say it did not show? A. I think I said it may not
show.

Q. Did you say it did not show? A. I don't think that I did say that.

Q. Have you examined the three Winchester bullets fired at Lowell? A.
I have.

Q. Do you find any evidence of a greater width at the top end of the
grooves made by the lands from the width of the same grooves at the bottom,
or base end? A. May I see the bullets again?

Q. I was going to save time. I thought you said you had seen them. A.
I have seen them.

Q. Do you need to examine them again? A. I would like to, yes, sir.

Q. (Mr. Katzmann hands bullets to the witness.) See if those are the
Winchesters, Exhibit 35. Look at the base. Take the one you marked with
the square or a square and a dot. A. This (indicating) has the square.

Q. Look at the one that has the square and two dots. No, the square is
the one with the letter "W" on it, I think. A. This (indicating) has the
square with the letter "W."

Q. Now, look at the groove that is at the base of the letter "W." Is the
upper end of that groove any wider than the lower end? A. At the base of
the "W"?

Q. That is the top. A. Of the land mark?

Q. Of the land mark, yes.

Q. Now, at the top of that land mark, directly under the letter "W," is
that land mark any wider at its bottom, or base end? A. You are speaking
now of that one mark on the left-hand side of the land mark?

Q. Yes. A. It is a clearly defined mark.

MR. KATZMANN. I ask that be stricken from the record, if your Honor
please.

THE COURT. That may be. It is not responsive. That is the reason, Mr.
Witness.

Q. Is it any wider, is the question: wider, that is the question. A. You
mean one side of the land mark?

Q. I do not mean any such thing. I am asking you if the groove made

directly under the letter "W" in the bullet which you hold in your hand, if
that groove mark is any wider at the top under the letter "W" than it is at
the base. That is the question. A. (Witness examines bullet.) I don't
understand what you mean by that one line being wider.

MR. KATZMANN. That is all, sir.

THE COURT. You may have a recess of five minutes. The mornings are
a little longer now than before.

(Short recess).

Q. (By Mr. Katzmann) Mr. Fitzgerald, you are employed, are you [1480]
not, at the Colt concern that makes the type of weapon shown as Exhibit 28?
A. I am.

Q. What is the standard width of the barrel in a "32"? A. Lands and
grooves. I have a blueprint of it here. You want the width of the lands?
You want the width of the lands?

Q. Did you hear the question? A. As I understood it, you want the width
of the grooves.

THE COURT. Read the question.

(The question is read).

THE WITNESS. .052.

Q. What is the standard width of a barrel groove in a Colt 32? A.
.105.

MR. KATZMANN. That is all.

REDIRECT EXAMINATION

Q. (By Mr. Jeremiah McAnarney) Will you kindly explain what you
mean when you say that cast does not show or may not show the pits? State
that fully so we may get it.

MR. KATZMANN. Does not or may not?

MR. JEREMIAH MCANARNEY. May not or does not. Perhaps his explanation will answer that.

THE WITNESS. When the sulphur is poured into the barrel hot it, of
course, finds the bottom of all the pits. When that is pushed out of the
barrel, the raised part or the part which goes into the pits, is, in a great
many cases, scraped off. Therefore, it is hard to find all the pits that are
in the barrel.

MR. JEREMIAH MCANARNEY. Is that plain to the jurymen? That is, the
sulphur sinks into the holes. Then when it is pulled out the sulphur that
was in the holes is broken off, but you have an imprint, do you not, of the—
pardon me, I am leading.

MR. KATZMANN. I wasn't thinking about that, but I was thinking about
your testifying.

MR. JEREMIAH MCANARNEY. Oh, well, I would like to have the chance.

Q. Now, looking at this sample again, what is the effect of what you have
just stated on the cast? Assume that the pit in its depth, as it sunk in, is
brushed off,—to what extent will it or does it show on the cast? Will it or
does it show on the cast? A. It would sometimes show a little mark on the
cast. If the cast was cool and the barrel was a trifle cool when it was taken
out, it would show a broken,—a little breakage. If it was warm it might have

a little elasticity and not show the pit in the same relation, but it usually shows a little stain or mark when it is taken out.

Q. Now, the matter you have been interrogated about, the flow-back on those shells, whether or not that is a common thing? A. It is common.

[1481] Q. In all automatic shells where the primer is of copper? A. It is a common thing.

Q. And what have you to say where the primer is of brass? I understand some of those Peters were brass primers, were they not, or were they all of them brass primers? A. Copper, I believe was what was shown me this morning, most of them. I guess they all were.

Q. I have 34 here. That is one or the other. The 34, they aren't copper, are they? A. Copper primer.

Q. That is the Winchester. Now, I direct your attention to 33, the so-called Peters shell. They are copper, are they? A. Copper primer.

Q. Now, with reference to the flow-back and the amount of flow-back you get, is it not dependent somewhat on the thickness of the metal in the primer? A. It is.

Q. That fullest amount of pressure developed by the explosion of the shell gives you the amount and the extent of the flow-back, does it not? A. Yes, sir.

MR. JEREMIAH MCANARNEY. That is all.

RE-CROSS-EXAMINATION

Q. (By Mr. Katzmann) Just one question. When I asked you the last two questions about the width of the barrel land and barrel groove, which you gave respectively as .052 and .105, did you not— A. I did.

Q. —that is the standard which you seek to attain, is it not, in the manufacture of the pistol? A. Yes, sir.

Q. Is it attained necessarily within one or two thousandths of an inch? A. Within a thousandth. We work within a thousandth.

Q. And that degree of difference from the standard is called "tolerance"? A. Tolerance.

Q. And you may run as high as two? A. I have never seen any.

Q. Did you hear Mr. Burns testify that the lower end of the land mark on the bullet was .050 yesterday in the Colt? A. I don't remember what he said.

Q. Did you hear him testify? A. I heard him testify.

Q. Don't you remember that? A. I do not remember that he said——

Q. Don't you remember that particular thing, it was .050 at the bottom and .060 at the top? A. I don't remember what he said about the particular question.

MR. KATZMANN. You don't remember. That is all.

REDIRECT EXAMINATION

Q. (By Mr. Jeremiah McAnarney) Just a minute. I want to show you —wait a minute, please. I show you two unmarked shells, unmarked. No, they are marked with letter "A." Will you examine those shells and state what they are? A. Peters, a Peters Cartridge Company shell.

[1482] Q. I ask you to examine the flow-back on those, particularly with reference to the flow-back in Exhibit 34. I desire you to be careful not to get those mixed up. Here (indicating) are your 34 exhibits. A. The up-settage, or flow-back, is different on all five shells.

Q. How do they compare with each other, the two there with any two or any one of those, with any one of those three? A. They all show flow-back.

Q. Is there any similarity, or are they different or can you tell? Is there enough of difference to tell? A. I couldn't tell. They are all slightly different. All five are slightly different.

MR. JEREMIAH MCANARNEY. I would like to have those put in an envelope and marked for identification.

(Two shells marked letter "A" are marked "Exhibit 25 for identification.")

Q. You said you did not understand what he meant by one question. Have you forgotten that question now? A. Something about the mark, wasn't it?

Q. If there was anything you didn't understand that you want to explain? A. I did not understand his last question, as he put it, I did not understand it.

MR. JEREMIAH MCANARNEY. Well, I have forgotten it. That is all.

THE COURT. Is this all the expert testimony to be offered by the defendants, Mr. McAnarney?

MR. JEREMIAH MCANARNEY. Well. I am inclined to think so.

THE COURT. (To Mr. Katzmann) Do you expect to offer evidence of experts in rebuttal?

MR. KATZMANN. Yes, your Honor.

THE COURT. Would it not be as well, if you are now prepared, to meet that at the present time in order that possible considerable expense might be saved?

MR. KATZMANN. Yes, it is.

THE COURT. It would be a saving of expense to the defendants as well as to the Commonwealth if you are now ready. If you are not ready, of course, I am willing to leave it to you to exercise—

MR. KATZMANN. I would like to have five minutes' conference.

THE COURT. Suppose you wait until afternoon.

MR. KATZMANN. I would be glad to.

THE COURT. You can confer then, and with the understanding you may offer that rebuttal after lunch. And then the experts on both sides can be excused. I should think that would be satisfactory to both sides. What do you say, Mr. McAnarney?

MR. JEREMIAH MCANARNEY. That is very agreeable to me.

THE COURT. All right. Then you put on some other evidence. They may confer with their experts during the noon hour, and at two o'clock they may go on with their expert testimony in rebuttal.

The testimony of the experts for the defense was utterly incompetent: it displayed an ignorance of any reliable method of proof employed in the identification of firearms. The lack of qualifications was manifested throughout the entire testimony. Both experts exhibited a lack of familiarity with the basic principles involved in the formation of the land engravings upon the surface of a bullet. Their testimony did not explain that the widening of the land mark is caused by the stripping of the bullet, and that a double land mark is produced when a bullet strikes the forcing cone with its axis inclined to the axis of the bore. The testimony did not reveal information as to the effects produced by variations in the diameters of bullets.

The tragic feature of the expert testimony for the defense was that the experts gave largely unresponsive answers to questions, and Burns, particularly, attempted to support his opinions with vague explanations. These vague explanations and indirections in meeting questions must have made a most unfavorable impression upon the jury, with the result that, despite the shortcomings of the testimony submitted by the experts for the commonwealth, this evidence for the prosecution, having been presented more sincerely and frankly, certainly carried the greater weight with the jury.

The first expert for the defense was completely discredited by his statement that the most important measurements in determining whether a bullet has been fired from a particular weapon were of class characteristics, that is, "The width and diameter of the lands and the same of the grooves." He contended that bullet III might have been fired from a Bayard despite his measurements of 0.050 inch for the Colt land width, 0.040 for the Bayard, and 0.050 to 0.060 inch for the width of the land mark on the fatal bullet No. III, and furthermore, he did not know anything about the pitch of rifling in a Bayard. His testimony was not intelligible; it lacked sincerity; and it may be characterized as absolute folly. He did not use the proper instruments, and he was not able to point out any characteristics to prove that neither bullet III nor the W.R.A. Fraher cartridge case had been fired in the Sacco pistol.

The testimony of the second expert for the defense manifested a knowledge of arms and ammunition, and he did point out that the "flow-back" depended "a great deal on the hardness of the primer" which tended to weaken Van Amburgh's testimony, but, like Burns, he failed to point out intelligently any characteristics to show that bullet III and the W.R.A. Fraher cartridge case had not been fired from the Sacco pistol. Instead of confining his examination to the markings on bullets and cartridge cases he foolishly stated, "I find no condition in this barrel that would cause a bullet to jump the lands as shown in bullet No. III."

From the examination of the testimony on the record it may be conservatively concluded that the jury was in no position to determine whether the No. III bullet or the Fraher W.R.A. cartridge case had been discharged in the Sacco Colt automatic pistol. The incompetent testimony did not throw any illumination on the science of firearms identification. The opinions that were offered must be considered valueless because they are not based upon a combination of class and accidental characteristics establishing an individual peculiarity and which serve to distinguish between the bullets and cartridge cases fired in two different weapons of the same class characteristics.

It should be noted that the experts at the trial had not based their opinions upon an examination under a comparison microscope. On the contrary, they employed low-powered single microscopes and magnifying glasses. Of the photographs that were taken, the experts themselves were "gun-shy." There was no display of a knowledge of precision of measurements.

The experts did not have either the requisite knowledge or equipment to arrive at reliable opinions. They did not explain or point out to the jury any significant markings in an intelligible manner. Certainly neither the experts nor the jury were qualified to form an opinion on the basis of the testimony given at the trial.

PART 4

THE SUMMARY OF COUNSEL AND JUDGE THAYER'S CHARGE TO THE JURY

MR. MOORE in closing for the defense referred briefly to the bullets and cartridge cases:

> What is the issue in this case? The primary issue and the only issue here is the issue of identification. The one issue is, Has the Commonwealth proved beyond reasonable doubt that the defendant Sacco fired the fatal shot that caused the death of one or the other of these men? Or did the defendant Sacco aid or abet or contribute in any wise to that crime?
>
> . . . .
>
> There are some other questions, many, that I would like to discuss. The issue of the government, you have heard the testimony of the experts pro and con, back and forth. Gentlemen, if the time has come when a microscope must be used to determine whether a human life is going to continue to function or not and when the users of the microscope themselves can't agree, when experts called by the Commonwealth and experts called by the defense are sharply defined in their disagreements, then I take it that ordinary men such as you and I should well hesitate to take a human life.
>
> Particularly is this true when you consider the fact of the ancestry of these bullets and these shells. Remember that the shells were found on the street. Mr. Bostock said he found three or four, he did not know how many, and he put them in somebody's desk up at Slater & Morrill's and then Mr. Fraher found them sometime later. He did not know where they came from, except that he found them in the desk, and so on.[10]

Mr. McAnarney for the defense handled the evidence on firearms identification as follows:

> Now, gentlemen, there were experts on revolvers here. You heard them testify. I am not going to call for an exhibit at all. I will take them as they come. I will take the Colt revolver. We had Van Amburgh. I will call him the "circles" man. He was put on here by the government. He testified, and I don't like

[10] "The Sacco-Vanzetti Case." Vol. II, pages 2125, 2147.

these fellows that do this when they are saying something pretty strong, but Van Amburgh said that that No. 3 shell, the fatal bullet that killed Berardelli, came from the 32 Colt. Now, mark that, gentlemen.

And he says that the bullet 3, the one the doctors say killed Berardelli came from the Colt revolver that was found on Sacco. That is a fearful statement to make. Now, I challenge the record and I will quote the record to you almost word for word. You know there must be some peculiar outstanding identifying thing about that revolver which would warrant any man in taking that fearful responsibility and that is the responsibility of saying that and making that statement, and what does he say when asked what is that condition? He says that there is a flare back, that the flare back shows on the firing primer and when you find the primer— and with the most absolute ease and perfect abandonment he says it is not an unusual thing to find in Colt revolvers. Well, good Heavens, he says, gentlemen, "this is the shell that comes from this revolver because of this condition that I find in it," and I want to read now. You may think I am not quoting right. Now, let this answer of the expert's question, let it go out where it came in.

"THE WITNESS. A set back on a primer is a little flowing back of the metal beyond the true surface. Here is the true surface of the metal, that portion of it which is bearing against the breech block.

Q. What is the breech block. A. That is the point from which the firing pin protrudes.

Q. The firing pin comes through the breech block? A. It does.

Q. And strikes the cartridge? A. It does.

Q. Similar to the way my finger is now indicating? A. Yes. The set back takes place around the firing pin sometimes. It did in this case.

Q. And what causes the set back? A. Largely, a little open-ing—in my experience I have found it to be usually a little opening in the mouth of the firing pin hole. Do I make that clear?

Q. Is that something that occurs in all guns? A. It is not an unusual thing, but it does not occur in all guns."

It is not an unusual thing. Now, something that is not unusual conversely may be a usual thing. It is not an unusual thing. He finds that on this shell, and he says that is the reason how he proves that this is the identical revolver that fired that shell. Will you see where that reason gets you? If he had said it is an unusual thing, it is an uncommon thing, it is something that rarely occurs, the percentage of chances of that occurring are very remote, "and I would feel that on that—and in my experience, my experi-ence it seldom occurs, I feel on that I feel warranted in saying in my opinion this shell came from that revolver." Well, what do

you say when the man says it is not an unusual thing to find it?
Now, just think what a jump that is?

I am putting that fair and square, and I want to look every
man in the eye on this panel when this case is through, and I
want him to know that I have tried to do my duty, and I want
every man here to do his, and I want—when I am giving you
that man's statement, I ask you can I do more fairly to him than
to ask you to discard such stretch of reasoning as that?

He finds another position, that he said there is a marking on
the side of the groove caused by a fouling of the barrel. I asked
him where he finds the fouling, and he says it is right at the shoulder where the groove and the lands meet. I asked him if that is
not where he usually gets fouling in a used revolver. "Yes." I
asked him if there is any condition in that revolver that is not usually to be found in a used revolver, and he says "No." For good-
ness sake, where do we get when men have that elasticity of con-
science that men of penetration of mind, that they will jump that
fairly large gap and say this is the revolver that fired that fatal
shell? [11]

Mr. Katzmann, the prosecutor, treated the evidence on firearms
identification at some length:

We say in Plain English that on the evidence we have proven
to you beyond any reasonable doubt that the defendant Sacco
fire a bullet from a Colt automatic that killed Allesandro Berardelli; that some other person whose name we do not know and
who is not under arrest, in custody or upon his trial, killed the
man Frederick A. Parmenter with a Savage automatic, and that
that was not the defendant Vanzetti.

That is as plain as I can make it in English. We say that the
defendant Vanzetti, although we have offered no evidence that is
controlling that he on the day in question fired any fire-arm that
took the life of either of these men, that he was there actively
aiding and abetting and assisting Sacco and the other man who
were actually doing the killing, and that his presense there for
the purpose of assisting, standing ready to stand by, to render
such assistance as might become necessary in the events that might
immediately follow the taking of life, makes him in the contempla-
tion of the law just as guilty as the defendand Sacco, who is the
man we say killed Alessandro Beradelli.

Who it was that with a Savage automatic fired three bullets
from a Savage automatic into Berardelli and two into Parmenter,
we do not know and we have offered no evidence on, except that
there was such man, so that the force of my brother's suggestion
that Vanzetti were foolish to give up a Savage automatic to take
a 38 revolver is of no force in the light of that claim, and I

say to you, gentlemen of the jury, it is substantiated on the evidence.

That we do say to you is that we expect you find upon all the evidence that the 28 Harrington & Richardson revolver that was found upon the defendant Vanzetti was the 38 Harrington & Richardson revolver that poor Beradelli tried to draw from his pocket to defend himself and before he sunk to his knees with the blood coming out of his mouth dying on that sidewalk that afternoon.

.   .   .   .

Bullet no. 3, and a bullet that had a left hand twist, a bullet fired from some gun that gave a left hand twist; and Mr. Burns for the defense says that a Bayard Belgium automatic 763, which is the same as 32 calibre, could possibly have fired that, and also a Colt could have fired it, and he wasn't sure which, gentlemen.

Why, he gave the widths of the lands in the barrel of a 32 Bayard as 40 thousandths of an inch. And he gave the lands, the land width of the barrel of a Colt 32, as begin, from the bottom to the top, a variable distance of 2050 to .060, and .060 is the measurement that was given by Capt. Van Amberg; .060, varying to a lesser, almost .050 is what Mr. Burns gave himself. Then on that one fact alone, the width of a barrel lands, the projection, how could a Bayard have fired it. The width of a Bayard is 40 one-thousandths.

But the most important measurement, because it is susceptible to measurement, agreed to I think by the several experts, is the distance on the inside of the barrel between any tow lands, and on a Bayard it is 120 thousandths of an inch and on a Colt, and on this Colt of Sacco's it is 107 thousandths of an inch. It could have been fired by a Bayard, and that is the only one other than a Colt that is suggested by the defense, and in giving the measurements which Mr. Burns gave honestly, he put a Bayard out of it.

Then gentlemen, this left handed twist bullet, No. 3, was fired by a Colt 32. Was it fired by this Colt 32? Some one of learned counsel for the defendant has said that it is coming to a pretty pass when the microscope is used to convict a man of murder. I say heaven speed the day when proof in any important case is dependent upon the magnifying glass and the scientist and is less dependent upon the untrained witness without the microscope. Those things can't be wrong in the hands of a skilled user of a microscope or a magnifying glass. . . .

I say to you on this vital matter of the No. 3 bullet, take the sulphur case of James E. Burns for the defense produced from the barrel of the Colt that belonged to Sacco. Take the No. 3 bullet. Take the three Winchester bullets that were fired by Capt. Van Amberg at Lowell and take the seven United States Bullets that were fired by Mr. Burns at Lowell, and lastly take the barrel itself which we will unhitch for you, and determine the fact for yourself, for yourselves.

Taking one other thing and keeping it carefully in mind, Take the photograph, the series of photographs produced by Mr. Burns that he made of the six bullets that were in the two bodies of these two decedents, fired, as we say, fired from a Savage, from the same Savage, because of the double marking that you will see on the bullets themselves, indicate of slippage as the bullet took the leads of the barrell, a personal peculiarity of the Savage that fired them that you will find on every one of the bullets, 1, 2, 4, 5, and X, not withstanding the fact that Mr. Burns could nor or would not see it on two of them.

Take the glass, gentlemen, and examine them for yourselves. If you choose take the word of nobody in that regard. Take the exhibits themselves. Can there be a fairer test than I ask you to submit yourselves to? Eliminate Mr. Fitzgerald from this case, and I will tell you why, I was pressing and had been pressing Mr. Burns, an exceedingly able man and undoubtedly an exceedingly estimabel gentleman of long training and experience, as to whether or not this widening of the nose, the end of the bullet channel or groove, was not something that was common, and then he was succeeded by Mr. Fitzgerald, and, gentlemen, I did not ask Mr. Fitzgerald I believe 15 minutes' worth of questions.

And when I asked him to look at two of the Lowell Winchesters,—you know what I mean by that, those that were fired by the experts in their experiments at Lowell on a Saturday during the curse of the trial, which Mr. Fitzgerald marked on the bottom with my knife a square and another with a square and two little points on it.

I said, "Take no. 3, which is the fatal bullet in Berardelli's body, the bullet which killed him and the only bullet which killed him, which has letter "W" up near the beginning of the nose of the bullet indicative of its manufacture, Winchester, and take those two that I am showing you of the Winchesters that were fired by Van Amberg at Lowell, and tell me if you don't find the same widening on those bullets." And I pressed him and pressed him on that question, and finally Mr. Fitzgerald says, "I don't understand what you mean by widening," and I said "That is all."

Any man who can't understand such plain question as that;— "Please look at bullet No. 3 with its widening and please look at those two Winchesters fired at Lowell by Capt. Van Amberg with their widenings"—which you saw, gentlemen—which were exhibited to you for the purpose of showing that to you right then and there—"and tell me if the widening isn't the same?" "I don't understand what you mean by you question." I wasn't going to waste any more time on him, gentlemen, and I would not blame you for wasting any more time on him. I consider it would be a waste of time.

Now, gentlemen, my good friend McAnarney,—and he is a good friend through seven weeks of hard trial. I am just as fond of him as I was when I started, and heaven knows I was fond of him then—my good friend McAnarney says in his argument, "Why, what does this circus man Van Amberg"—I don't know what he means by "Circus man Van Amberg, "But that is the way he chose to described him—"What does this circus man Van Amberg say is the identification of bullet No. 3, as being fired from Sacco's Colt?"

Why, he says because of the back flow and the primer of the discharged shells. Absolutely not the fact. Capt. Van Amberg did say it was indicative of a common property, but that it was not of importance. He ascribed as the reason why bullet No. 3, was fired by Sacco's Colt to a wholly different cause, and the answer to it, if you will just examine the exhibits, is within the discernment of your own eyes, and you don't have to trust Van Amberg or anybody else. Use your own eyes and you will see it.

But before I disclose what he gave as the reason why bullet No. 3, was fired by the Colt that belonged to Sacco and not by any other Colt, I want to tell you the reason that Mr. Burns gave and the answer to it. Mr. Burns said, "I am of the opinion"—and he showed you the photographs and told you to look at bullet No. 3— "I am of the opinion that bullet No. 3 was not fired by the Sacco Colt because if you will look at this groove in the bullet"—It was on one of the plaes, I think, the third or fourth plate—"you will see instead of the sides being parellel in the bullet groove, which is the part of the bullet that was marked by the land of the barrel, the projection of the barrel at the top or nose end, you will see it flares out a little bit," and he said, "the reason why I know that bullet No. 3 was not fired by Sacco's Colt is because the United States Bullets which I fired"—he fired—"at Lowell, showed absolutely parellel side, with no splaying or widening of the nose end of the groove."

He is right, gentlemen, they do not show any widening of the groove on the bullet. You can examine them to your heart's content and you will see they are absolutely parallel, and there is lies the skill and the value of James E. Burns to his defense. He knew that bullet No. 3, the fatal bullet, was a Winchester. He used seven United States Bullets, and he gave as the reason the fact that the bullet no. 3, was of old manufacture by the Remington (Repeating) Arms, the Winchester (Repeating) Remington, and that he could not procure one just like it, and that the nearest to it was the United States bullet.

Well, gentlemen, the answer to that expert's basis of claim that bullet No. 3 was not fired by the Colt Sacco is to be found with your eyes, gentlemen, your eyes, not my eyes, not Van Amberg's eyes, not Burns denial, not Fitzgerald's denial, but with your eyes, and in two of the three bullets of the Winchester make that were fired by Van Amberg at Lowell out of Sacco's Colt, out of the same gun that Burns fired the seven United States you will find, three of the Winchesters fired by Van Amberg at Lowell, on the base of two of those—and if you have a strong enough glass—

what I saw about these two that are marked is equally true as to the one which is unmarked, and I will explain what I mean in a moment.

On the base of the two Winchesters, and you can so identify them, fired by Van Amberg at Lowell out of Sacco's gun, (first a little square mark and the other one has a little square with two pits put on by Mr. Fitzgerald.) Those Winchesters were fired by Sacco's gun, the Seven United States were fired by Sacco's gun. The same seven United States bullet grooves show no widening at the top. The Winchesters fired by the same gun, which Mr. Burns says is the identification mark on which he eliminates the Sacco Colt, show the same widening, gentlemen, as does the bullet No. 3. Don't you forget that.

And it is equally true of the third bullet when on close examination, on a very close examination, you will find a little piece of metal here which gives the effect of making the line parallel. If you examine it closely you will see 3, an unmarked one, is widened similarly to bullet No. 3, and that is what Mr. Burns says is his means of identification, and his failure to find it on the seven United States is the reason that he says bullet No. 3 was not fired by the Colt of Sacco.

What is the reason Capt. Van Amberg gives for saying that bullet No. 3 was fired by the Colt of Sacco. A short statement of his reason is this: Not flow back, gentlemen.—That was a matter of minor detail, common in the primers of cartridges, not the reason that Capt. Van Amberg gave for saying that Sacco's Colt fired bullet No. 3, but the fact that while fouling from rust is a peculiarity of any gun that is not cleaned properly after it is fired and more or less common, the pitting on the inside of the barrel of the Sacco revolver, when you heard speak of "gear" marks, if I was to speak of them at 9 o'clock by the clock and look at 6 o'clock, that is so marked one inch in from the muzzle end at the right hand side of the land that you would there see at 6 o'clock identically the same to what Mr. Burns says was a ten-thousandth of an inch on that sulphur case, confirmatory absolutely, if you will only use your eyes to see them, of the scoring that would come from the pits caused by rust.

Now, what did Capt. Van Amberg say would be the effect of rust pits in a barrel? He said it would cause scoring along the edge of the groove in the bullet, Look, gentlemen, at bullet No. 3, Look, gentlemen of the jury, at the three Winchesters fired at Lowell through the Sacco Colt, and see if you do not find pronounced scorings on the edge. It is just like you took a plane and went along the edge of the groove, and you will find pronounced scoring on one of the bullet grooves common to those four bullets, No. 3, the fatal bullet, and the three Winchesters.

More than that, I asked you to eliminate Mr. Fitzgerald. I will ask you to bring him back, not eliminate Mr. Fitzgerald for the moment. The reason is this: He said that he found a great deal of pitting at three of the lands as shown by the sulphur case from the inside of the Colt barrel more pronounced in the case of two than in the case of one.

I ask you to look again, gentlemen, at that sulphur cast, and what we are talking about as pits show as a sort of pebble eruption on the cast so you would understand that, because if there is an actual pit on the barrel itself, when the sulphur went in it filled it, and when removed it had left a raised portion.

I say to you, gentlemen, that there is something still more to be discerned by your eyes as to which no expert has testified, and that is that there are four of the six rifling marks, which are called lands, on the inside of the Sacco Colt that show pitting, one more than the other three, that you will find scoring on the edge of the bullet grooves of the three Winchesters at Lowell and the fatal bullet No. 3, and you will find it more pronounced as to one of the lands, but measure up four, look it over gentlemen, and you can come to but one conclusion with respect to bullet No. 3.

May I have a recess of five minutes out of my time? [12]

On this subject the Court charged the jury in the following color-less and indefinite fashion:

Now, the Commonwealth claims that there are several distinct pieces of testimony that must be considered upon the question of personal identification. Let us see what they are. First, that the fatal Winchester bullet, marked Exhibit 3, which killed Berardelli was fired through the barrel of the Colt automatic pistol found upon the defendant Sacco at the time of his arrest. If that is true, that is evidence tending to corroborate the testimony of the witnesses of the Commonwealth that the defendant Sacco was at South Braintree on the 15th day of April, 1920, and it was his pistol that fired the bullet that caused the death of Berardelli. To this effect the Commonwealth introduced the testimony of two witnesses, Messrs. Proctor and Van Amburg. And on the other hand, the defendants offered testimony of two experts, Messrs. Burns and Fitzgerald, to the effect that the Sacco pistol did not fire the bullet that caused the death of Berardelli.

Now, gentlemen, what is the fact, for you must determine this question of fact, and when determined it may be of assistance to you in determining the ultimate fact. Of course, this evidence cannot be considered by you in any manner whatsoever against the defendant Vanzetti unless you find as a fact that he, too, was present aiding and assisting Sacco and the other conspirators in the shooting and killing of Berardelli.[13]

[12] "The Sacco-Vanzetti Case," Vol. II, pages 2183, 2223-2228.
[13] "The Sacco-Vanzetti Case," Vol. II, pages 2254, 2255.

Clearly, there is no doubt that counsel failed by a wide margin to handle properly the evidence on the all-important question of whether Sacco's pistol had been discharged at the *situs* of the murder and whether any of the fatal bullets had been fired therefrom. The summary of the defendants' counsel was pitifully inadequate. By subjecting the evidence offered by the prosecution to a painstaking analysis, the defense could have substantially weakened the commonwealth's proof on this issue, irrespective of the competence or incompetence of the expert testimony. Whatever the sincerity of Mr. Katzmann, the prosecutor, his excess enthusiasm produced a brilliant and erroneous analysis by his clever emphasis upon the significance of the markings on the exhibits observable to the naked eye and under a low-powered magnifying glass. His treatment of the evidence must have misled the jury.

Mr. Moore touched the question but lightly in his brief comment upon the use of microscopes in determining whether a human life shall continue to function. As evidence of the unreliability of the microscopes, he particularly called attention to the sharply defined disagreements among the users thereof. Mr. Moore's protest against the use of microscopes could have been justified upon the appropriate ground that the experts neither possessed nor were qualified to manipulate the proper instruments.

Mr. McAnarney, on whose shoulders rested the burden of meeting the commonwealth's proof on this issue, was guilty of stupidity. He had neglected to cross-examine the experts for the prosecution with respect to the mental reservations underlying their opinions; and he likewise failed in his summary to expose their true states of mind. Proctor had testified that "my opinion is that" bullet III "is consistent with being fired by that pistol." When asked whether the marks on the six test cartridge cases and the W.R.A. Fraher cartridge case were consistent with being fired from the same weapon, he replied, "I think so, the same make of weapon." Surely, the omission to explain the language of these statements allowed the opinions to convey far more certainty to the jury than that warranted by the personal convictions of the expert.

Mr. McAnarney's treatment of the language of the Van Amburgh opinions was far worse than a mere neglect to point out the under lying mental reservations. The expert had testified that "I am inclined to believe that it was fired, No. 3 bullet was fired, from this Colt automatic pistol." Upon being interrogated as to whether there was a similarity between the test cartridge cases and the so-called Fraher cartridge cases, he answered, "There is one of these so-called Fraher shells, the one marked 'W. R. A. Co.,' meaning Winchester Repeating Arms Company, and three that were fired in Lowell, W.R.A. Co., shells, a very strong similarity." Despite the doubtful state of mind evidenced by the language of the opinions, Mr. McAnarney represented that Van Amburgh had stated positively that bullet III and the W.R.A. Fraher cartridge case had been discharged in Sacco's pistol.

Aside from neglecting to make an exhaustive analysis of the language of the opinions, the defendants' counsel failed to call attention to the discrepancies and weaknesses in the testimony submitted by the experts for the commonwealth. Proctor identified the W.R.A. Fraher cartridge case with all the test cartridge cases, whereas Van Amburgh identified the W.R.A. Fraher cartridge case with only the W.R.A. test cartridge cases. The two experts disagreed on measurements. Proctor testified that the land width of a caliber .32 Savage automatic pistol was .035 of an inch and that this was the measurement of the width of the land mark on bullets Nos. 1, 2, 4, 5, and 6. The corresponding measurements of Van Amburgh were, for bullet 1, between .035 and .036 of an inch; and for bullets Nos. 2, 4, 5, and 6, about .036 of an inch. Both experts admitted that they did not know of any caliber .32 automatic pistol, other than a Colt, having a left-hand twist. One of the experts for the defense, Burns, pointed out that a caliber .32 Bayard automatic pistol had such a twist. Van Amburgh testified that the flow-back appeared on the W.R.A. test cartridge cases and not on the Peters test cartridge cases, because of the greater pressure generated by the discharge of a W.R.A. cartridge. The defense expert Fitzgerald testified that the existence of a flow-back depended largely upon the hardness of the primer metal. Proctor acknowledged that he was not familiar with the component parts of either firearms or ammunition.

The cleverly misplaced emphasis of Mr. Katzmann was manifested by his continually urging the jurors to determine the identity of the fatal bullets and the Fraher cartridge cases on the basis of phenomena visible to their eyes. Examine the exhibits and decide for yourselves, was the everlasting plea of the prosecutor. He stressed the marks on the bullets produced by scoring as being an individual peculiarity and instructed the jury to examine the pits which caused the scoring, both in the barrel of Sacco's pistol and the corresponding eruptions on the sulphur cast. Obviously, Mr. Katzmann should have realized that the jury was not properly equipped to reach an intelligent conclusion on this issue by such an examination of the exhibits.

Throughout his closing argument, the prosecutor assumed that Van Amburgh had given positive opinions that bullet III and the W.R.A. Fraher cartridge case had been fired in Sacco's pistol. Mr. Katzmann attempted to demolish the defense by indicating that the markings on the U. S. test bullets of Burns matched the engraving on the No. III bullet within the limits of two bullets of different manufacture, that is, all discrepancies could be reconciled upon the ground that bullet III was a W.R.A. product whereas the said test bullets were of U. S. make. This unwarranted inference must have unduly caused the jury to discount discrepancies.

The effect of the summary for the commonwealth was that the jurors, at the close of the trial, were impressed with the thought that they were qualified to decide the issue by a mere examination of the observable data. The brevity of Mr. McAnarney and the clever emphasis of Mr. Katzmann induced in the jurors an attitude such that they were likely to decide the issue in favor of the commonwealth on the basis of superficial consistencies perceivable to the naked eye upon an examination of the exhibits.

The trial judge, in his capacity of a fair and impartial commentator upon the evidence, could have mitigated the confusion arising from the expert testimony. But instead of assisting the jury to marshal the evidence clearly and concisely, the Court, in the space of a few sentences, contributed the shallow comment that the testimony of Proctor and Van Amburgh indicated that bullet III was fired from Sacco's pistol, and that the testimony of Burns and Fitzgerald was offered to rebut this contention.

Undoubtedly, the issue was sadly mistried—it was neither fairly nor intelligently handled. The jury retired to the jury room with the exhibits and a mass of incongruous information relating to the primer indentation, the flow-back, the widening of the land marks, and the scoring marks produced by pits in the barrel. Counsel had failed to dissect the evidence and to make a thorough analysis of the complicated items. They had neglected to assort the contending elements of proof and to weigh one against the other. The opinions of the experts were not clearly separated from the facts on which they were predicated, and the jury was deceived by all counsel with regard to the language of the opinions. The charge of the trial judge was worse than useless.

The incompetence of the expert testimony was not brought to the attention of the jurors. In dealing with this evidence they were forced to rely upon the most convincing, irrespective of its merits. The experts for the commonwealth, being more frank, responsive, and conservative, made a better impression upon the minds of the jury than did those for the defense, one of whom created a suspicion of insincerity.

The inevitable conclusion is that the defendants did not receive a fair trial on this all important and vital issue. The jurors were not properly equipped to make an intelligent determination. They had to rely upon incompetent expert testimony which was not, within the narrow limits of information set by the experts, either intelligently or fairly explained during the presentation and summary; and the observable data which the jurors were advised to consider had but little, if any, significance to them.

PART 5

MOTIONS FOR NEW TRIAL

JUDGE THAYER filed his decision on December 24, 1921, denying the motions which each defendant had made upon the ground that the verdict was contrary to the weight of the evidence. In discussing the evidence on firearms identification, the learned trial judge used the following language:

Now, Capt. Van Amburgh testified that in his judgment the bullet that killed Berardelli was fired through the Sacco pistol. Both experts for the defendants, testified that in their judgment the fatal bullet was not fired through the Sacco pistol. In this state of evidence, the important question was raised of whether or not Sacco's pistol fired the bullet that killed Berardelli. On this question there is no dispute but that marks in the barrels of pistols leave their identifying marks upon the bullets fired through them. Sometimes rust and other marks on the inside of the barrels of pistols make impressions and identifying marks upon the bullets as they pass through them.

. . . .

The Court further said that the jury

had the benefit of personal observation, with the naked eye and by the use of magnifying glasses, of comparing the identifying marks on the fatal bullet with the marks upon the other bullets fired through the Sacco pistol.

When counsel on both sides asked the jurors to make these comparisons of identifying marks on the various bullets was it not a recognition on their part that these various identifying marks were disputed question of fact which might according to the judgment of the jury go a long way in convicting or acquitting the defendant Sacco? This was obviously an important question of fact for the determination of the jury. The jury had a right to find that these marks were the "tell tales" of guilt and not marks indicating innocence. As proof of the importance of this evidence let me refer to Com. v. Best, 180 Mass. p. 492.

And in conclusion the Court said:

To determine this question correctly between the experts on both sides, it would depend upon what identifying marks the jurors saw when making the comparison between the different bullets. This being true, it is not for me to decide what the jurors did or did not see with their own eyes.[14]

The motions did not call upon the Court to take the place of the jury and to pass upon the question of the reliability of the witnesses and the credence to be given their testimony. And the mere fact that more than one inference could be drawn from conflicting items of evidence did not empower the Court to find that the verdict was unsupported by the weight of the evidence. The precise question raised by the motions was whether, taking into account the undisputed deficiencies of the state's case, there was still such a preponderating weight of evidence, which had been presented to the jury in a usable and intelligible form, as would justify the jurors in finding that the proof of the defendants' guilt was so unclouded by the shadow of any reasonable doubt that the state would be justified in taking their lives.

The decision of Judge Thayer with respect to the trial of the firearms issue was untenable. Actually the jury had no data upon which to base an intelligent finding. It is plausible to contend as a general proposition that the trial judge, lacking superior knowledge in the field of firearms identification, believed that the testimony of the four firearms experts was competent evidence. But it is difficult to understand how the learned judge, after examining the record and noting the obvious and hopeless confusion throughout the expert testimony, could have concluded that the jury had been properly equipped with the information necessary to justify a finding against the defendants. Furthermore, a careful survey of the record would have disclosed that the jury's attention had not been called to the flimsiness of the state's proof, with regard to the identity of bullet III, which was manifested by Proctor's answer of "consistent with," and by Van Amburgh's answer of "inclined to believe." Surely, then, the Court failed in its duty to give thorough consideration to the firearms issue because, as already pointed out, the personal identity testimony was exceedingly weak and the witnesses giving it had been badly impeached at the trial.

The Fifth Supplementary motion for a new trial was filed on behalf of the defendant Vanzetti on April 30, 1923, and on November 5, 1923, a supplement to this motion was filed on behalf of each defendant. These motions were based upon the discovery of new

---

[14] "The Sacco-Vanzetti Case," Vol. IV, page 3542.

evidence relative to the firearms issue. The supplement involved the two questions of whether any of the Fraher cartridge cases had been discharged in the Sacco pistol, and whether bullet III had been fired from Sacco's pistol.

AFFIDAVIT OF WILLIAM PROCTOR

Captain Proctor made this affidavit at the request of Mr. William G. Thompson, one of the counsel for one of the defendants. The Captain testified in the affidavit as to the real state of his mind existent at the trial. The introduction of this testimony raised squarely the question of the misconduct of the prosecuting attorneys with reference to the testimony given by Proctor at the trial. The affidavit read in part as follows:

> During the preparation for the trial, my attention was repeatedly called by the District Attorney and his assistants to the the question: whether I could find any evidence which would justify the opinion that the particular bullet taken from the body of Berardelli, which came from a Colt automatic pistol, came from the particular Colt automatic pistol taken from Sacco. I used every means available to me for forming an opinion on this subject. I conducted, with Captain Van Amberg, certain tests at Lowell, about which I testified, consisting in firing certain cartridges through Sacco's pistol. At no time was I able to find any evidence whatever which tended to convince me that the particular model bullet found in Berardelli's body, which came from a Colt automatic pistol, which I think was numbered 3 and had some other exhibit number, came from Sacco's pistol and I so informed the District Attorney and his assistant before the trial. This bullet was what is commonly called a full metal-patch bullet and although I repeatedly talked over with Captain Van Amberg the scratch or scratches which he claimed tended to identify this bullet as one that must have gone through Sacco's pistol, his statements concerning the identifying marks seemed to me entirely unconvincing.
>
> At the trial, the District Attorney did not ask me whether I had found any evidence that the so-called mortal bullet which I have referred to as number 3 passed through Sacco's pistol, nor was I asked that question on cross-examination. The District Attorney desired to ask me that question, but I had repeatedly told him that if he did I should be obliged to answer in the negative; consequently, he put to me this question: Q. Have you an opinion as to whether bullet number 3 was fired from the Colt automatic which is in evidence? To which I answered, "I have." He then proceeded. Q. And what is your opinion? A. My opinion is that it is consistent with being fired by that pistol.

That is still my opinion for the reason that bullet number 3, in my judgment, passed through some Colt automatic pistol, but I do not intend by that answer to imply that I had found any evidence that the so-called mortal bullet had passed through this particular Colt automatic pistol and the District Attorney well knew that I did not so intend and framed his question accordingly. Had I been asked the direct question: whether I had found any affirmative evidence whatever that this so-called mortal bullet had passed through this particular Sacco's pistol, I should have answered then, as I do now without hesitation, in the negative.[15]

Mr. Katzmann, the prosecutor, retaliated with an affidavit to the effect:

that I did not repeatedly ask him whether he had found any evidence that the mortal bullet had passed through the Sacco pistol, nor did he repeatedly tell me that if I did ask him that question he would be obliged to reply in the negative.[16]

Mr. Williams, who conducted Proctor's examination at the trial, averred:

I conducted the direct examination of Captain Proctor at the trial and asked him the question quoted in his affidavit, ''Have you an opinion as to whether bullet number 3 was fired from the Colt automatic which is in evidence?''

This question was suggested by Captain Proctor himself as best calculated to give him an opportunity to tell what opinion he had respecting the mortal bullet and its connection with the Sacco pistol. His answer in court was the same answer he had given me personally before.[17]

Proctor's own testimony appears on the record, and his affidavit exposes the true nature of his opinion. Clearly, then, his evidence at the trial exhibited only a misleading fragment of his actual state of mind. Considering his affidavit with those of Mr. Katzmann and Mr. Williams, it is possible for a lawyer to draw only the inference that the prosecuting attorneys knew the precise limitations of Proctor's opinion, and that the form of the question was probably determined by prearrangement to permit the giving of a deceptive answer. However, aside from the affidavits and the acts of misconduct set forth therein, the whole record bristles with circumstances which negate the possibility that the district attorney and his assistants were

15 ''The Sacco-Vanzetti Case,'' Vol. IV, pages 3642, 3643.
16 ''The Sacco-Vanzetti Case,'' Vol. IV, page 3681.
17 ''The Sacco-Vanzetti Case,'' Vol. IV, page 3682.

ignorant of the deceptive nature of Proctor's testimony. If they knew his entire opinion, they were guilty of misconduct which should be far beyond the bounds of the discretionary power of the trial judge to conceal, permit, or condone.

Undoubtedly Proctor's testimony at the trial had been interpreted by the defense and by the trial judge as the expression of an unqualified opinion that bullet III had been fired from Sacco's pistol.[18] This was the only interpretation submitted to the jurors, and it is therefore reasonable to conclude that the jury so construed the incongruous Proctor testimony. It is possible that the Court's misapprehension at the trial, like that of the defense, was due to a bona fide lack of understanding. But when Judge Thayer received the affidavits he should have realized at least that the prosecuting attorneys had knowingly permitted Proctor to express a half-truth which had the effect of a lie. The sufficiency of the affidavits for the purpose of directing attention to the gross mistrial of the vital firearms issue cannot be questioned. The jury had been deceived, and it is evident from the verdict that the deception was inexcusably and substantially prejudicial to the defendants. The decision of the learned judge with respect to the affidavits read in part as follows:

> In my judgment, the questions propounded by Mr. Williams were clearly put, fairly expressed, and easily understood; they have been so commonly used by experienced trial lawyers throughout the Commonwealth for so many years that they have become almost stereotyped questions. It cannot be said that Captain Proctor did not have time to fully understand and appreciate the full meaning of the questions propounded to him, because the first one was put twice to him before he answered.
>
> Let us now carefully consider the questions and answers by applying to them the usual tests of ascertaining facts. Were any of these questions "framed" by Mr. Williams on account of any unworthy or improper motive? Study the fairness of the questions propounded by Mr. Williams, with the view of determining whether or not Captain Proctor did not have the fullest opportunity to express his honest opinion. Is there any hidden mystery or unworthy concealed motive in the first question, which is as follows: "Have you an opinion as to whether bullet number 3 was fired from the Colt automatic which is in evidence?" Bullet No. 3 was the mortal bullet, and the only Colt automatic which was in evidence was the Sacco pistol.
>
> Is there anything about that question that any man who has been a member of the State Police for over thirty years could not easily

18 See, *supra*, summary of defendants' counsel and the judge's charge to the jury.

and readily understand, if he was then in full possession of his mental faculties and was desirous of expressing his true opinon? Was there any power under the sun that could have prevented him from expressing his honest conviction at that time?

. . . .

Now, let us proceed further with this inquiry, with the further object of ascertaining what was then the real situation. Counsel for the motion, with great earnestness and force, argued that the true interpretation of this affidavit is as follows: that the District Attorney, knowing that Captain Proctor honestly believed that the mortal bullet was not fired through the Sacco pistol, by prearrangement with Captain Proctor prevailed on him to compromise the truth, in that Captain Proctor should testify that it was his opinion that it was consistent with its having been fired through the Sacco pistol. In other words, that Captain Proctor, by prearrangement (which means intentional) compromised the truth with the District Attorney by his (Captain Proctor's) testifying knowingly to something that was false.

I do not believe that the interpretation of counsel for the motion is the true one. . . .

With my findings of facts, it is unnecessary for me to state at length and in detail the contents of the counter-affidavits of Messrs. Katzmann and Williams,—they are clear and convincing. I cannot, however, close this phase of the matter without saying that I carefully watched the conduct of both Messrs. Katzmann and Williams in the trial of these cases for nearly seven weeks, and I never observed anything on their part but what was consistent with the highest standard of professional conduct.[19]

HAMILTON MOTION

Albert H. Hamilton made the affidavit in support of this motion for a new trial, and the affidavits in opposition to the motion were made by Charles J. Van Amburgh and Merton A. Robinson. The motion was based upon the following broad grounds:

1. A Bausch and Lomb professional compound microscope equipped with a Filar micrometer, measuring to 1-100,000 of an inch, reveals a different knurling in the cannelure of the No. III bullet than the knurling of the six W.R.A. cartridges found in the possession of Sacco at his arrest.

2. New and independent evidence since the trial.

a. Above microscope shows discrepancies between the markings on the test bullets from the Sacco gun and the markings on the fatal bullet.

19 "The Sacco-Vanzetti Case," Vol. IV, pages 3699-3701, 3703, 3704.

b. Measurements under the above microscope reveal a difference between the six lands and six grooves of the test bullets and the six lands and six grooves of the fatal bullet.

c. Under this microscope the measurements of the land and groove widths of the barrel of the Sacco Colt agree with the land and groove widths on the test bullets.

d. Microscopic examination of the rim of the barrel of the Sacco gun reveals that it does not conform with each land and groove mark on the number 3, fatal bullet.

3. A microscopic examination of the four Fraher cartridge cases and the six test cartridge cases indisputably eliminates the possibility of any one or more Fraher cartridge cases being fired from the Sacco gun.

a. The character of the ejector marks on each of the Fraher cartridge cases differs from the character of the ejector marks on the test cartridge cases.

b. Character of firing-pin marks on the Fraher cartridge cases differs from the character of firing-pin marks on the test cartridge cases.

c. Flow-back on the Fraher cartridge cases differs from the flow-back on the test cartridge cases.

d. Relative position of firing-pin dent not the same on the Fraher cartridge cases and on the test cartridge cases.

e. Condition of interior of firing-pin dent is not the same in the test cartridge cases and the Fraher cartridge cases.

f. There is a difference between the test cartridge cases and fatal cartridge cases in "the embossing appearing on the primer around the dent made by the impact of the primer as against the bushing of the firing pin at the time of the explosion of the shell." The file marks on the bushing of the Sacco Colt conform to the embossing on the test cartridge case primers but not the embossing on the primers of the Fraher cartridge cases.

HAMILTON AFFIDAVIT

Affiant's Qualifications: A graduate of pharmacy, chemistry, and microscopy of the former New York College of Pharmacy, which is now a department of Columbia University; for 37 years his profession has been that of a micro-chemical investigator and criminologist in connection with investigations of crime; as an investigator he originated a scientific micro-chemical method of examination of exhibits in suspected homicide and suicide cases; called as an expert in 165 homicide cases and for the last fifteen years he had devoted his entire time to this work.

Hamilton contended that the number III bullet was not manu- factured at the same time as the six W.R.A. cartridges found on Sacco at his arrest. The basis of the contention was a difference in the knurlings in the cannelures of the fatal bullet and the six car- tridges. The discrepancy was in the angles of the teeth in the knurl- ings, i.e., the teeth in the knurling of the six cartridges were perpen- dicular whereas the teeth in the knurling of the fatal bullet were three degrees off perpendicularity.

This contention was completely discredited by the rebutting affi- davit of Robinson, a former employee of the Winchester Repeating Arms Co., who testified that the teeth in the knurling of all W.R.A. bullets were parallel to the axis. The fatal bullet was deformed in striking the bones of the deceased so that it would have been diffi- cult to ascertain the true axis of the teeth in the knurling under any conditions.

Hamilton attempted to prove that the fatal bullet was not fired from the Sacco Colt. The basis of the proof was discrepancies be- tween the measurements of the rifling in the Sacco Colt barrel and measurements of the rifling marks on the number III bullet. He ren- dered an unqualified opinion that bullet number III was not from the Sacco pistol because: "If the twelve land marks on the mortal bullet did fit the lands and grooves consecutively and simultaneously on the Sacco pistol, it would be proof positive that the mortal bullet came from the Sacco pistol, but that not being the case, it is equally proof positive that the mortal bullet did not come from the Sacco pistol."

When consideration is given to the differences existing in the co- efficients of expansion of steel, lead, copper, zinc, tin, and antimony; the plasticity and elasticity of metals and alloys; the shock effect pro- duced when a bullet comes in contact with a resisting medium,—one can readily understand that such a bullet is not an appropriate object for precision measurements. As a matter of fact, if it had been shown that the measurements referred to by Hamilton were in perfect agreement, this would have furnished a sound basis for proving that bullet III could not have been fired in Sacco's pistol. The very es- sence of Hamilton's contention reduces itself to an absurdity.

Hamilton explained the individuality of a particular barrel in the following terms:

The cutting process starts with a cutting blade fashioned to the predetermined standard width of groove blade measuring .105 of an inch and leaving the unremoved space between the two grooves as a land approximately .050 of an inch. Theoretically, each groove and each land so cut should be uniform to the above measurements, but due to the high friction under which the cutting tool works it rapidly wears away the right and left edges of the tool, constantly making it narrower and narrower until there comes a time, after a few hours, when the cutting tool has to be removed and a new one installed, due to the fact that it has worn to an extent that it is cutting a groove too narrow, leaving a land too wide to conform to the specification of the ballistic engineer. As a result of this constant wear upon the edges of the cutting tool, and the frequent changing of a worn tool to one less worn, the microscopic width of the groove and the land are constantly changing between certain limits, and the twelve measurements in their sequence, extending around the interior of the barrel, six lands and six grooves, have an ever changing, combination of twelve measurements. That it is this constant change of the cutting blade that gives the individuality to the new pistol, fresh from the factory; that subsequently as these pistols go out upon the market, are fired, become filled with nitrous corroding fumes from exploded smokeless powder and are improperly cleaned by the use of improper tools and emery flour, which has a tendency to again wear away the sides and corners of the lands, a new individuality is given to these pistols. That this latter process of improperly cleaning a weapon will not affect the individuality of a pistol to such an extent as to affect the relationship between the mortal bullet fired through it and test bullets fired at a later date where the mortal pistol has been constantly in the hands of some officer of the court and out of use.[29]

Hamilton's testimony grossly exaggerates the formation and extent of constructional variations in the dimensions of the bore. That there are dimensional variations is admitted. If barrels of the same class characteristics are subjected to precision measurements (at constant temperature) to one hundred thousandths of an inch, such measurements will show variations in the dimensions of the cross-sections of the bores of the barrels from breech to muzzle; on the other hand, such dimensional variations can not be determined from bullets fired through the bores of these same barrels. Bullets fired from a number of caliber .32 Colt automatic pistols have been compared under the comparison microscope with a magnification of 30 diameters and these comparisons indicated perfect agreement in the land and groove widths.

Furthermore, the rifling tool is passed through the barrel and

20 ''The Sacco-Vanzetti Case,'' Vol. IV, pages 3628, 3629.

back through one groove after which the barrel is rotated one-sixth of a turn in the case of a six-groove barrel, so that on the next pass of the rifling tool, it passes through the next groove, and so on, mitigating the wearing effect of the cutter as causing a dimensional divergence in any one groove.

Another example of Hamilton's useless measurements is as follows:

> Microscopic examination of the most clearly defined of the two double land-marks appearing on the mortal bullet, Exhibit 18, shows that said double land mark consists of four identifying scratches arranged parallel as follows:—Two at the left 4/100ths of an inch apart; 7/100ths of an inch to the right are two more parallel scratches 6/100ths of an inch apart, and 5/100ths of an inch to the right of these is a parallel, deep scratch or cut. While upon each of the six Proctor-Van Amberg Lowell bullets appears a double land-mark, and upon each of said double land marks appear eight parallel fine scratches 3/100ths of an inch apart, except that the sixth and seventh lines counting from the left are 6/100ths of an inch apart.

> Assuming from the above uniformity of measurements of the scratches upon the double land-marks of the Lowell test bullets, that one land in the Sacco pistol made this double land mark upon each of the Lowell bullets, it is conclusively demonstrated that the double land-mark upon the mortal bullet with its different set of measurements of the scratches on the double land-mark, and with the extra wide land at the top, shows that a pistol other than the Sacco pistol fired the mortal bullet.[21]

It is utter nonsense to attempt to reach any definite conclusion by measuring marks on bullets unless the examiner knows exactly what he is measuring, because the relative position of certain marks on one bullet will not exactly agree with the corresponding relative position of these marks on another bullet from the same gun. As a matter of fact, research has yet to find two bullets fired from the same weapon that are exact duplicates of each other so that, if placed under a comparison microscope, any part of the surface of one bullet will exactly match the corresponding surface of the other bullet in all details.

Aside from the ordinary difficulties of measuring marks on the cylindrical surface of a bullet, the number III fatal bullet was distorted because of contact with a resisting medium.

Hamilton's theory of identification will not survive the test of severe criticism. The only scientific method of determining whether or not two bullets were fired in the same weapon, is by a comparison

21 "The Sacco-Vanzetti Case," Vol. IV, page 3631.

under the comparison microscope. (It should be remembered that Carl Zeiss, Jena, has been *manufacturing the comparison eyepiece since 1920.*) To base the identification upon measurements alone was not just an impossibility—it was absolute folly. The best evidence of the unreliability of measurements lies in an examination and comparison of the measurements submitted by the experts for and against the Hamilton motion. These measurements are tabulated on pages 226 and 227.

Augustus H. Gill, professor of technical-chemical analysis at the Massachusetts Institute of Technology, made his measurements for Hamilton. Professor Gill tried to account for the divergence in his own measurements, made at two different times, by the following explanation:

> The reason is, I suppose, because my August measurements were made under different conditions from the September measurments; and to speak frankly, my August measurements were the first time that I had ever measured a bullet.[22]

As to the causes of the discrepancy between the August and September measurements, Gill cited different temperatures, different points on the bullet, and the personal equation. On the basis of either measurement Gill stated that number III bullet did not come from the Sacco Colt.

Gill attempted to reconcile his data with the Hamilton measurements on the ground that he made his measurements across the base of each land and groove in the Sacco barrel whereas Hamilton measured across the tops of the lands and grooves. Gill believed the bottom of the groove to be wider than the top and that the difference in the method of measuring would account for the smaller dimensions of Hamilton. He said:

> I made my measurements on the bullet in the same way that I did on the pistol, as contrasted with Hamilton's method, the average of my measurements of the protuberances or so-called lands of the bullet made by the grooves in the pistol, and of the grooves of the bullet made by the lands of the pistol, should also be slightly larger than Hamilton's.[23]

The affidavit of Robinson stated that the grooves of the barrel are wider at the top than at the bottom; the grooves are wedge shaped to facilitate cleaning and to obtain a perfect groove cut. Gill's measure-

22 ''The Sacco-Vanzetti Case,'' Vol. IV, page 3650.
23 ''The Sacco-Vanzetti Case,'' Vol. IV, page 3649.

HAMILTON'S MEASUREMENTS

Bullet III

Land mark		1	0.050	inch	Groove mark		2	0.1000	inch
"	"	3	.0512	"	"	"	4	.1025	"
"	"	5	.0525	"	"	"	6	.1050	"
"	"	7	.050	"	"	"	8	.1000	"
"	"	9	.050	"	"	"	10	.1025	"
"	"	11	.050	"	"	"	12	.1025	"

Sacco Pistol

Land mark		1	0.050	inch	Groove mark		2	0.1050	inch
"	"	3	.045	"	"	"	4	.1050	"
"	"	5	.050	"	"	"	6	.1025	"
"	"	7	.0475	"	"	"	8	.1025	"
"	"	9	.050	"	"	"	10	.1025	"
"	"	11	.0475	"	"	"	12	.1050	"

GILL'S MEASUREMENTS

Bullet III: Lands and Grooves

August 14, 1923	September 25, 1923	
0.0487 inch	0.0507 inch	0.0520 inch
.1000	.1038	.1040
.0500	.0512	.0512
.0975	.1020	.1040
.0500	.0515	.0504
.0987	.1040	.1024
.0487	.0512	.0512
.0987	.1032	.1032
.0500	.0512	.0512
.0962	.1040	.1040
.0492	.0512	.0512
.0962	.1040	.1040
———	———	———
0.8839 inch	0.9280 inch	0.9288 inch

GILL'S MEASUREMENTS (*Continued*)

Sacco Pistol: Lands and Grooves

August 14, 1923	September 25, 1923	
0.0500 inch	0.0562 inch	0.0562 inch
.1075	.1048	.1056
.0500	.0568	.0568
.1075	.1056	.1064
.0515	.0562	.0562
.1080	.1064	.1040
.0510	.0562	.0568
.1075	.1064	.1056
.0500	.0568	.0562
.1075	.1075	.1064
.0492	.0562	.0562
.1075	.1056	.1069
0.9472 inch	.9747 inch	.9733 inch

VAN AMBURGH'S MEASUREMENTS

Bullet III

Land width	Groove width
0.0533 inch	0.1039 inch
.0544	.1045
.0543	.1036
.0534	.1050
.0524	.1032
.0532	.1047

Sacco Pistol

Land width	Groove width
0.0527 inch	0.1056 inch
.0505	.1050
.0525	.1058
.0507	.1050
.0519	.1058
.0523	.1059

ments of the grooves at the bottom should have been smaller than the Hamilton measurements across the top.

Gill's credibility is seriously weakened by his assertion, "I am absolutely convinced from my own measurements that the so-called mortal bullet never passed through the Sacco gun," in the face of his admission of lacking skill and experience in this line.

As to the scoring on the number III bullet, Hamilton summarized:

> That the scoring or scraping of the metal jacket on the mortal bullet possesses no evidentiary value at all as indicating or tending to prove in any wise whatsoever that the mortal bullet was fired from the Sacco automatic pistol, inasmuch as the same kind and character of scoring or scraping of the metal jacket appears on bullets fired from new Colt automatic pistols; all as appears from the photographs on pages fifteen, seventeen and eighteen of the album, and as likewise appears on the bullets fired from a new Harrington & Richardson and a new Savage pistol, all as appears on page six of the album. That the scoring or scraping of the metal jacket on a bullet possesses no mark of individuality that enables one to determine, at all, from what pistol a given bullet was fired, or even to determine, from what manufacture of pistol the given bullet was fired.
>
> The last above statement is subject to the qualification that where there is a rust pit or other injury immediately at or upon the muzzle edge of the rifling of the pistol, same may register itself upon a metal jacketed bullet and will register itself upon a soft lead bullet. Again attention is directed to the fact that the Sacco pistol shows no rust pit or other injury at or near the muzzle.[24]

If the bullets referred to in Hamilton's summary had been compared under the comparison microscope, they would have revealed considerable difference in the character of the scorings as well as the effects of the accidental characteristics of the barrels from which they were fired, and by these characteristics the barrels could be identified. Hamilton's testimony forces the conclusion that he was trifling with the truth.[25]

Hamilton gave an absolute answer on the question of the Fraher cartridge cases, i. e.,

> as a result of his examination he gives it as his unqualified opinion that Fraher shells F1, F2 and F3, as same appear photographed on page one of the album, were not fired from an American made automatic pistol but were fired from some automatic

---

[24] "The Sacco-Vanzetti Case," Vol. IV, page 3634.

[25] See, New York *Herald-Tribune*, March 11, 1934, "Strewl Defense Expert Admits Deceiving State."

pistol of foreign manufacture; that Fraher shell F4, the Winchester shell, was fired from a Colt automatic pistol, 32 calibre, but was not fired from the so-called Sacco pistol.

From the size of the firing-pin dent he concluded that no one of the Fraher cartridge cases was fired from a Savage or Harrington and Richardson. The ejector marks on F1, F2, and F3 were not those of a Colt, Infallible, or Savage, so the expert gave his unreserved opinion that they were discharged from a weapon of foreign manufacture.

Hamilton contended that F4 was not from the particular Colt in evidence because the firing-pin dent was in the exact center in F4 but approximately 23 per cent off center in the test cartridge cases; because of the presence of an imperfection at the right interior of the firing-pin dent 0.6 inch long, which was absent on the test cartridge cases; and because the file marks on the base of the F4 cartridge cases were coarser than the file marks of the bushing and breechblock of the Sacco Colt.

To recapitulate the reasons why the Fraher cartridge case F4 could not have been fired from the Sacco pistol, Hamilton stated:

(a) That the firing pin dent on Fraher shell F4 is substantially in the exact centre of the primer. While the firing pin dent upon all of the test shells is off centre approximately twenty-three (23) per cent, same being visually demonstrated by the photographs appearing upon page three of the album and more clearly defined in the enlarged photographs appearing on pages twelve and fifteen of the album.

(b) That there is an imperfection upon the firing pin of the Sacco pistol which registers upon the Van Amberg test shells but which does not register upon the Fraher shell F4, which is visually demonstrated by the photographs appearing on pages twelve and sixteen of the album.

(c) That there are file marks upon the face of the breech block and bushing of the Sacco pistol made by the mechanic at the time of the assembling and fitting of the Sacco pistol in the factory, which give individuality to the pistol and which register upon shells discharged from that particular pistol; that such imperfections are registered upon the Van Amberg shells as same appear on pages twelve and sixteen of the album, but same are not registered upon Fraher shell F4. That Fraher shell F4 does have upon its face markings indicating the form and structure of the breech block and bushing with which it came in contact at the time it was discharged, but said markings are entirely different, both as to their relative form and relationship one to the other and as to their fineness, from the markings appearing upon the test shells. All of which is visually demonstrated by photographs appearing on pages twelve and sixteen of the album.

(d) That the flow-back in the firing pin dent of the Fraher shell F4 is an entirely different flow-back, both as to size, form and location, from that appearing on the Van Amberg test shells all of which is visually demonstrated by the photographs appearing on pages twelve and sixteen of the album.

(e) That the ejector marks appearing upon the Fraher shell F4 is an entirely different ejector mark, both as to size and form, from the ejector marks appearing upon all three of the test shells; which differences are visually demonstrated by the photographs appearing on pages twelve and sixteen of the album.[26]

Hamilton's method of tracing the identity of cartridge cases, like the method he employed to identify bullets, was utterly unreliable. The glaring weakness, that is, a measurement of particular characteristics, was in his basic premise. The examiner of cartridge cases fired in automatic pistols is primarily interested in the relative position of the characteristics one to the other in a combination, such as the impression of the breeching face with reference to the impression of the circumference of the hole in the breechblock through which the firing pin strikes; the relative positions of the marks made by the extractor and ejector with reference to the mark made on the head of the cartridge case by the lower surface of the breeching face in the process of forcing the cartridge from the magazine into the chamber. The firing pin may not always strike the same position in relation to the center of the primer, but the impression of the tool-mark pattern of the breeching face will always occupy the same relative position with reference to the impression made by the circumference of the firing-pin hole in the breechblock. The fact that the firing-pin indentation was in the center of the primer in F4 and 23 per cent off center in the test cartridge case does not in itself prove anything.

The Fraher and test cartridge cases should have been examined under a comparison microscope to determine the agreement or disagreement of the characteristics as a combination. The breeching face impression on the primer cup is influenced by the relative hardness of the primer and the pressure developed in the barrel. With the comparison microscope the tool-mark pattern is matched as the indicia. An imperfection on the firing pin of one pistol may turn up in different dimensions on numerous primers owing to the variations in the pressure developed and the constituency of primers of same manufacture, but it will be in the same relative position with reference to the other characteristics.

AFFIDAVIT OF CHARLES J. VAN AMBURGH

There was a remarkable change in the state of this expert's mind since the trial, at which time he retained a mental reservation in rendering the requested opinions.

The affidavit opinions ran as follows:

I am absolutely certain that the Fraher shell F4 was fired in the Sacco pistol. The score of the firing pin in the pistol is registered in the firing pin indentations in the Lowell Winchester shells and in the Fraher shell F4. The impressions of the breech block of the Sacco pistol is clearly shown on the primer surfaces of the Lowell Winchester shells and of Fraher shell F4. Not only do the photographs taken of these marks on the shells demonstrate these facts, but microscopic measurements show the same relation of the lines on the primer surface of Fraher shell F4 to each other as the relation of the lines to each other on the primer surfaces of the Lowell Winchester shells.

I am also positive that the mortal bullet was fired in the Sacco pistol. The marks in the groove cut, to which I testified at the trial, and which I then showed to the jury, were caused by pitting and corrosion in the barrel of the Sacco pistol. Since the trial I have made an exhaustive microscopic examination of the mortal bullet, of the three Lowell Winchester bullets, and of the barrel of the Sacco pistol. These examinations were made, as I have above stated, with the aid of a compound microscope with a Filer micrometer attachment. This is a much more powerful microscope than the smaller one which I used at the time of the trial. The microscope measurements which I have made of the land and groove widths in the pistol and on the bullets are extremely significant. The facts which I have found from my entire investigation are so clear that, in my opinion, they amount to proof.[27]

Van Amburgh gave the following reasons for variations in the location of the firing-pin indentations on cartridge cases fired in the same pistol:

    1. The tolerance between the firing pin and the firing pin hole.
    2. The difference in the diameter of the cartridge and the chamber in which it is placed which difference results in a certain looseness of the fitting of such cartridge in such chamber.
    3. The looseness of the fit of the slide, that is, that part of the pistol which houses or covers the barrel.

[27] "The Sacco-Vanzetti Case," Vol. IV, pages 3666, 3667.

4. The construction of the cartridge itself. Under the centre of the primer surface is a hard anvil made of brass. When the firing pin strikes the primer surface over this anvil there is a tendency of the firing pin to be deflected to one side. In the Sacco pistol the space between the firing pin and the firing pin hole is .007 inch, which means that the firing pin can move a distance of .007 of an inch within the circular rim of the firing pin hole.[28]

Van Amburgh measured the ridge in the firing-pin indentation caused by an imperfection of the firing pin. He used a method different from Hamilton's, with the following results:

	Length	*Width*
Lowell Winchester No. 1	0.040 inch	0.0043 inch
Lowell Winchester No. 2	.041	.0047
Lowell Winchester No. 3	.041	.0050
Fraher F4	.040	.0050

Depths of Firing-Pin Indentation

Lowell No. 1	0.028 inch
Lowell No. 2	.025
Lowell No. 3	.028
Fraher F4	.027 [29]

Hamilton measured the ridge as 0.60 inch long on F4 and 0.25 inch long on test cartridge cases, and Van Amburgh gave as a reason for this discrepancy: "In fairness to Mr. Hamilton I must assume that said measurements were made not from the original shells but from the photographs appearing on either pages 12 or 16 of his photograph album."

In the last analysis Van Amburgh's basis of identification was a comparison of measurements of certain features which he finds to agree most remarkably. No great credence can be attached to any part of his affidavit because after all, assuming that he was capable of making precise measurements and that he had the necessary equipment, his measurements nevertheless were of relatively insignificant value from the standpoint of identifying the particular pistol from which a given bullet was fired or in which a given cartridge case was discharged. Van Amburgh should have obtained ten Colt caliber .32 automatic pistols with barrels in about the same condition as the barrel in Sacco's Colt; fired four test shots from each, and then subjected these barrels, test cartridge cases, and test bullets to the same series of measurements. A tabulation of these results, when compared

[28] "The Sacco-Vanzetti Case," Vol. IV, pages 3653, 3654.
[29] "The Sacco-Vanzetti Case," Vol. IV, pages 3655, 3656.

with the results of corresponding measurements made with Sacco's pistol, would probably have added to the confusion as well as to the examiner's experience. And certainly, only by an exhaustive examination with a comparison microscope can it be determined whether two cartridge cases were discharged in the same pistol, in the face of the many variations that can sometimes be found in two cartridge cases fired consecutively from one pistol even if the ammunition is taken from the same original package of ammunition. It is to be noted that Van Amburgh also measured the lines of the breechblock impression of fatal and test cartridge cases.

Van Amburgh rebutted Hamilton's theory of rifling a barrel:

> Uniformity in groove and land dimensions in a given barrel is absolutely necessary to obtain accurate shooting. I know it to be a fact that the Colt Company produces weapons all of which are noted for their accuracy. The system of rifling described by Mr. Hamilton would admittedly produce a barrel with non uniform grooves, would destroy its accurate shooting qualities, and is not the method employed by said company.

Van Amburgh described the method of rifling a barrel:

> The degree of pitch, sometimes known as "twist" having been determined, the machine is set to the desired point and the operation is carried on through the medium of a rifling rod which moved forward and back, and which has on its forward end a rifling head containing a rifling cutter. The rifling cutter is ground and honed to the desired shape, which shape determines the contour of the grooves which are out in the pistol barrel. It is not the practice to complete one groove in the pistol barrel before moving to and commencing the next one, but rather to commence the cutting of each groove with a very light cut and to move progressively with similar light cuts to all the grooves in the barrel, cutting each groove deeper and deeper and so continuing round and round until sufficient groove depth is reached, after which the rifling tool is withdrawn. The method of rifling the barrel which I have described is one which produces a groove of uniform width throughout a given barrel.[30]

Van Amburgh minimized the accidental variations in the dimensions of the bore caused by rifling the barrel. Yet, despite the alleged uniform groove width, he purported to identify the Sacco Colt by measuring the rifling dimensions on the bullets fired therein. It is interesting to note that his data of measurements of the Sacco barrel, the mortal bullet, and the test bullet did not check exactly:

30 ''The Sacco-Vanzetti Case,'' Vol. IV, pages 3662, 3663.

I have carefully made measurements of the lands and grooves of the mortal bullet and submit them as follows:

Land width	Groove width
.0533	.1039
.0544	.1045
.0543	.1036
.0534	.1050
.0524	.1032
.0532	.1047

I have also made careful measurements of the width of the lands and grooves of one of the Lowell Winchester Bullets, and submit them as follows:

Land width	Groove width
.0529	.1058
.0531	.1047
.0535	.1047
.0537	.1050
.0522	.1068
.0507	.1054

I have carefully measured the width of the lands and grooves in the Sacco pistol which I submit as follows:

Land width	Groove width
.0527	.1056
.0505	.1050
.0525	.1058
.0507	.1050
.0519	.1058
.0523	.1059

Average width of mortal bullet land is.................. .0535
Average width of Lowell Winchester land............... .0527
Average width of Sacco Pistol land................... .0517

Average width of mortal bullet groove................. .1041
Average width of Lowell Winchester groove............ .1054
Average width of Sacco Pistol groove................. .1055

Total width of lands and grooves in mortal bullet......... .9459
Total width of lands and grooves in Lowell Winchester
    bullet .... .9485
Total width of lands and grooves in Sacco Pistol........ .9437

It is apparent again that my measurements of the lands and grooves in the Lowell Winchester bullet, known to have been fired from the Sacco pistol, do not absolutely agree with my measurements of the Sacco pistol. Nor do my measurements of the mortal bullet absolutely agree either with my measurements of the Lowell Winchester bullet or with my measurements of the Sacco pistol. The closeness of my measurements of the mortal bullet, however, to both the measurements of the Sacco pistol and the measurements of the Lowell Winchester bullets are significant in identifying the said mortal bullet, first, as having been fired in the Sacco pistol, and second as being alike in reference to width of lands and grooves with the Lowell Winchester bullet known to have been fired in the Sacco pistol. Measurements of land and groove widths through a microscope depend for their accuracy, first, upon the eye sight of the examiner; second, on his manipulation of the examining instrument or microscope; and, third, on the examiner's familiarity with the construction of bullets and pistol barrels, with particular reference to the land walls of such barrels and the groove contours; and, fourth, on the condition of the pistol barrel and bullets when measured. The land walls of a pistol barrel, which also form the boundaries of the grooves, do not have parallel sides, but slope from the top of the lands to the base of the grooves, this condition being necessary to give the rifling cutter a clearance. The land walls at their base are also slightly rounded where they intersect the grooves. These sloping land walls necessarily impress their character on the bullet fired through the barrel. In order to obtain as correct a measurement as possible of a land or groove width either in barrel or on bullet, measurements should be taken from a half way point on the opposite land wall. If the person making the measurements does not accurately obtain the proper point on the land wall from which or to which to measure the resulting measurement will be either too large or too small. The person measuring also encounters defects in the shape of rust, pitting and fouling, which prevents, in many instances, the obtaining of a true measurement. I am now referring to measurements made with a microscope and Filer micrometer eye piece.

The most accurate method of obtaining the bore diameter of a barrel and from the bore diameter computing the bore circumference is by the use of a plug gauge. This is a purely mechanical method of measurement and is absolutely accurate. Measuring the Sacco pistol with such plug gauge, the true bore diameter is found to be .3045 inch. Multiplying this bore diameter by the usual constant, the true circumference of the bore is obtained, which is .9556 inch. Comparing this figure so obtained with the total width of lands and grooves in the Lowell Winchester bullet, obtained by microscopic measurement, there appears to be a difference of .0071. Comparing again this figure with the total widths of lands and grooves obtained by microscopic measurement of the mortal bullet, there appears to be a difference of .0027. Comparing again the total width of lands and grooves of the Lowell Winchester bullet, obtained by microscopic measurement, and the total width of lands and grooves on the mortal bullet, obtained by such measurement, the difference is .0026.[31]

The data and language of Van Amburgh based upon measurements were subject to the same objection as was the substance of Hamilton's affidavit. The individual peculiarity of the Sacco pistol could not be established by such measurements. The accidental variations in the bore dimensions were not susceptible of precision measurements on the surfaces of the bullets fired therein for the purpose of establishing an individual peculiarity of the Sacco Colt. A reasonable amount of research would have demonstrated the uselessness of such measurements.

Van Amburgh discussed the double land mark:

> This double land is caused by what is termed "slippage." The slippage in each instance is shown at the right of the true land cut. Where a pistol barrel has a left twist as has the Colt automatic pistol, slippage when it appears on the bullet fired in such pistol is always to the right of the true land. It is caused by the bullet jumping ahead when the explosion occurs, and momentarily failing to take the rifling. The bullet "staggers" as it is sometimes called. As the rifling takes effect the bullet proceeds in its normal course to the left and with the twist of the rifling, but the initial impression of its jump ahead is registered at the top of the bullet and slightly to the right of the true land mark. This slippage or registering or a double land mark commonly occurs in the firing of automatic pistols. The appearance of such mark on bullets so fired is not significant of this case.
>
> This double land mark is made by the land in the pistol. The top surface of the land is formed by the action of a reamer which rotates in its journey through the barrel, and its action leaves a series of fine machine tool marks running transversely across the surface of the land. These tool marks permit the collection or metallic fouling or scraping from the surface of the bullet jacket through abrasion as the bullet proceeds through the barrel. The condition of the land varies accordingly from shot to shot, and scratches or marks made by the land upon the bullet, as are referred to by Mr. Hamilton in this case, have, in my opinion, no consequence.[32]

Such a discussion of tool marks seriously impeaches the credibility of Van Amburgh's testimony as it demonstrates a characteristic lack of understanding of the subject. As a matter of fact, the "scratches or marks made by the land upon the bullet" may be very

[31] "The Sacco-Vanzetti Case," Vol. IV, pages 3663-3665.
[32] "The Sacco-Vanzetti Case," Vol. IV, pages 3665, 3666.

valuable in identifying the firearm from which a bullet was fired if they appear on every bullet fired from the weapon. Generally speaking, the land impressions on a bullet are capable of furnishing more reliable evidence for the purpose of identification than are the marks left by the grooves of the barrel on the surface of the bullet. The surface of the bullet which comes in contact with the surfaces of the lands in the barrel is subjected to the greatest radial compressive stresses.

Nils Ekman of Bridgeport, Connecticut, at the request of Van Amburgh, took photographs of the mortal bullet and one test bullet. These photographs were taken through a microscopic arrangement magnifying to approximately 13 diameters. Ekman took twelve pictures of each bullet, to represent each land and groove. On the strength of these photographs Van Amburgh concluded: "By the Ekman photographs of all the land and groove impressions on the mortal bullet and the same of a Lowell Winchester bullet, the Court is afforded an opportunity of making comparisons between the two bullets, not only of one impression, but of the whole twelve impressions. Such comparison demonstrates absolutely the truth of the conclusion stated by me in my previous affidavit that the so-called mortal bullet was fired in the Sacco pistol.

"The photographs of the mortal bullet" were "numbered from one to twelve consecutively, and the photographs of the Lowell Winchester bullet" were "numbered from 'A' to 'L' consecutively. . . . For purposes of comparison photographs 1 and A, 2 and B, 3 and C, etc., should be examined together."

The photographs were taken consecutively moving from right to left around the bullet as the bullet appears setting on its base, starting with a certain groove impression which shows two significant scores and scratches.

Examples of the agreement in the markings as pointed out by Van Amburgh were:

> In photograph 3, in the left center of the groove impressions are four parallel lines extending downward from the knurled groove to a point near the base of the bullet. The same lines appear in photograph C. Although this latter photograph is slightly blurred they are, nevertheless, visible.
> In photograph 4 a land impression is shown which on its right border clearly indicates that some slippage took place during the travel of the bullet through the bore of the barrel. All lines within this land impression are not parallel, indicating that a change in direction took place when slippage occurred. In the

right center of the land impression, extending downward from the knurled groove, are two lines close together and parallel to each other. Two other lines to the left of those just described stand out prominently. Photograph D has in it the same lines as described in 4, appearing in the same locations and of the same character.

. . . .

Photograph 11 is a groove impression which registered lightly against the bore surface. The main bearing surface which is represented by the dark patch just above, and extending to, a point just below the knurled groove to a "V" shaped point below its center, is largely occupied by the area of the knurled groove itself. In this main portion of the bearing surface should be found the best registration of rifling marks. The limited area left to the observer is the "V" shaped portion directly under the knurled groove. In the "V" shaped portion are found two parallel lines extending upward and touching the lower part of the knurled groove. Photograph K is a groove impression. The dark patch in the center of the groove is that portion of the bullet which was most effected by friction with the bore surface. In the center of the dark patch are two parallel lines visible throughout the length of the patch and lightly but clearly, registered. They occur at the same distance with reference to the right and left borders and are of the same appearance and character as the mark found in 11.[33]

Van Amburgh advocated a comparison by means of these photographs of the mortal and test bullets: the examiner surveys together two pictures purporting to be images of the engraving of the corresponding land or groove of the barrel on two bullets, and the pictures may be superimposed one on the other to bring them into alignment. The record is not very explicit in its description of the method by which these photographs were obtained.

These photographs were not reproduced in the record as printed. It is to be inferred that the bullets were mounted on a base free to rotate about a vertical axis, and provided with means for accurately measuring the angle of rotation. In order that the photographs can be accurately superimposed, accurate phase relations must be maintained, and it is necessary that each bullet be mounted on the base with its axis of symmetry in exact coincidence with the axis of rotation of the base. Obviously, this can not be accomplished with a bullet that is deformed. On the other hand, photomicrographs can be made through the comparison microscope, comparing the purported

[33] "The Sacco-Vanzetti Case," Vol. IV, pages 3669, 3670.

corresponding surfaces of two bullets, even if they are deformed. Such photomicrographs have the added advantage in that the image photographed can always be reproduced at will in court by placing the bullets under the microscopes in the same position they occupied at the time the photomicrograph was made. In photographing the entire bullet, as was done in the Ekman photographs, illumination is an all-important factor; by improper illumination it is possible entirely to eliminate important details in the photograph. It is also to be noted that a photographic enlargement is quite different from a photomicrograph. A photographic enlargement has no more resolution of detail than that in the photograph from which it is made.

MERTON A. ROBINSON AFFIDAVIT

Affiant's Qualifications: In 1904 he entered the ballistic laboratory of the Winchester Repeating Arms Co., and after six years became foreman of that department. He held that position for five years when he was transferred to the Central Engineering department and given the title of ballistic engineer, still retaining the supervision of the ballistic department, which he then held.

He rebutted Hamilton's contention that the W.R.A. fatal bullet was not of the same manufacturing period as the W.R.A. cartridges found on Sacco at his arrest. Hamilton claimed the angle of the knurls to be different. Robinson stated:

> The Winchester Repeating Arms Company, which manufactured the bullet known as Exhibit 18 has always grooved the bullet with a knurl parallel to the axis of the bullet. If any individual bullet is found with a knurling which is slightly off from the parallel, it must be caused by stretch or distortion of the surface when engaging the lands and not by the process of manufacture.
>
> This bullet [fatal] is badly deformed, one side is flattened or caved in, and the base is mis-shaped. Under these conditions it is impossible to determine the true axis of the bullet, and it is impossible to determine whether the knurling is at right angles with the true axis or not.[34]

Robinson contended that the groove of a barrel is wedge shaped, being wider at the top than at the bottom. This upset the Gill attempt to reconcile measurements on the grounds that the groove was wider at the bottom than at the top.

[34] "The Sacco-Vanzetti Case," Vol. IV, pages 3674, 3675.

The expert rendered the following opinions:

From my careful examination and measurements, I am satisfied that the Fraher shell F4, Exhibit 30, was fired from Colt automatic pistol No. 219722, Exhibit 28.

From my careful examination and measurements, of the fired Winchester bullet known as the mortal bullet, Exhibit 18, I am satisfied that this bullet was fired from Colt automatic pistol calibre 32, No. 219722, Exhibit 28.[35]

Robinson merely supported Van Amburgh; he did not contribute new material. Without the aid of a comparison microscope, Robinson purported to identify the F4 shell with the test shell:

The firing pin dent shows an imperfection on the side, and the face of the flattened metal of the primer where it is set back against the breech block shows clearly defined marks and characters from the imprint of the breech block. I have observed the relation as to position of these significant markings with respect to the ejector mark, extractor mark, and the imperfection in the primer indent.

I have examined the Colt automatic pistol calibre 32 No. 219722 Exhibit 28, and find that the firing pin has a defect or gouge on its side. I find that the face of the breech block shows a number of machine or tool marks and other groups of characters, and that these agree as to number and shape with the markings on the fired primer in the Fraher shell F4 Exhibit 30. I find further that the location of these various tool marks or machine marks with respect to the extractor and the ejector clearance agrees as to location of the imprints shown on the primer of the Fraher shell F4 Exhibit 30.[36]

Robinson compared the Ekman photographs and said, for example:

Photographs 3 and C show the next adjacent groove cut to the land impression shown on Photographs 2 and B. At a location between the second and third knurl from the left hand side of the groove cut is a characteristic scratch or score and on either side are two or more scores or scratches showing on both photographs.

Photographs 6 and F show the land impression adjacent to groove cut shown on photographs 5 and E. Near the right hand edge is a characteristic score or scratch shown on both photographs. Near the left hand side, there is a set of three scores or scratches on both photographs. The turn-over to the right hand edge of land impression shows the same characteristics in both photographs.[37]

---

[35] ''The Sacco-Vanzetti Case,'' Vol. IV, pages 3676, 3677.
[36] ''The Sacco-Vanzetti Case,'' Vol. IV, page 3675.
[37] ''The Sacco-Vanzetti Case,'' Vol. IV, pages 3678, 3679.

Hamilton attempted to attack the Ekman photographs, which the photographer testified were of *approximately* 13 diameters' magnification. Hamilton made measurements on the photographs and then divided these measurements by 13. For example, he stated that on the photographs of the mortal bullet, all six groove marks measured 1.32 inches which divided by 13 gave as the actual width of all six groove marks 0.1015 inch. Van Amburgh, in his original affidavit, gave 0.1039, 0.1045, 0.1036, 0.1050, 0.1032, and 0.1047 inch as the actual widths of the groove marks on the mortal bullet, whereas at the trial he gave these widths as 0.107 and 0.108 inch.

This was another of Hamilton's silly attempts to cloud the issue by adding to the already existing confusion on measurements. Hamilton measured the photographs in hundredths of an inch, and therefore his division by 13, which was only approximate, would not produce results accurate to ten thousandths of an inch, as he would like one to infer.

These photographs gave the prosecution an advantage because, without a comparison microscope, they were a step in the right direction, and irrespective of the merits thereof, they must have weighed heavily with the trial judge. He only knew what the experts told him; he was in no position to criticize all the experts' testimony by virtue of his superior knowledge.

DECISION OF JUDGE THAYER ON THE MOTION FOR A NEW TRIAL

The opinions of the experts who submitted affidavits were accurately summarized by the learned judge: "Both Mr. Hamilton and Prof. Gill have stated in their opinion that the mortal bullet undoubtedly was not fired through the Sacco pistol; while Messrs. Van Amburgh and Robinson have affirmed with equal positiveness that said mortal bullet was fired by the Sacco pistol." His impression was that the experts were unable to agree upon any single material and important issue.

The affidavits were built around the theory of identification of bullets by measurements, and the judge's reaction thereto was: "The jury had at the trial the fullest opportunity to pass judgment upon this testimony introduced by both sides." This comment is absurd when the inaccuracy of the trial measurements and their significance is considered.

The judge placed too much faith in the trial testimony and the ability of the jury to track down the truth when he commented:

"This question of whether or not the mortal bullet was fired by the Sacco pistol was tried out at the trial most thoroughly and ably by counsel on both sides and was submitted to the jury for their determination as to the existence or non-existence of said rifling marks and whether or not said rifling marks were caused by the Sacco pistol. . . . The jury had the advantage of the use of a cast of the interior of the barrel of the Sacco pistol." In the light of previous explanation, the unreasonableness of the judge's contention requires no further illumination.

To be concise, the trial judge denied this motion for a new trial for the reason "that the defendants have not maintained by a fair preponderance of the evidence that there is any individuality in the location of the firing pin indentation on the primer or in the flow-back of the metal around the firing pin indentation, as affirmed by Mr. Hamilton." As to the individuality of the gouge near the end of the firing pin in Sacco's pistol he was "not convinced that the claim of Mr. Hamilton is correct. Therefore, on this particular issue I find as a fact that the defendants have not maintained the burden of proof that the law requires."

The judge reasoned similarly with respect to Hamilton's material regarding the ejector and breechblock marks.

The granting of a motion for a new trial, based upon newly discovered evidence, under the Massachusetts rule, rests in the sound judicial discretion of the trial judge. The *raison d'être* for the rule is that the trial judge is familiar with the witnesses and the evidence offered at the trial. The Supreme Court of Massachusetts has said that "it is enough if the newly discovered evidence appears to be so grave, material and relevant as to afford a probability that it would be a real factor with the jury in reaching a decision. . . . It is not essential in all cases that the judge must be convinced that the verdict at a new trial would inevitably be changed by the new evidence." It appears that the learned judge, in the exercise of his sound and impartial judicial discretion, should have applied the practical test of whether, if this evidence had been submitted to the jury, it would have made any difference in their conclusions with respect to the firearms issue. Furthermore, considering the confusion manifested in the evidence on firearms identification, the judge should have entertained the question of whether the new evidence reasonably demonstrated the doubt shadowing the proof of the state.

In behalf of Judge Thayer's decision it may be said that the credibilities of Hamilton and Gill were far from impressive, and seriously impaired the credence attributable to their affidavits. Gill, who positively stated that the fatal bullet "never passed through the barrel of Sacco's pistol," used the following language: "The reason is, I suppose, because my August measurements were made under different conditions from the September measurements; and to speak frankly, my August measurements were the first time that I had ever measured a bullet."

Hamilton is justly criticized by the Superior Court judge:

This motion depends very largely upon the affidavits of Mr. Hamilton and I have therefore given them my best consideration, and this consideration forces me to the conclusion that his greatest weakness is found in his overenthusiastic willingness to adapt himself to almost any condition that may arise within his line of professional inquiry. To show what I mean by this expression of "ever-enthusiastic willingness" let me refer, by way of illustration only to a portion of his affidavit of the Ripley motion.

Of course it was important on that motion to establish that certain ink marks were placed upon the Vanzetti bullets by some of the jurors, although there was no evidence of that fact, but Mr. Hamilton under oath gave his opinion on this subject as follows: That the affiant has examined the entire trial record in the above entitled case insofar as the same bears upon the history and introduction of the said so-called Vanzetti cartridges, Exhibit 32 and his examination of said four ink marks and is of opinion that said ink marks in each of them were placed upon said bullets after said cartridges went into the jury-room and before they were returned by the jury to the sheriff. The ink marks as a matter of fact must have been on the bullets for more than a year and a half before Mr. Hamilton saw them, and yet, notwithstanding this fact, he made oath that he could tell the day and the place that said shells were marked and the persons (the jury) who marked them. This is to give an illustration of what I mean by his "over-enthusiastic willingness" in his cause. [38]

In Hamilton's affidavit it was strongly contended that the angle of the knurling of the fatal bullet was three degrees off from being parallel to the axis, when in reality, this bullet was badly distorted, and according to an employee of the Winchester Repeating Arms Co., all W.R.A. ammunition has knurlings parallel to the axis of the bullet.

Hamilton foolishly stated that a new hammer had not been put in Berardelli's revolver because an essential screw did not show the marks of having been removed. It was quite possible that a new screw might have been inserted or the old screw removed and replaced by an expert with no appreciable addition to the marks on the screw head made by the original fitting of the screw.

[38] "The Sacco-Vanzetti Case," Vol. IV, page 3712.

Hamilton was guilty of using all possible arguments in his client's behalf whether founded in fact or in fiction, and his testimony creates a suspicion of charlatanism. Such an expert was not responsible and could not command the respect of the Court. His qualifications indicated a diversity of interests with no special training in the field of firearms identification.

Throughout the trial and the subsequent proceedings for a new trial, the expert testimony on firearms identification was crucial—it was the natural determinant of the guilt or innocence of Sacco. And until Sacco was proved guilty there was no case against Vanzetti. However, all the experts failed to use the essential instrument in firearms identification, the comparison microscope. Without this device it is impossible to obtain reliable results through an accurate comparison of the significant markings appearing upon the surfaces of the bullets and cartridge cases. The expert testimony at the trial was hopelessly incompetent; the experts did not have any helpful instruments, and their testimony evinced a characteristic lack of understanding of the method of firearms identification. The testimony embodied in the affidavits on the motion for a new trial showed an improvement, but the method employed therein, being fundamentally a comparison of the measurements of particular characteristics, was utterly unreliable, and the useless affidavit opinions should not have been admitted in evidence. The measurement of single characteristics on the cartridge cases, without taking into account the relative positions of these characteristics as a combination under the comparison microscope, was meaningless. No thought was given to the true individual peculiarity of a firearm which is constituted by a combination of its accidental characteristics. All the experts erred by not confining their examinations to the markings found on the surfaces of fired cartridge cases and bullets. They were not aware of the fact that identification of any kind is based upon a comparison of evidence of *like kind*—signature on bullet with signature on bullet and *not* signature on bullet with visual or other characteristics of a firearm.

Both the jury and judge at the trial were misled by the expert testimony presented for the state. The qualified opinions were treated as positive statements on the question of the identity of bullet III and the W.R.A. Fraher cartridge case. The unresponsive manner of the defense experts made those for the prosecution appear, by contrast, reliable and competent. Sadly enough, the experts for the state on the motion for a new trial created a better impression than did those for the defense, who made such absurd statements that no weight could honestly be given to their affidavits.

The inevitable conclusion to be drawn from an analysis of all the expert testimony is that what might have been almost indubitable evidence was in fact rendered more than useless by the bungling of the experts. And after the trial and the submitting of the affidavits on the motion, the trial judge was unable to evaluate the conflicting items of testimony; he lacked the necessary capacity in that he did not possess superior knowledge in the field of firearms identification. However, the affidavit testimony should have convinced the trial judge that the jury had not been properly equipped to make an intelligent finding on the firearms issue. And from a reading of the affidavits, which contained remarkable disagreements in measurements as between those of the various experts and between the measurements of a single expert, Judge Thayer, like any other reasonable person, should have realized the weakness in the state's proof upon the identity of bullet III and the W.R.A. Fraher cartridge case. He should have appreciated the doubt shadowing the prosecution's testimony, and, on the basis of this doubt, granted a new trial.[39]

[39] The Supreme Judicial Court of Massachusetts sustained the rulings of the trial judge (255 Mass. 369; 151 N. E. 839). The Lowell Committee likewise upheld the conviction. In connection with the results of an examination of the firearms exhibits submitted to the Lowell Committee, see the *Nation* (December 7, 1927), Vol. 125, No. 3257, page 625, ''A Sacco Revolver Expert Revealed.''

An explanation of the actions of the trial judge, as well as the affirmance by the Supreme Judicial Court, is found in an excellent discussion by Professor Edmund M. Morgan in 47 *Harvard Law Review*, 538.

CHAPTER III

ANALYSIS OF THE AUTHORITIES ON FIREARMS IDENTIFICATION

INTRODUCTORY

THE scientific methods of proof requisite to the identification of a firearm by means of its **signature** [1] have but recently become of sufficient reliability to merit judicial sanction. However, during the past half-century, the courts have been confronted with the question of the admissibility in evidence of testimony and objects offered as proof that the victim of a crime had been struck by a bullet or pellet discharged from a particular firearm. [2]

The courts, in determining the admissibility of material on firearms identification, have looked to the settled exclusionary rules of evidence

---

[1] It should be noted that ''Ballistics is the science that treats of the motion of projectiles, and is a particular branch of applied mechanics. Interior ballistics treats of the motion of the projectile while still in the bore of the gun and exterior ballistics treats of the motion of the projectile after it leaves the muzzle. Interior ballistics includes a study of the mode of combustion of the powder, the pressure developed, the velocity of the projectile along the bore; and the calculation of the dimensions of the powder chamber and of the powder which, for any particular design of gun and projectile, will give the required muzzle velocity while not exceeding the allowable interior pressure. Having determined the powder-pressure curve for the gun, the thickness of wall of the gun to withstand the expected pressure at each point may be determined by the principles of gun construction.'' Exterior ballistics includes ''Solution of the simplest problem of the trajectory— that of the trajectory *in vacuo*, the hypothetical case of a projectile passing through space without air resistance. Method of computing a trajectory in air— considering both air resistance and the force of gravity. Discussion of ballistic tables and firing tables,'' McFarland, ''Ordnance and Gunnery,'' pages 56, 411. (See also, Tschappat, ''Ordnance and Gunnery.'') To the foregoing may be added ''ballistics of penetration'' which deals with the effect of the projectile upon the target. A knowledge of ballistics furnishes a valuable background for the field of firearms identification; however, it is clear that an expert ballistician may be utterly unable to identify a particular firearm by means of its signature.

[2] Note, 66 A.L.R. 373; note, 66 U.S.L. Rev. (1932) 180. See also Wigmore, Principles Of Judicial Proof (2d ed.) section 62.

rather than to new criteria.[3] The greatest difficulty encountered by
the courts has been in the application of the **opinion rule**. Under this
rule, generally speaking, the opinion of a witness is excluded from evidence wherever the facts can be placed before the jury and where
the facts are of such a nature that the jurors are as capable as the
witness of predicating opinions thereon.[4] One jurist has stated in
effect that the policy behind the opinion rule is that, whenever the
jurors are as capable as the witnesses of drawing inferences and conclusions from the facts, the opinions of the witnesses are excluded
upon the theory that such opinion testimony is superfluous.[5] Another writer has said in effect that the opinion rule has been enunci-
ated to preserve the right to a trial by jury; the reasoning of the
jurors with respect to facts must not be supplanted by that of the
witnesses, unless the jurors are, under the circumstances, incapable
of reasoning in a rational manner without the assistance of opinion
testimony.[6]

In the application of the opinion rule the courts will exclude the
opinion of a lay witness unless he testifies with respect to personally
observed data which cannot be adequately reproduced for the jury.[7]
The opinion of a **specially trained witness** will be admissible only
wherever the drawing of an inference from fact to conclusion requires
specialized knowledge.[8] The opinions of witnesses are not binding
upon the jurors, who must draw their own conclusions with the
assistance of such opinions, and therefore opinion testimony does not
usurp the function of the jury.[9] The jury may accept or reject the

[3] "Developments in the Law, Evidence," 1932, 46 Harv. L. Rev., 1138, 1167;
"Progress in scientific methods of proof necessitates for the most part only the
application of old tools to new materials."

[4] Pearce v. Stace (1913) 207 N. Y. 506, 511, 101 N. E. 434, 436.

[5] 4 Wigmore, Evidence (2d ed.) sections 1917, 1918.

[6] 3 Chamberlayne, Evidence, sections 1796-1799, 1807.

[7] State v. Pruett (1916) 22 N.M. 223, 160 Pac. 362 (a lay witness was permitted to give his opinion that a certain mark observed by him was a knee-print).
See the note following this case in L.R.A. 1918A 662, on the "right of witness
to express an opinion on non-technical subject because of impossibility or difficulty
of reproducing the data"; 4 Wigmore, Evidence (2d ed.) section 1924.

[8] Recent Cases, 43 Harv. L. Rev. 497: "Opinion testimony is admissible
wherever the court feels that specialized knowledge will be of aid to the jury in
interpreting facts from which the lay mind cannot readily draw the correct conclusions. People v. Derrick, 85 Cal. App. 406, 259 Pac. 481 (1927); Pridgen v.
Gibson, 194 N.C. 289, 139 S.E. 443 (1927); 4 Wigmore, Evidence (2d ed. 1923)
section 1923"; Dougherty v. Milliken (1900) 163 N.Y. 527, 57 N.E. 757.

[9] 4 Wigmore, Evidence (2d ed.) sections 1920, 1921.

reasoning of the witnesses. And furthermore, "to say that the testimony of any witness is conclusive upon the jury clearly invades its province to pass upon the credibility of the witness." [10]

No hard and fast rule can be applied to determine upon what subjects the jury need the assistance of specially trained witnesses.[11] Of course, the subjects must not be those of common knowledge. The subjects must be those which are not understood by jurors generally, and it may be assumed that jurors are usually persons of common education engaged in the common occupations of life.

Before the opinion of any specially trained witness can be admitted in evidence, his **qualifications** must be established by showing that he possesses the required specialized skill, and the decision with respect to qualifications rests in the discretion of the trial judge.[12] Depending upon the subjects there are all grades of knowledge which must be possessed by the specially trained witnesses. The capabilities of the witnesses must be proportionate to the specialized nature of the inferences which they offer to state. The qualifying experience may be derived from two sources: first, experience in some occupation, such as banking, which is called occupational experience; second, experience in some science, such as chemistry, which is called scientific experience.

The term **expert** is commonly used to include all witnesses whose opinions are admitted on the ground of specialized knowledge.[13] However, there are two types of specially trained witnesses. The first class, strictly speaking, is composed of experts who render opinions upon data which are observed by others and are presented to them in hypothetical form. Here the ability to observe may be possessed by the layman. The second class includes the **skilled witnesses,** who are skilled in the observation of data which cannot be intelligently observed without such skill, and in addition possess special skill in

[10] Commonwealth Life Ins. Co. v. Harmon (Ala. 1934) 153 So. 755, 757.

[11] 4 Wigmore, Evidence (2d ed.) section 1923.

[12] General Paint Corporation v. Kramer (1933) C.C.A. 10th Circ. 68 Fed. (2d) 40, 42: "Whether a witness possesses the requisite qualifications to testify as an expert is a matter resting largely in the discretion of the trial court, and its decision will not be disturbed unless it is manifest that the trial court has fallen into error or abused its discretion. Clarke v. Hot Springs E.L.&P. Co. (C.C.A. 10) 55 F. (2d) 612, 615; Gila Valley, G.&N.R. Co. v. Lyon, 203 U.S. 465, 27 S. Ct. 145, 51 L. Ed. 276; People v McCarthy, 115 Cal. 255, 46 P. 1073''; Masocco v. Shaaf (1931) 234 App. Div. (N.Y.) 181, 254 N.Y.S. 439; Dougherty v. State (Ind. 1934) 191 N.E. 84; 1 Wigmore, Evidence (2d ed.) sections 560, 561.

[13] 22 Corpus Juris, section 624; 1 Wigmore, Evidence (2d ed.) sections 672-686.

drawing deductions from the data. It has been stated that the experts who render conclusions in connection with hypothetically stated premises occupy a unique position; their opinions are based upon facts which are known to others. The skilled witnesses who observe the data occupy essentially the same position as any other witness; they state, either by way of fact or inference, that which they claim to know.[14] Obviously, a specially trained witness may testify as either an expert or a skilled witness.

Specially trained witnesses in the field of firearms identication are skilled witnesses; the layman cannot intelligently either observe or draw proper conclusions from the data. The testimony of these skilled witnesses as to what they have observed through their specially trained faculties is a statement of fact rather than a matter which is usually deemed opinion, and it is within the usual province of the witness. Substantially the same facts will be observed by the trained witnesses using proper instruments and reliable methods which reduce to a minimum the part played by the imagination of the observer. The testimony of the skilled witnesses concerning the analysis of the data as to their origin and significance is properly a matter of opinion which, with an ordinary witness, is in the province of the jury. As all the opinion testimony of specially trained witnesses is admitted upon the theory that the jury without assistance cannot draw rational inferences from the facts, there is no danger in the prevailing use of the term expert. And a competent witness in drawing deductions from data on firearms identification will by necessity be skilled in the observation of the data. Osborn has said that "the problem of identification of any kind has two principal phases: first, accurate observation and mental recognition or perception of the evidence; and second and more important, correct interpretation of the meaning or significance of the evidence." [15]

In the identification of firearms, in which it was assumed that all the data could be reproduced for the jury, the trial judges, when a party sought to introduce the opinion of an expert, had to make the preliminary determination as to whether the subject matter of the identification was so complicated for the average juror that the specialized skill of an expert was necessary to draw a correct inference from fact to conclusion. If the court found that the opinion of an expert would be of assistance to the trier of fact, it was faced with the further problem of deciding whether the witnesses introduced

[14] 3 Chamberlayne, Evidence, sections 1805, 1950.
[15] "The Problem of Proof," page 58.

as experts were sufficiently skilled to qualify as such; and this, of course, involved the question of whether the witnesses' methods of proof were dependable enough to permit their use in evidence.

Many rulings of the courts on these questions are found to be erroneous when tested by the present understanding of the highly technical science of firearms identification from ammunition fired therein.[16] The serious errors were made by permitting the juries to draw their own inferences from the markings on the firearms exhibits without the aid of expert opinions, and by allowing experts to testify who employed unreliable methods of proof predicated upon incomplete data. However, the decisions of the judges should not be criticized without taking into consideration some of the factors underlying their judgment. In the first place there was no helpful written material on the subject prior to the last few years. Furthermore, the relevant authorities indicate that the testimony of the witnesses who purported to be experts can be subjected to criticism for having misled the courts as to the extent of the development of the science of firearms identification. Obviously, the judges could not have been expected to undertake scientific research in this field of identification; they looked to the testimony of the witnesses for guidance. The burden was on the shoulders of the witnesses to perform the necessary research and to be familiar with the shortcomings that arose from the inadequacy of the accumulated data.

It appears that the witnesses, whether testifying as experts or merely pointing out the observable facts concerning the firearms exhibits, assumed an unwarranted air of experience and on many occasions gave testimony which a modicum of research would have proved to be fallacious in numerous aspects. The early witnesses lacked the proper scientific equipment; their emphasis upon the plainly observable data prompted the courts to allow the juries to form conclusions without the assistance of expert opinions. No consideration was allocated to the necessity of having a competent witness draw an inference predicated upon a skilled observation and interpretation of the data. With the development of scientific methods, facilitated by the application of helpful instruments, and the more general recognition by the courts that expert testimony was essential,

---

[16] Firearms identification from ammunition fired therein is demonstrated as a science in Chapter I, *supra*. A working knowledge thereof cannot be acquired through occupational experience. It requires scientific judgment based upon scientific experience which necessitates deliberate study and investigation directed along scientific lines.

the witnesses did not continue to develop the science thoroughly. Many of their present theories, which are now the subject of disproof, were originated at the beginning of the scientific era (when the comparison microscope was applied intelligently in the systematic acquisition of pertinent data). The testimony failed to interpret successfully the differences found in the respective signatures of a single firearm and the similarities found in the signatures of different firearms. The failure of the courts to detect the outstanding deficiency of ignoring differences is due apparently to the erroneous belief that it was certain that the respective signatures of a particular firearm would be identical. (The courts realized that the respective signatures of any person would not be identical. Firearms were distinguished on the theory that they were inanimate objects lacking the personal element existing in humans.) Further evidence of the lack of foundation for the testimony is found in the language of the experts who qualified as ''ballistic experts'' although they were purporting to identify firearms by means of the markings on ammunition fired therein.

The attitude of the witnesses, in general, has been manifested by the faulty treatment which the subject of firearms identification has received from the courts. The judges were on the alert to learn of a theory which would logically and reasonably explain the proposition that the signature of each firearm is individual. In this approach the courts were likely to approve of the theory that appeared most plausible at the time, and consequently, the methods of proof accepted on the basis of this relative standard were not always, on their own merits, of sufficient dependability to justify their admission in evidence. A great deal of the testimony appeared to be very accurate because it was for the most part consistent with the outstanding circumstances of the cases which created strong probabilities that particular weapons had been employed. The apparent accuracy of the testimony creates a suspicion that the experts were often apprised of the circumstances of the cases previous to the trials. This was a vicious practice, and it is reasonable to conclude that, if the experts had confined their conclusions strictly to examinations of the markings on bullets and cartridge cases, they would not have been so readily accepted by the courts; the judges were more critical of expert testimony inconsistent with the other evidence in the cases. In fairness to the experts it should be stated that unquestionably many of them actually believed that they were competent and were unaware that they were basing their opinions on utterly insufficient data.

ANALYSIS OF AUTHORITIES

As early as 1881 [17] a defendant, having been convicted of murder, maintained that the trial court had erred in not permitting an expert to examine and make experiments with the pistols exhibited at the trial to see if he could ascertain from which one the fatal bullet had been discharged. The expert had stated that the proposed experiments would change the condition of the pistols. The ruling of the trial judge was sustained on the ground that it was within his discretion to determine whether the pistols should go to the jury in the condition which existed at the time they were brought into the custody of the court.

The stage of the development of firearms identification, prior to the year 1900, was for the most part analogous to the early tracing of printed material by means of the class characteristics of the type.[18] The relatively few firearms in existence, as well as the comparatively small amount of ammunition, created the probability that for the purpose of a particular case the class characteristics could be treated as being individual. The procedure, in linking the defendant's firearm with the homicide or assault, was to ascertain whether the defendant possessed a weapon adapted to ammunition of the same size and shape as that found at the *situs* of the crime, and whether he owned ammunition of that certain size and shape. A factor which contributed to the success of this early method was that some of the ammunition was homemade and the class characteristics of the product of any maker were likely to be individual.

In 1876,[19] at a trial on an indictment for assault with intent to commit murder, the court allowed the prosecution to introduce the testimony of a witness, as a circumstance indicative of the defendant's guilt, to the effect that pellets taken from the body of the victim corresponded in size with those drawn from the barrel of the defendant's shotgun, the witness having made the comparison.[20] In

[17] State *v.* Smith, 49 Conn. 376.

[18] See Introduction, *supra.*

[19] Moughon *v.* State, 57 Ga. 102. The defendant was convicted of assault by the trial court, but the judgment was reversed on the ground of error in the charge to the jury.

[20] A witness, after stating that he had been familiar with the use of guns all his life, was allowed to express his opinion that an examination of the defendant's shotgun on the day following the assault revealed that it had recently been fired.

Evidence on the question of how recently a firearm has been fired may be properly a part of the identification of a particular firearm. For example, it is sometimes possible to prove that the firearm had not been fired as recently as the time of the perpetration of the crime, and this evidence would corroborate the

State *v.* Outerbridge (1880) [21] the defendant was convicted of murder. The state was permitted to prove by the opinion of a witness and by a demonstration to the jury that a bullet taken from an oak tree near which the body of the deceased was found, and another extracted from his body, fitted into bullet molds which had been found in a pocket of the defendant's clothing. The court ruled that the opinion of the witness had been properly admitted on the ground that the fitting of the bullets into the molds was pertinent evidence and that the defendant did not take a timely exception thereto.[22] A trial court of California (1892)[23] allowed a gunsmith, called as an expert, to compare a bullet which had been extracted from the head of the deceased with another discharged from a pistol belonging to a companion of the defendant, and to state his opinion that they were not similar. The Supreme Court held that this was not a proper subject for expert testimony as it did not require special skill to determine whether "the end of one bullet was feather-edged, and another not so, or was flat or concave at the end." [24]

---

opinion that the fatal bullet had not been fired from the particular weapon. However, evidence relating to how recently a firearm has been discharged must not be confused with the science of identifying a firearm by means of the ammunition fired therein.

[21] 82 N. C. 617.

[22] There was evidence to show that the defendant had recently molded bullets and that soon after the murder his gun bore fresh marks of having been discharged.

[23] People *v.* Mitchell, 94 Cal. 550, 29 Pac. 1106. The conviction of murder in the second degree was reversed by the Supreme Court.

[24] 94 Cal. 550, 555, 29 Pac. 1106, 1107. The Court similarly treated the opinion of the expert that a certain cartridge had never been in a pistol because there were no marks on it.

Other cases dealing with the comparisons of ammunition are: Meyers *v.* State (1883) 14 Tex. App. 35 (a comparison of the character of the gun wadding found at the *situs* of the crime with that taken from the defendant's shotgun); Granger *v.* State (Tex. 1895) 31 S. W. 671 (a witness gave his opinion that pellets found in the defendant's shotgun were of the same size as those taken from the planks of the victim's house, and as there were no samples of the shot in court, the opinion was thought to be the only way of putting this evidence before the jury).

Cases involving opinions as to how recently a firearm has been discharged are: Meyers *v.* State, *supra* (persons experienced in the use and handling of firearms are competent to give expert opinions on whether or not a firearm has been recently discharged); State *v.* Davis (1899) 55 S. C. 339, 33 S. E. 449 (a witness familiar with firearms testified that a firearm had not been recently fired); Pemberton *v.* State (1909) 55 Tex. Crim. Rep. 464, 117 S. W. 837 (after qualifying, a witness may state his conclusion as to when a firearm was last discharged); Holder *v.* State (1917) 81 Tex. Crim. Rep. 194, 194 S. W. 162 (a deputy sheriff, after qualifying, stated that in his opinion the pistol had been fired only once);

The substantial increase in the number of firearms possessing the same class characteristics, as well as in the quantity of ammunition, created a need for the identification of a firearm by means of its accidental characteristics. The common duplication of class characteristics rendered the former method of identification grossly inadequate for most purposes.[25]

Gravette v. State (Ala. 1933) 147 So. 641 (a sheriff, after stating that he had had considerable experience with firearms, gave his opinion that the pistol had been recently fired); see also, L.R.A. 1918A 670.

Another type of evidence which has some probative value in firearms identification is the "carrying distance" of a particular firearm (Smith v. State (1913) 8 Ala. App. 187, 62 So. 575). See, also, Bearden v. State (1903) 44 Tex. Crim. Rep. 578, 73 S. W. 17 (the extent to which shot will scatter at various distances). This type of evidence concerns the science of ballistics and not the science of firearms identification from ammunition fired therein.

[25] However, class characteristics may still be of importance. In Collins v. State (1918) 15 Okla. Crim. Rep. 96, 175 Pac. 124, the defendant possessed a caliber .41 firearm and an issue was raised at the trial as to whether the deceased was killed by a caliber .38 or caliber .41 bullet. A sheriff, testifying as an expert, stated his opinion that the fatal bullet and a bullet taken from the defendant's firearm were of the same caliber. The appellate court held that "the subject of firearms and the size of bullets shot by different kinds of pistols is not a subject upon which the ordinary layman would be qualified to express an opinion." This view of the court can be reconciled with the ruling in People v. Mitchell, supra. The court in the Mitchell case, in excluding the expert's opinion, focused its attention upon the conspicuous differences between the shapes of two bullets exhibited at the trial. In the Collins case, the court admitted the expert's opinion on the theory that a familiarity with firearms and ammunition was necessary in distinguishing between the relatively inconspicuous differences of caliber .38 and caliber .41 bullets.

In State v. Rusnak (1931) 108 N.J.L. 84, 154 Atl. 754, a policeman was murdered and the most plausible theory for the defense to propound under the circumstances was that the fatal bullet had been discharged from the revolver of a fellow police officer who was at the scene of the crime. The defendant had a Colt revolver adapted to the caliber .38 S.&W. Sp'l cartridge, whereas the police officer carried a Colt revolver adapted to the caliber .38 S.&W. cartridge. The expert demonstrated that the revolver of the police officer could not have fired the fatal bullet which was a caliber .38 S.&W. Sp'l with two grease cannelures. Fig. 138 is a photomicrograph of the bullet which was extracted from the body of the victim.

In People v. Sonoqui (Cal. 1934) 31 Pac. (2d) 783, 787, the court held that the prosecutor was not required to call a "ballistic expert" although "he introduced the cartridges recovered from the defendant's home and the empty shells found at the scene of the homicide" to show that they were of the same size and caliber. The court held that "it was the province of the jury to determine as a question of fact whether or not the cartridges or shells corresponded to the shells or bullet which it was shown produced death."

An attempt was made in Commonwealth v. Best (1902) [26] to identify a firearm through its accidental characteristics. The trial court admitted in evidence two fatal bullets, a test bullet of the same cali-

Fig. 138.

ber which had been pushed through the barrel of the defendant's rifle, and photographs thereof,[27] in order to show that the marks from

[26] 180 Mass. 492, 62 N.E. 748. The defendant was convicted of first degree murder.

[27] The markings on test bullets and cartridge cases are standards of comparison and they are in this respect analogous to a specimen of typewriting. In People v. Storrs (1912) 207 N. Y. 147, 100 N.E. 730, the court admitted a specimen of typewritten material, not otherwise relevant to the issue on trial, ''upon the principle that, where an impression is made upon paper, wood, leather, or any other plastic material by an instrument or mechanical contrivance having or possessing a defect or peculiarity, the identity of the instrument may be established by proving the identity of the defects or peculiarities which it impresses on different papers. . . . Inasmuch as its work affords the readiest means of identification, no valid reason is perceived why admitted or established samples of that

the rifle on the test and fatal bullets coincided so closely as to prove that all three had passed through the same barrel.[28] The defendant excepted to this evidence on the theory that the conditions of the actual shooting had not been reproduced in the performance of the experiment. The exceptions specifically mentioned the difference in the forces used to impel the test and fatal bullets, the rusting of the barrel during the interval between the murder and the experiment, and that three shots fired during this interval "increased the leading of the barrel."[29] The Supreme Court of Massachusetts held that the suggested sources of error were trifling. The evidence was thought to be the best attainable and the only means of informing the jury on the question of how a lead bullet is marked by the barrel through which it is fired.[30]

The defendant objected to the photographs on the further ground that they were arranged to bring out the likeness in the markings of

work should not be received in evidence for purposes of comparison with other typewritten matter alleged to have been produced upon the same machine.'' It is within the discretion of the trial judge to determine whether or not a standard of comparison is genuine. If there is doubt, test bullets and cartridge cases may be obtained by discharging the firearm in the presence of the court as in the case of State of N.Y. *v.* Small, Kings County, Dec. 17, 1930.

The courts have intelligently ruled upon the question of the admissibility in evidence of test bullets and cartridge cases as standards of comparison. They have not applied the technical restrictions which have emanated from the historical background of handwriting identification.

The use of photographs and photomicrographs must be governed with caution owing to the possibilities of deception by manipulation of lighting effects. See Chapter I, *supra*. Photomicrographs are invaluable because they present to the jury a permanent reproduction of the image visible under the microscope. Each juror may receive a copy to be studied at his leisure in the jury room. The photomicrographs can be verified for the trial judge by placing the purported object under the microscope.

See generally, Osborn, ''Questioned Documents'' (2d ed.); Osborn, ''The Problem of Proof'' (2d ed.); Wigmore, ''Principles of Judicial Proof'' (2d ed.).

[28] The Supreme Court, in affirming the conviction, used the words, ''the marks from the rifle in the two cases coincided so closely as to prove that all three bullets had passed through the same rifle barrel.'' The question raised by ''so closely,'' when identifying a particular firearm from all others, illustrates the necessity of having a competent expert skillfully interpret the observable markings.

[29] 180 Mass. 492, 495, 62 N.E. 750.

[30] The Court admitted as expert testimony the opinion of a witness that the fatal bullets were marked by rust in the same way that they would have been, if they had been fired through the evidence rifle, and that it took at least several months for the rust in the rifle to form. Surely, such testimony emanated from an over-zealous witness.

the bullets. The Court manifested its belief that the jurors were qualified to detect the significant markings and to draw their own conclusions therefrom when it said of the photographs, "But the jury could correct them by inspection of the originals, if there were other aspects more favorable to the defence."[31]   As a matter of fact the jurors were as competent as any of the witnesses at that time. The proper equipment was not being utilized, and there is nothing to show that constructive research had been performed. The test bullet admitted in evidence by the Massachusetts Court was submitted to the jury under the erroneous belief that the same markings are produced whether a bullet is pushed or fired through a barrel.[32]

In People *v.* Weber (1906)[33] the defendant was convicted for the murder of four members of his family. A firearm was traced to his possession, and a witness at the trial fired test bullets therefrom. The trial judge originally admitted the opinion of the witness, as expert testimony, that the test and fatal bullets had been discharged from the same weapon. On the basis of subsequent testimony that the opinion was predicated upon a comparison of the marks on the bullets, the trial judge concluded to strike the opinion from the record. He decided that a comparison of the marks was not a matter of expert testimony, but one within the ordinary capacities of the jurors, who could inspect the bullets and determine for themselves whether the test and fatal bullets had been emitted from the same firearm.[34] The Supreme Court of California sustained the decision of the trial judge.

The marks found upon the cartridge cases discharged in the defendant's rifle were used as the standard of comparison in State *v.* Clark (1921).[35] The defendant was convicted of manslaughter on the finding of the jury that Clark had negligently shot and caused the death of his companion during the course of a deer hunt. The evidence was purely circumstantial, and the identification of the defen-

---

[31] 180 Mass. 492, 496, 62 N.E. 750.

[32] The use of markings on a test bullet produced by "pushing," as a standard of comparison with markings on a fatal bullet which has been fired, is reliable only for certain characteristics. See Chapter I, *supra*, Figs. 120-122.

[33] 149 Cal. 325, 86 Pac. 671.

[34] The defendant contended that the trial court had erred in admitting the test bullets on the ground that the fatal bullet was not fired under the same conditions. The Supreme Court decided that the only unsimilarity related to exterior blood marks. The Court also determined that it was not error to allow the prosecution to present the bullets to the jury in a pre-arranged sequence, because the jury would doubtless, at their convenience, change the order as they saw fit.

[35] 99 Or. 629, 196 Pac. 360.

dant's rifle was a significant factor leading to the conviction. The substance of the testimony on firearms identification was discussed by the Supreme Court of Oregon in upholding the admissibility of such evidence. The Court said:

It will be remembered that defendant and deceased were both carrying guns of the same make and caliber, 30-30 Winchester rifles, and that they were both using the same kind of ammunition, 30-30 cartridges with steel jacketed soft-nosed lead bullets, the cartridges perhaps most commonly used by hunters, and ordinarily it would have been impossible to identify by the cartridge the gun from which it was fired. In the case at bar that difficulty was surmounted by certain physical facts which cannot be disputed. At the boulder heretofore mentioned the witness found at the base of it an empty 30-30 shell, and another empty shell was taken from the barrel of the rifle of deceased, both of which bore on the brass part of the primer a peculiar mark evidently caused by a flaw in the breech-block of the gun from which they had been fired. The primer of a Winchester shell is composed of two parts, one, the outer rim of the primer, being brass, and the center portion where the firing pin strikes being copper. The copper center is approximately the diameter of an ordinary pin head, while the brass portion would measure 3/16 of an inch across, including the copper center. The brass rim from the exterior of the copper center to the exterior of the brass portion will measure something less than 1/16 of an inch. When the firing-pin strikes the primer in discharging the gun the primer is driven slightly in and below the general surface of the butt of the cartridge, but the explosion, which causes a pressure of about 32,000 pounds to the square inch, drives back the brass part of the primer against the breech-block, leaving it flush with the butt of the shell. But the copper portion, being held by the firing-pin, is not so pressed back. The result of this enormous backward pressure upon the primer is to stamp upon it the mark of any flaw or protuberance that may be upon the breech-block. If there is a depression, the primer will show a protuberance, and vice versa. In this case the flaw had caused a very slight, almost microscopic protuberance in the primer of the shell, which enlarged photographs make very clear to the naked eye, although it is difficult to discern the mark without the aid of a microscope or such photographic enlargements. A careful examination of the shell found at the base of the boulder disclosed the mark upon the brass portion of the primer and this was compared with two shells found where defendant had shot the porcupine on the evening when he and the deceased had started out to hunt, and identical marks were found upon these. Later several shells were fired from defendant's gun and that weapon was found to produce the same mark. The extractor of a Winchester rifle makes a little scratch upon the inside of the rim of the shell in the process of throwing the cartridge from the barrel. The extractor of defendant's gun was found to make a sort of double scratch, while the rifle of deceased made only a single scratch, thus negativing the theory that deceased might have been accidentally shot with his own gun. Another singular circumstance developed: When the body of deceased was found, his rifle was lying by his side and in the barrel was an exploded shell. An examination of this shell disclosed the fact that it had the same flaw mark upon it that was produced by defendant's gun, and subsequent tests disclosed the fact that the gun of deceased produced no such mark in the process of firing. More than this, an examination of the shell disclosed that the inner rim of the cartridge showed that it bore, not only the extractor mark made in taking it out of the barrel when it was found, but another extractor mark, indicating that it had been taken from the barrel of some other, gun before being placed in the barrel of defendant's gun.[36]

The Court also said:

> The peculiarities of the shells found were pointed out to them [jurors] by expert witnesses and made clearer by enlarged photographs.[37]

A fair inference to be drawn from the language of the Court is that the jury was led to believe that the existence of the "flaw" on the test and boulder cartridge cases proved conclusively that all had been fired in the same firearm. Such a belief was prejudicial to the defendant. The evidence did tend to show that the boulder cartridge case had not been fired in the deceased's firearm and that it might have been fired in the defendant's. But to prove that the boulder cartridge case had been discharged in the defendant's rifle, it would have been necessary to undertake a more comprehensive investigation, taking into consideration the character of all the marks on the cartridge cases produced by contact with the firearm and the relative position of these marks, in order to establish the individual peculiarity of the defendant's firearm.[38] Furthermore, there are differences in the markings on cartridge cases discharged in the same firearm, (see Fig. 139), and it is very possible that the identification evidence in this case was of no probative value on the question of whether the boulder cartridge case had been discharged in the defendant's firearm rather than some other firearm.

The Court held that it was proper to admit in evidence enlarged photographs of the cartridge cases on the theory that they presented

---

[36] 99 Or. 629, 649-651, 196 Pac. 360, 367.

[37] 99 Or. 629, 654, 196 Pac. 360, 368.

[38] See Chapter I, *supra,* page 98.

to the jury a clearer view of the condition of the cartridge cases than that afforded by the naked eye. It overruled the objections made to the introduction of testimony relating to the experiments performed with the rifles and stated that "the admission of testimony concerning tests of this character rests very largely within the sound discretion of the court."[39] It pointed out that the possession of the rifles had been carefully traced; that there had been no appreciable

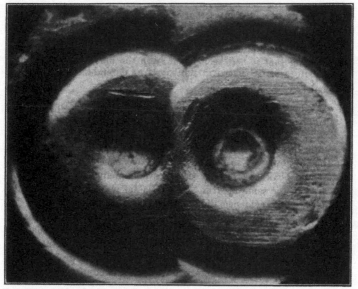

Fig. 139.—Comparison of Primers of Two Cartridges Fired in the Same Revolver.

change in the condition of the rifles because the "hard steel of a breech-block is not liable to sudden changes when in disuse, unless it is intentionally disfigured";[40] and that the testimony tended to prove that the cartridge case found near the boulder had been fired in the defendant's firearm.[41]

[39] 99 Or. 629, 665, 196 Pac. 360, 372.

[40] *Id.*

[41] On the question of the admissibility of testimony relating to experiments performed by the witness: State *v.* Nagle (1903) 25 R. I. 105, 54 Atl. 1063. (A doctor was permitted to testify as to the nature of a wound. His testimony, pertaining to experiments performed with a firearm to determine the character and

In State v. Vuckovich (1921)[42] the defendant was convicted of first degree murder. At the time of his arrest a caliber .32 Colt automatic pistol and ammunition therefor were taken from his possession. The trial judge admitted expert testimony to show that a cartridge case picked up at the scene of the homicide, and test cartridge cases produced by discharging the defendant's ammunition in his pistol, bore a mark characterized by the court as "a peculiar crimp or mark at the open end," [43] and that this mark did not appear when the same cartridges were fired in another pistol of similar make.[44] There was further expert testimony to show that the fatal bullet was of the same weight as those taken from the defendant and that the same marks were "made by the lands and grooves in the barrel of the pistol"[45] on the test and fatal bullets.

The Supreme Court of Montana approved of this evidence. The main objection of the defendant was directed to the evidence of the experiments made by the expert. The Court in disposing of the objection said:

nature of powder burns, was admitted on the theory that whenever the opinion of a witness is deemed relevant, the grounds on which such opinion is based are also relevant.)   People v. Ferdinand (1924) 194 Cal. 555, 229 Pac. 341 (an expert gave testimony on the subject of experiments which he had performed to determine the nature of powder marks produced on a garment by discharging a firearm at stated distances therefrom; the court held that it was not error to admit such testimony because substantially similar conditions were shown to have existed between the experiments and the circumstances under which the victim was killed); Epperson v. Commonwealth (1929) 227 Ky. 404, 409, 13 S.W. (2d) 247, 249 ("in order that an experiment shall possess sufficient probative value to warrant the admission of evidence thereof, it is necessary that the circumstances under which the experiment is made shall have been similar to the circumstances prevailing at the time of the occurrence involved in the controversy, and that the materials and instrumentality used shall be substantially the same"); Prudential Ins. Co. of America v. Tuggle's Adm'r (Ky. 1934) 72 S.W. (2d) 440; State v. Hendel (1894) 4 Idaho 88, 35 Pac. 836 (the court refused the defendant's request that a bullet be subjected to a microscopical and chemical examination, as well as the substances adhering thereto, for the purpose of determining whether this bullet, found in the clothing of deceased, had caused the victim's death; the main objection raised was that the performance of the experiments would require over two weeks); see State v. Allison (Mo. 1932) 51 S.W. (2d) 51, 85 A.L.R. 471, and the annotation following this case in 85 A.L.R. 479, entitled: "experimental evidence as affected by similarity or dissimilarity of conditions."

[42] 61 Mont. 480, 203 Pac. 491.

[43] 61 Mont. 480, 494, 203 Pac. 491, 494.

[44] Clearly, the identification was based upon incomplete data.

[45] 61 Mont. 480, 494, 203 Pac. 491, 495.

It seems to be a well-established rule that it is largely within the discretion of the trial court to permit experiments to be made, and that caution should be exercised in receiving such evidence. It should be admitted only where it is obvious to the court from the nature of the experiments that the jury will be enlightened, rather than confused. Such evidence should not be excluded merely because it is not necessary in establishing the facts sought to be shown by the prosecution, if it tends to corroborate the position taken by the expert witness whose evidence has been received; for whenever the opinion of a person is admitted to be relevant the grounds on which it is based are also relevant. Evidence of experiments made out of court and not in the presence of the jury are admissible upon the same principle as the experiments which are conducted in the jury's presence.[46]

The courts admitted in evidence testimony relating to the identification of firearms from ammunition fired therein on the theory that it was evidence concerning experiments. At the time of these early cases no scientific research had been undertaken. Consequently, in each instance, the witness experimented to ascertain whether the firearm to be identified produced markings different from those produced by other firearms to which the witness had access. Of course, these identifications were for the most part worse than useless because they were based upon utterly incomplete data. After some research had been accomplished the witnesses did not perform such experiments in each case; they made primarily a comparison of the signatures on the test ammunition with the signatures on the ammunition used in the crime. The background of research obviated the necessity of performing specific experiments, and the courts for the most part no longer referred to the testimony as "evidence of experiments."

The Supreme Court of Oregon again ruled upon the admissibility of evidence on firearms identification in State v. Casey (1923).[47] The defendant, jointly indicted with one Burns but given a separate trial at his request, was sentenced to death for murder. The victim of the murder was a special agent guarding railroad cars. He was armed with a caliber .38 special, Smith and Wesson revolver. There was evidence to show that the bullets found in and about his body were fired from a caliber .38 firearm. Two exhibits at the trial were caliber .38 "Colt's army special revolvers," one having been traced to Casey's possession and the other having been found in a bed belonging to Burns under which Casey was hiding when arrested. The defen-

[46] Id.
[47] 108 Or. 386, 213 Pac. 771, 217 Pac. 632.

dant objected to certain exhibits composed of bullets which had been "driven" through the two revolvers by a witness. The defendant likewise objected to the testimony of this witness who compared "rifling marks" on the test and fatal bullets and stated that, in his opinion, the bullets found in and on the body of the deceased had each been fired from a "Colt's army special revolver."

The expert was dealing with class characteristics, and it should have been a simple matter to ascertain whether a bullet might have been fired from a Colt or a Smith and Wesson revolver because the direction of the twist of the rifling is different in the two cases. But to state positively that a given bullet has been fired from a caliber .38 Colt army special revolver requires a careful comparison to establish definitely the existence of the class characteristics of a caliber .38 Colt army special revolver.[48]

The Court held that it was proper to admit the test bullets and the opinion of the witness, the weight of evidence of experiments being exclusively for the jury.[49] It sanctioned the use of enlarged photographs on the theory that they aided the jury to visualize the evidence and that they were true representations, "the sufficiency of the verification is a preliminary question for the trial court."[50]

The Supreme Court of Illinois in People v. Berkman (1923)[51] characterized as absurd and unfounded the testimony of a witness which purported to prove that a particular bullet had been discharged from the caliber .32 Colt automatic pistol introduced in evidence. The Court said:

> The State undertook to prove by the opinion evidence of Officer Dickson that this revolver was the identical revolver from which the bullet introduced in evidence was fired on the night Rahn was shot. The State sought to qualify him for such remarkable evidence by having him testify that he had had charge of the inspection of firearms for the last five years of their department; that he was a small-arms inspector in the National Guard for a period of nine years, and that he was a sergeant in the service

[48] See Chapter I, *supra*, page 74.

[49] The Court also said: "Recent cases involving experiments presented to the jury are: Kohlhagen v. Cardwell, 93 Or. 610, 619 (184 Pac. 261, 8 A.L.R. 11). State v. Holbrook, 98 Or. 43, 61 (188 Pac. 947, 192 Pac. 640, 193 Pac. 434); State v. Clark, 99 Or. 629, 665 (196 Pac. 360)."

[50] 108 Or. 386, 406, 213 Pac. 771, 778.

[51] 307 Ill. 492, 139 N. E. 91. The Court reversed the judgment of the trial court which convicted the defendant of assault with intent to commit murder, and the cause was remanded.

in the field artillery, where the pistol is the only weapon the men have, outside of the large guns or cannon. He was then asked to examine the Colt automatic 32 aforesaid, and gave it as his opinion that the bullet introduced in evidence was fired from the Colt automatic revolver in evidence. He even stated positively that he knew that the bullet came out of the barrel of that revolver, because the rifling marks on the bullet fitted into the rifling of the revolver in question, and that the markings on that particular bullet were peculiar, because they came clear up on the steel of the bullet. There is no evidence in the case by which this officer claims to be an expert that shows that he knew anything about how Colt automatic revolvers are made and how they are rifled. There is no testimony in the record showing that the revolver in question was rifled in a manner different from all others of its model, and we feel very sure that no such evidence could be produced. The evidence of this officer is clearly absurd, besides not being based upon any known rule that would make it admissible. If the real facts were brought out, it would undoubtedly show that all Colt revolvers of the same model and of the same caliber are rifled precisely in the same manner, and the statement that one can know that a certain bullet was fired out of a 32-caliber revolver, when there are hundreds and perhaps thousands of others rifled in precisely the same manner and of precisely the same character, is preposterous.

There are many instances in which opinion evidence is admissible on the part of both expert and non-expert witnesses. The speed of trains, of automobiles, of horses, values of property, sanity or insanity, intoxication of individuals, physical condition of a person, size and color and weight of objects, and many other such facts may be shown by opinion evidence when such opinion is based upon proper facts and opportunity of the witness to observe the things or persons and where it is impossible for the witness to detail all pertinent facts in such a manner as to enable the jury to form a conclusion without the opinion of the witness. The general rule that facts, and not conclusions, should be stated is a wise and safe one and cannot be too strictly followed. Mere opportunity does not change an ordinary observer into an expert, and special skill does not entitle a witness to give an opinion when the subject is one where the opinion of an ordinary observer is admissible or where the jury are capable of forming their own conclusions from the pertinent facts susceptible of proof in common form. [Jones' Commentaries on Evidence, Secs. 359, 360.] If it were possible in this case to determine whether or not the bullet in question was fired from the gun in question, it must have been by the peculiar rifling or condition of the gun that made what are called the peculiar markings on the bullet aforesaid. If any facts pertaining to the gun and its rifling existed by which such fact could be known, it would have been proper for the witness to have stated such facts and let the jury draw their own conclusions.[52]

The Court was justified in condemning this testimony because it did not show that the witness possessed special skill or that specialized knowledge was essential to the drawing of a proper conclusion from the observable data. The testimony did not explain the basis of the proposition that the signature of each firearm is individual.[53] The Supreme Court of Kentucky (1928)[54] followed the example established by the Illinois Court and held that a trial court had committed error by admitting as expert testimony the opinions of two witnesses for the Commonwealth that the fatal bullet had been discharged from the defendant's pistol. One of the witnesses was a former deputy sheriff and jailer, the other was the jailer at the time of the trial. They testified that they had gained some experience in the handling of firearms and had made a study of catalogues issued by the various firearms companies, and that the pistol introduced in evidence was a special make and carried a 32-20 Winchester cartridge. The witnesses made an examination of the pistol barrel and the fatal bullet under an ordinary magnifying glass and stated that "from the rifles in the pistol and the marks on the ball they could identify that particular ball as having been fired from that pistol."[55] They also fired a test bullet into a sack of bran and testified that an examination of the test and fatal bullets under the magnifying glass revealed that "the two corresponded."[56]

The Court held that this evidence was important if competent and highly prejudicial if incompetent, demanding more than a casual notice. It commented that the witnesses were not prepared "to measure the width nor depth of the rifles nor the distance between them, nor to make such measurements upon the bullets, nor to make any scientific test nor comparison between them."[57] The basis of the

[52] 307 Ill. 492, 500, 501, 139 N.E. 91, 94, 95.

[53] During the same year the Court of Appeals of the District of Columbia in Laney v. United States (1923) 54 App. D. C. 56, 294 Fed. 412, upheld the admissibility of expert testimony which tended to establish the fact that the fatal bullet was fired from the defendant's pistol. The defendant had confessed to the firing of a number of shots at a crowd assembled in the street.

[54] Jack v. Commonwealth 222 Ky. 546, 1 S.W. (2d) 961. The conviction of murder in the trial court was reversed and the cause remanded.

[55] 222 Ky. 546, 549, 1 S.W. (2d) 961, 963.

[56] Id.

[57] Id.

Court's conclusion that firearms identification had become a science was predicated upon a knowledge of an article which appeared in *Popular Science Monthly*, November, 1927. The article was discussed as follows:

As described by this writer, the first step in the manufacture of pistols and rifles is to bore a hole through a cylindrical iron bar and to ream it smooth, after which certain spiral grooves or rifles of uniform width, depth, and space are cut into the inner surface of the barrel for the purpose of giving the necessary spin to the bullet. In some makes five grooves are used and in some six. Also the twist in some makes is to the left, while in others it is to the right. The grooves are cut with a "rifle cutter," a sharp-edged tool with fine sawlike teeth. These teeth are visible only under a microscope, and in cutting out the hard steel of the barrel, make minute scratches on the inside of the barrel; also, the "cutter" to some extent is worn or dulled in rifling a single barrel, and as in the case of finger prints no two finished barrels are identical, though each barrel of the same pattern will show the same width and depth of grooves and width of lands between the grooves. When a bullet is fired through the barrel, an impression is made upon its sides by both the grooves and the other marks mentioned, and the same marks will appear upon the side of every bullet fired through that particular barrel, and these can be seen plainly under a properly constructed microscope. In making a test to determine whether or not a bullet which is known to have produced death was fired through the barrel of defendant's pistol, there is in use a special microscope consisting of two barrels so arranged that both are brought together in one eyepiece. The fatal bullet is placed under one of these barrels, and a test bullet that has been fired through defendant's pistol is placed under the other barrel, and this brings the sides of the two bullets together and causes them to fuse into one object. If the grooves and the other distinguishing marks on both bullets correspond, it is said to show that both balls were fired from the same pistol. If the grooves correspond, but the other distinguishing marks do not, it is thought the two balls were discharged from different pistols of the same make. If neither the grooves nor marks blend, it is evident that the balls were fired from different makes of pistols. Other microscopic instruments, such as the helixometer and the micrometer, have been developed, by which the depth of the grooves and the width of the grooves and lands in the pistol barrel may be measured and compared with similar marks on balls fired from such weapons, and other instruments of like character are in use.[58]

This article is the forerunner ot voluminous testimony based upon

[58] 222 Ky. 546, 549, 550, 1 S.W. (2d) 961, 963, 964.

the same general lines.[59]   The testimony may be characterized as lacking in a proper consideration of the character of all the marks appearing upon the surface of a bullet or cartridge case produced by contact with the surfaces of the firearm in which fired, and the relative position of these marks with respect to each other.   The testimony evidences an overemphasis of the effect of the tool-mark patterns of the grooves of a barrel as produced upon the surface of a bullet.   It has been pointed out that the tool-mark pattern in a groove of a barrel may never be duplicated in its entirety, but the resultant effect produced upon the surface of a bullet, the groove engraving, may be reproduced.[60]   Generally speaking, the effect produced upon the cylindrical surface of a bullet by its contact with the surfaces of the grooves in the barrel is not in itself sufficient to determine the identity of the particular firearm from which the bullet was fired; the groove engravings, instead of being an individual peculiarity, furnish corroborative evidence as to the identity of the firearm.   The surface of the bullet which comes in contact with the lands of the bore is subjected to the greatest radial compressive stresses, and it is the accidental characteristics of the lands of the bore which engrave themselves upon the surface of the bullet, the land engravings, which, in general, are the important elements in the identification of a particular firearm.[61]   The testimony attempts to explain the individuality of a new barrel in terms of the wearing away of the tools used in its manufacture.   The rifling operation is a metal-cutting operation.   The result of a metal-cutting operation is a *tearing* rather than a true shearing action, and the tool-marks are produced by this tearing of the metal and not by "fine sawlike teeth." [62]

The testimony places undue reliance upon the "fine lines" appearing upon the cylindrical surfaces of bullets.   The most reliable standards of comparison are the conspicuous elements, such as two or more prominent striae in a combination.   The fine striae which in themselves may be reproduced are used in conjunction with the conspicuous elements and not by themselves.   Aside from allocating undue weight to a single element without an intelligent orientation of that element to all the others, as is essential in establishing the individual peculiarity of a firearm, measurements are unnecessarily stressed

[59] See State *v.* Boccadoro, *infra*; Evans *v.* Commonwealth, *infra*; People *v.* Fisher, *infra*; State *v.* Campbell, *infra*.

[60] See Chapter 1, *supra*, page 71.

[61] *Id.* page 93.

[62] *Id.* Figs. 27-29.

in the face of the more reliable and adequate results obtained by the use of a comparison microscope. The identifications are not precisely limited to a comparison of the signatures found upon the surfaces of bullets and cartridge cases.

In State v. Boccadoro (1929) [63] the evidence relating to the personal identity of the murderer was far from convincing. It appeared that the defendant had in his possession on the night of the crime a revolver which had been previously stolen from a Doctor Black. The revolver was not found, but a bullet discharged therefrom by the Doctor some two or three years before the murder was dug out of the ground alongside his house. The experts for the state testified that the fatal bullet had been discharged from the same firearm as the unearthed bullet. The testimony of the witness, which received the approval of the Court of Errors and Appeals of New Jersey, demonstrates both the overemphasis of the importance of the tool-mark patterns in the grooves of the barrel and the undue reliance placed upon ''the little tiny barrel scratches'':

Q. By the way, have you seen a Smith and Wesson 38 calibre manufactured? A. I have.

Q. And have you specimens you could show us of this rifling and so forth? A. Yes. This long rod, considerably longer than the full length of the barrel, is passed backward and forward. In a slot on that rod, this cutting edge is situated. Now, that edge which is shown in this upper picture, much magnified in this especial case, as I say, cuts a groove of its own width through that barrel. The rod is passed backwards and forward, turning as it goes, so that the grooves shall have a spiral twist and not be perfectly straight. The fact is that this spiral twist gives the bullet a spinning action and makes it fly further and faster than if it were not so spun. The barrel is turned. The position of the cutting edge is not changed beyond the spiral that it goes through. When cutting our grooves, instead of moving the cutter around the whole barrel is moved around. For instance, we start one here. The first operation shaves out a very tiny shaving of metal from the interior of that barrel. Then the whole barrel is turned around and a second groove is begun, another fine shaving turned out. Then the barrel is turned again and another and until five shallow gutters situated an equally distance apart have been started. After those five have been begun, this cutting edge is jacked up by a wedge underneath it so that on the next draw it will cut deeper, so that it makes five little deeper cuts in the same channels and deeper cuts on the next turn, and so forth until the grooves are cut to the required depths, to the predetermined depth

[63] 105 N. J. L. 352, 144 Atl. 612.

for that arm, which as I recall, in five-thousandths of an inch.
Now, this cutting edge, this rifle edge which is shown in section
here, this tiny cutting edge and this groove which it cuts, this edge
of the cutter, and which to the naked eye is a perfect arc, is actually interrupted by little iregularities, teeth practically, which
exist on any fine cutting edge when you magnify it when you look
at it under magnification under a high-powered microscope or under a microscope of moderate power. This edge instead of having
this appearance (indicating) has that appearance. As a result of
that instead of the bottom of the grooves having nice beautiful
contours, perfect arcs in the bottom of the grooves, they have
these little hills and valleys, a perfect row of tiny little depres-
sions, and elevations which correspond with other tiny teeth on
the edge of the cutter. Now, when you draw one piece of hard
metal against another piece of hard metal wear occurs, and wear
occurs so rapidly and so steadily in this rifling operation that
these teeth edges are constantly crumbling and new teeth are constantly forming. As a result, the little tiny barrel scratches left
in this groove are not the same as are left in this groove or in
this (indicating), with the result that the two grooves are different
in the same barrel, with reference to the same fine scratches as
they run down in the bottom of the groove. Now, after the rifling
operation a rod with a washer on it is put in the barrel and molten
lead is poured into it, and oil and emery is splashed on to the
molten lead, and after the lead has solidified it seeps down in the
cracks, and this slug of lead is worked back and forward to take
out what it can of the coarser scars left in this operation. It also
produces longitudinal scars if tiny bits of emery are caught in the
lead of the barrel and rubbed up and down, so these elevated surfaces also are not perfectly free from scratches, although the majority of the scratches are present in the grooves. That carries us
through the manufacture of the actual bore of the barrel.

. . . .

Q. That is of the comparison microscope? A. Yes. . . . In due
time I found a point where there were markings of identity common to both bullets. I do not mean they had broad groove marks,
but I mean certain finer markings, the origin of which I have already explained in the manufacture of a gun barrel.[64]

It is evident from the description of the manufacturing process
that the important accidental characteristics arising in manufacture
were thought to be the tool-mark patterns of the grooves of the barrel and that the individuality of the tool-mark pattern depended upon

[64] ''State of the Case,'' pages 226-228, 255-256. The witness briefly explained
the ''drilling'' and ''reaming'' processes in the manufacture of the barrel. However, in this connection he did not point out any significant accidental variations
as he did in the case of the rifling cutter.

the wearing away of the rifling cutter. The witness also testified as to measurements, and in this respect it should be noted that the difficulties of making precision measurements on deformed fatal bullets are often obviated by the use of the comparison microscope, under which the widths and directions of rifling can be carefully and accurately compared:

Q. Is that the Filar micrometer microscope? A. Yes, sir.

Q. And how accurate is that microscope? A. It is graduated down to five-thousandths of an inch.

Q. And what is it used for? A. It is used for measuring accurately short distances on any object you wish to measure.

Q. Is it used for bullets? A. Yes, sir.

Q. Now, this microscope differs from any other microscope only in that on the eye piece it has a device by which the measurement can be carried out. A. . . . If they have the same number of grooves going in opposite directions they can never be confused, but when two makers have the same number of grooves inclined in the same direction, then, naturally, it is difficult to tell whether the bullet has been fired from one or the other, but no two makers use the groove the same width or of the same inclination. If they get to the same width one may be inclined to the perpendicular and the other not, but no two makers use the same width, just as they do not in bullets, so by measuring the width of the groove and the insulation which this groove has made by the use of this microscope, by which we can rotate our cross hairs through the angle of the groove, we can determine the width of these grooves and the angle that they make. Now, engineering tables have been worked out and are available, and I have a volume containing such tables, which shows the rate of twist of a given spiral over a given angle of a given diameter of that spiral, so knowing the diameter of the bullet and the diameter of the angle by measuring these angles, we can determine what it is, and I determined this to be fired through a Smith & Wesson. That was my first test. Then I took a bullet which I knew to be fired through a 38 calibre Smith & Wesson and compared it with this under the microscope and the grooves were the same, and that was a practical check on the theoretical determination.[65]

Fig. 140 is a photomicrograph of the fatal bullet, Fig. 141 is a photomicrograph of the bullet dug out of the ground in the garden alongside the Doctor's house, and Fig. 142 is a photomicrograph of a comparison of the portions of the two bullets which show the greatest degree of similarity.[66] It is interesting to note that the circumstances

[65] "State of the Case," pages 251-253. See, also, Fig. 81.

[66] In 1930 the above photomicrographs were made by the authors through the courtesy of Lt. Joseph Linarducci, Jr., of the County Detectives.

surrounding this case were well known to the authorities before the trial and that the circumstances created a strong probability that the crime had been committed with the same revolver that had been stolen from Doctor Black. Among the loot found in the possession of the defendant was a maple leaf pin which was identified as the pin stolen from the Doctor at the time his revolver was taken. A witness gave a rough description of a revolver which had been in the possession of the defendant shortly before the murder, and this description

FIG. 140.   FIG. 141.

tallied with that of the revolver stolen from the Doctor. When the revolver was stolen it was loaded with four cartridges of Peters ammunition of an old type. The witness who described the revolver in the defendant's possession stated that four cartridges were therein, and the fatal bullet was found to be of the same make and type of old ammunition with which the Doctor had loaded his revolver.

In 1929 the Supreme Court of California [67] affirmed a conviction of murder and thereby approved of testimony on firearms identification. The basis of the identification was referred to in the follow-

[67] People v. Beitzel, 207 Cal. 73, 276 Pac. 1006.

ing unscientific language: "Because of peculiar 'spurs' in the nozzle of the gun and 'gouges' on the bullets found in the body of the deceased, a firearms expert testified that these bullets were fired from this particular gun." [68] The Supreme Court of Wisconsin,[69] during the same year, affirmed a conviction of assault with intent to murder and stated that proof was made by competent evidence that the bullet extracted from the body of the victim had been discharged from the defendant's firearm.

The testimony on firearms identification was made the center of attack in the appeal taken from a conviction of manslaughter in

Fig. 142.

Evans v. Commonwealth (1929).[70] The chief of police of Pineville, Kentucky, was slain at 10:30 P.M. near the town depot. A number of people were in the neighborhood at the time, but no one was able to establish the identity of the slayer. Circumstances indicated that the defendant had committed the crime: his presence and conduct in Pineville during the evening evoked suspicion; he was noticed wearing a dark suit of clothes similar to the garb

described by the witnesses who caught a glimpse of the criminal leaving the scene of the homicide; and the evidence showed that the slayer was a left-handed person, as was Evans. However, the personal identity testimony was not convincing and, consequently, the tracing of the defendant's firearm was a significant factor leading to the conviction. The trial court had before it the caliber .45 Colt automatic pistol which belonged to Evans, the seven cartridges extracted therefrom, six cartridge cases found where the shooting took place and a bullet which had been dug out of the earth in that vicinity. It appears that the witness who testified as an expert on firearms identification, after having been examined with respect to his qualifications, was given the pertinent exhibits and excused by the court while he performed the necessary experiments. He was subsequently recalled

[68] 207 Cal. 73, 76, 77, 276 Pac. 1006, 1007.
[69] Galenis v. State, 198 Wis. 313, 223 N. W. 790.
[70] 230 Ky. 411, 19 S. W. (2d) 1091, 66 A.L.R. 360.

and allowed to express his opinion that the bullet dug out of the earth had passed through the barrel of the defendant's pistol and that certain of the cartridge cases retrieved at the scene of the crime had been discharged in that same pistol.

The Supreme Court of Kentucky held that the testimony on firearms identification was competent, and it affirmed the conviction.

The witness, describing the method he employed in establishing the identity of the firearm from which the unearthed bullet had been fired, said in part:

I fired a test cartridge through this pistol after having examined the evidence bullet to determine if possible what company made that bullet, I brought with me sample unfired bullets made by the different American cartridge companies, I compared this fired evidence bullet with those unfired samples, and the evidence bullet matched the product of the Peters Cartridge Company. I therefore selected a cartridge of the same type having a 230 gr. metal jacket bullet, the evidence bullet being of the 230 gr. type; the cartridge I selected being a Peters make and fired this through the evidence pistol; before firing, the shell of this cartridge was marked by me in the presence of several witnesses "T-1" to mean test No. 1; the bullet was fired into a basket filled with cottonwaste and recovered by me from that waste in the presence of the various witnesses and immediately marked on the base "T-1." The evidence bullet was now placed under one microscope of a double microscope which I use known as a comparison microscope; the test bullet was now placed under the second microscope of this pair; . . . By keeping one bullet fixed, set, and rotating the other bullet they can be matched, by trying one bullet against the fixed portion of the other bullet, if these two bullets were fired from the same pistol they will match, that is through the same barrel; after you have tried all the way around that bullet you will find points left on these two bullets by the pistol barrel, fine lines left by the wearing of the tool will match across the dividing line. After we find such a point then we can move the bullets together all the way around and you will find more similar points all the way around both bullets. Unless they are through the same particular gun these vary, and the similarity ends; if through the same particular gun the distinguishing marks on both bullets correspond. That is the condition I found when I compared the evidence bullet with the test bullet which I fired from the evidence pistol. If they are through the same particular gun all of these fine marks, not all of them, but a great per cent. will match.[71]

The witness explained the basis of his conclusion that certain of

[71] 230 Ky. 411, 418, 419, 19 S.W. (2d) 1091, 1094, 1095.

the retrieved cartridge cases had been discharged in the defendant's pistol:

> I examined these shells and find that there were three .45 caliber automatic shells of Remington make, there was one of United States Government make, and two Peters make; one of these two had a shiny area, polish on the head of the shell right next to the primer, one of these two Peters shells, I noticed that particularly, because all of the Peters shells given to me as taken out of the magazine of this evidence pistol were polished in that same fashion, inasmuch as the evidence bullet matched the product of the Peters Cartridge Company, as I have already stated the test shells which I fired were Peters make, I fired three test cartridges of Peters make, one through the evidence pistol, number 376,281, one through another similar pistol, number 15509 and one through my own pistol of the same type, number 70. The shells were marked "T-1" "T-2" and "T-3" indicating tests 1, 2 and 3; I compared the shells recovered after test one with the two Peters shells in evidence, the two empty. Peters shells in evidence. . . . I compared these shells through this microscope which I have described. I find that the impression left by the firing pin on these shells, on the test shells, was identical with the one left on the two empty Peters shells picked up at the scene. That impression had a peculiarity about it which is quite pronounced in that on the edge of the circular hole there is an additional little lip such as you see on the edge of an oil can, a little lip on the edge of the depression left by the firing pin; that lip is noticeable on the shell picked up and on the other shell which I fired in the evidence pistol. On the shells which I fired in the other two pistols the impression left by the firing pin is quite different, the depression, the hole made by the firing pin was different and the character of the hole was different from that presented on the evidence shells picked up at the scene. They in no way bore any degree of similarity at all.[72]

The "little lip on the edge of the depression left by the firing pin," termed a "peculiarity" by the witness, is caused by the protrusion of the firing pin during the process of ejecting the cartridge case. This "lip" is of relatively small value because it may or may not occur on cartridge cases that are discharged consecutively from the same firearm as shown in Figs. 65, 66.

The Supreme Court recapitulated the testimony given by the witness with respect to the proposition that the signature of each firearm is individual:

> He then testified that, when a pistol is fired, there is developed within the shell a pressure of approximately 1,500 pounds to the

[72] 230 Ky. 411, 420, 19 S.W. (2d) 1091, 1095.

square inch, that it is this pressure which drives the bullet out of the barrel, and that the shell is driven back against the breech of the pistol with an equal force. He testified that all the little scratches, markings, and irregularities on the hard breech of the pistol are thereby stamped on soft butt of the shell. He said that, in the making of the barrel of a pistol, after the hole had been drilled through the bar out of which the barrel is to be made, and the hole has been polished and finished, a rifle cutter is then passed through the barrel with a twisting motion so that it cuts and scrapes away a thin shaving of metal as it goes, thus leaving in the barrel a series of twisted grooves which are known as rifles. The spaces between these grooves are called lands. He testified that, when the pistol is fired, these grooves and lands leave their marks upon the bullet, and that, owing to the constant and rapid wearing of the tools used in this rifling, the rifles of no two pistols are exactly alike, even though made by the same company by the same machine. He testified that these rifles and lands will leave their marks upon every bullet that is fired through the pistol barrel, and that all bullets through the same barrel will have the same markings, the same tiny scratches, the same width of lands between the scratches, and that those will not be the same as those made on bullets fired through another barrel of the same make.[73]

The impression which the Court received from the witness was at variance with scientific developments. The Court was unaware of the complex problem raised by the differences in the signatures of the same firearm and the similarities existing in the signatures of different firearms.[74]

The defense presented two witnesses to weaken the effect of the testimony. One Pickard testified that although his caliber .45 Colt automatic pistol No. 14692 was practically new five cartridge cases discharged therein revealed that the firing pin did not strike the center of the cartridge cases. One Steward testified that eight cartridge cases fired in his caliber .45 Colt automatic No. C-128134 showed that the "primer prints varied" despite the good condition of the firearm. The state's expert was recalled and given the pistols. He was able to pick out the eight cartridge cases fired in the Steward pistol and the five fired in the Pickard pistol. He put two bullets fired from the Steward pistol under the comparison microscope so that the jury could see the similarity. The witness then placed under the microscope a bullet discharged from the Steward pistol and one retrieved from the scene of the crime and allowed the jury to see the dissimilar-

[73] 230 Ky. 411, 421, 19 S.W. (2d) 1091, 1095, 1096.
[74] See Chapter I, *supra,* Figs. 82-87, 126, 128.

ity. The same demonstration was made with cartridge cases. The Court held that this was proper rebutting evidence.

The defendant contended that the evidence of the witness was highly technical, unreasonable, extremely doubtful, and therefore inadmissible. The Court dismissed the argument with the comment that the same objection could apply to the evidence of finger prints which is commonly held to be admissible. In connection with this ruling by the Court that the evidence was of sufficient reliability, the following testimony of the witness is important. He stated that a certain factory manufactured pistol barrels in pairs, that is, the barrels are made in double lengths and after completion they are cut in two. The witness testified that two bullets were fired from each of six barrels comprising three such pairs and that he was able to identify the particular barrel from which each of the twelve bullets had been fired.

The experiment was described to the Court for the purpose of illustrating the skill of the witness. The fact that he could distinguish between bullets fired from two barrels manufactured as a pair would naturally create a favorable impression upon the Court and jury. However, the significance of this experiment, not being precisely understood, is often exaggerated.

These barrels were manufactured at Springfield Armory. It has been explained in Chapter I that the interior surfaces of two barrels manufactured as a pair will be quite different from each other, and that bullets passing through two such barrels, when separated, will rotate in opposite directions with reference to the axis of the double barrel, and the driving edges of corresponding lands will be reversed.[75]

The Court held that the experiment disposed of the defendant's objection that the making of barrels in pairs contradicted the statement of the witness that no two barrels are exactly alike. It is clear that the experiment does not answer this objection because the bullets could be distinguished by the reversal of the position of identical elements, if such did exist.

The defense vigorously urged that the witness had been permitted to answer many questions in which he invaded the province of the jury by stating conclusions, instead of confining himself to opinions and the facts in support thereof. The objection was specifically directed at the following statements: "I am convinced as a result of

[75] *Id.*, pages 50-53.

this test that the bullet in evidence was fired through the pistol in evidence,'' and, referring to cartridge cases, ''I am satisfied they were fired through the pistol in evidence.'' [76] The Court properly held that these conclusions, predicated upon the personal observations of the witness, were not submitted as facts but rather as opinions, and that the data underlying these opinions had been explained to the jury.

In support of its decision that the conclusions of the witness had been properly admitted, the Court mentioned that ordinarily an expert gives his opinion with respect to certain facts which have been observed by someone else and are presented to him in hypothetical form, ''whereas a man skilled in a particular business, who makes his own observations, and testifies to what he has observed and his conclusions therefrom, is regarded as a skilled witness. He occupies the same position as any other witness, except that it is recognized that within certain lines he possesses a superior knowledge which enables him to understand, as one without such special knowledge could not, what he has observed. Thus, in litigation about a horse, an experienced horseman, who has seen and examined the horse, would be permitted to state whether he was a saddle or a draft horse.'' [77] The Court likewise pointed out that witnesses must necessarily give conclusions when it is impossible to reproduce the relevant data for the jury, such as in the case of the presence of certain odors, and that wherever this situation arises the right of cross-examination can be exercised to test the correctness of the conclusions. The liberal attitude of the Court is indicated by the following statement:

> For the discovery of truth and the establishment of justice, certain rules have been formulated governing the admissibility and competency of evidence. These rules should usually be followed, but, when a rigid adherence to them would be subversive of the ends for which they were adopted, they should no longer be rigidly adhered to.[78]

The trial judge permitted the witness to supplement his verbal explanation of the facts relevant to his opinion by a detailed drawing upon a blackboard. The Supreme Court found no merit in the defendant's objection thereto because the illustration afforded the jury a clear understanding of matters difficult to describe. The Court

76 230 Ky. 411, 423, 19 S.W. (2d) 1091, 1096.
77 230 Ky. 411, 422, 19 S.W. (2d) 1091, 1096.
78 230 Ky. 411, 424, 19 S.W. (2d) 1091, 1097.

likewise approved of the jurors' looking through the comparison microscope at bullets and cartridge cases on the ground that they could see better than with the naked eye. The aid of microscopes was thought to be analogous to a situation in which a juror put on spectacles to help his sight during the examination of a bullet.

The Supreme Court of Kentucky cited some of the earlier cases dealing with firearms identification as authority to substantiate its decision that the testimony of the witness was properly admitted. The comments made in this regard manifest the importance of giving the courts an opportunity to become acquainted with the principles of firearms identification in order that they will not accept testimony in the future on the strength of its being similar to testimony which has been previously received by another court. Many of the early experts were in fact incompetent, and the judicial sanction of their testimony must not be used as a basis for allowing incompetent witnesses to testify in the future. For example, the Court said of the Sacco-Vanzetti case:

> In the case of Com. v. Sacco et al. 255 Mass. 369, 151 N. E. 839, a .32 caliber Colt automatic pistol and some cartridges therefore were found in the possession of Sacco. A .32 caliber bullet was taken from the body of Beradelli, the man who was slain. The commonwealth introduced two witnesses in that case, Charles H. Proctor and Charles J. Van Amberg, and their testimony was along the same line, and very similar to the testimony of Goddard in this case. That testimony was admitted. The admission of it was approved, and the judgment affirmed.[79]

That justice will be expedited by enlightening the courts upon the subject of firearms identification is made clear by the following comment on the Weber case in which the California Court held that the jury was competent to determine the identity of the firearm from which a bullet had been fired by an examination of the markings on the exhibited bullets:

> In People v. Weber, 149 Cal. 325, 86 P. 671, Weber was convicted of murder and sentenced to death. In that case bullets taken from the bodies of Mary Weber and Bertha Weber were exhibited to the jury, together with other bullets which had been fired from the pistol which a pawn broker identified as one he had sold defendant. The admission of this evidence was approved, and that judgment was affirmed, though it carried with it a death sentence. There was a dissenting opinion in that case, but the

[79] 230 Ky. 411, 425, 19 S.W. (2d) 1091, 1097.

judge dissenting did not base his dissent on any error in the admission of this evidence.[80]

The Court of Appeals of Ohio (1930) [81] sustained the admissibility of the following testimony on the ground that the ordinary juror was not familiar with the subject matter of firearms identification and that the finding of the trial court on the question of the expert's qualifications should not be reversed unless predicated upon an error of law or a serious mistake or abuse of judicial discretion. The testimony shows that a witness may possess experience and familiarity beyond that of the jury and yet not be equipped with the skill necessary for expressing a reliable opinion which will be of sufficient assistance to the trier of fact to warrant its admission in evidence. The testimony was in part as follows:

Q. Tell the jury whether or not you have made any examination of the rifles in that particular gun, State's Exhibit B? A. I did.

Q. Tell the jury how you made an examination of that? A. In the first place I poured some heavy oil in the barrel and then made a paper tube to fit that to show the impression of the rifling on this tube.

. . . .

Q. Just tell the jury what you did. A. After I filled this with oil and got that impression I measured the distance of the rifling of the impression on the paper and the distance of the rifling on this bullet and they compare exactly, showing that the cuts on this bullet are the same width as the rifling in the barrel.

. . . .

Q. Of the many thousands of guns there are in the world none of them marks a ball exactly alike? A. Every gun has its own in-

[80] *Id.*

[81] Burchett v. State, 35 Ohio App. 463, 172 N. E. 555. The defendant was convicted of second degree murder on strong evidence. The Court said: ''The possibility of identifying a bullet that has been fired with the firearm from which it was projected is now receiving intensive study by engineers. The Engineers' Foundation of New York is promoting such an investigation by Major Gunther, a professor of Stevens Institute of Technology, who has a highly technical paper on the subject in *Mechanical Engineering* for February, 1930. . . . In principle its admission as legal evidence is based upon the theory that the evolution in practical affairs of life, whereby the progressive and scientific tendencies of the age are manifest in every other department of human endeavor, cannot be ignored in legal procedure, but that the law, in its efforts to enforce justice by demonstrating a fact in issue, will allow evidence of those scientific processes which are the work of educated and skillful men in their various departments, and apply them to the demonstration of a fact, leaving the weight and effect to be given to the effort and its results entirely to the consideration of the jury.''

dividual marks, and any real gun expert could tell you the ball that was fired from any gun; in other words, he could take 100 guns and 100 balls and tell you which ball was fired from which gun.

Q. In the millions of guns in the world there are no two alike?
A. The rifling is not alike in any two guns.

Q. I am to understand you that the rifling in that gun can not be duplicated, or is not duplicated by any other gun in the world?
A. No, sir, it is not.[82]

The Supreme Court of Illinois approved of expert testimony on firearms identification in People v. Fisher (1930).[83] The Court had previously excluded such testimony in People v. Berkman, *supra*, but the two decisions are reconcilable. It did not appear from a reading of the testimony in the Berkman case that the witness possessed skill above that of the jury or that the drawing of an inference from fact to conclusion required specialized knowledge. The Court held in the Fisher case that the opinions of the expert were based upon specialized knowledge and that he possessed the requisite qualifications. The situation presented by these two decisions may be attributed to the witnesses. The early witness purported to identify firearms by a method which was equally available to the jury and, consequently, the Court was not aware that unusual skill or experience was necessary. The later expert demonstrated the requirement of specialized training.

Fisher and three other defendants were convicted of murder and all were sentenced to death with the exception of one Jenkins who was sentenced to life imprisonment. The defendants, together with one Hare who was given a separate trial and one Dixon who was not apprehended, had robbed a Chicago bank during business hours. Five of the robbers had entered the bank armed with pistols, revolvers, and a sawed-off shotgun. While four of the bandits aimed their weapons at the patrons, officers, and employees of the bank, the fifth robber climbed over the top of the teller's cage and drew his gun. Before he could reach any money a special police officer for the bank pulled out his gun and, being joined by a number of employees and officers of the bank, exchanged shots with the bandits. The robber in the teller's cage hurriedly seized a handful of bills and ran from the building. His confederates followed, and they all drove away in their automobile.

About twenty shots had been fired during the brief interval of the

[82] Defendant's Bill of Exceptions, pages 222, 223, 226.
[83] 340 Ill. 216, 172 N. E. 743.

robbery. The police officer for the bank was mortally wounded by two loads of shot discharged from a shotgun. The hospital authorities also found in his body the wadding from the shotgun cartridge and a caliber .38 bullet which in itself had not caused a dangerous wound. The confessions written by the defendants alleged that the shotgun had been carried by Dixon, the bandit at large. The defendants stated in their confessions that they had been armed with two caliber .38 revolvers, a caliber .32 revolver, a caliber .32 Savage automatic pistol, and a caliber .45 Colt automatic pistol. The police located these firearms by information elicited from the defendants. The sawed-off shotgun and two cartridge cases of that same gage were found in the abandoned bandit car.

Certain of the witnesses were able to submit convincing personal identity testimony. The evidence of the witness on firearms identification was introduced to prove that the bullet extracted from the body of the bank guard had been discharged from a caliber .38 revolver which had been carried by one of the defendants. He also testified that the retrieved cartridge cases had been fired in the sawed-off shotgun found in the car and that the wadding taken from the victim's body was made and used by the same company which made the cartridge cases.

The Court recapitulated as follows the testimony concerning the caliber .38 bullet:

> He testified that by examining a bullet it was possible to determine what company manufactured it; that it was also possible to determine from the examination of a fired bullet what make of gun fired it, and also to tell the exact gun which fired the bullet if the gun was available for examination. He explained how it was possible to tell, by examining a fired bullet, what company had made the bullet and the make of gun from which it was fired. He explained that he had fired a bullet from the .38-calibre Smith & Wesson revolver, in evidence as exhibit 17, into a barrel filled with waste; that by placing the test bullet, together with the one introduced in evidence, under a microscope, he was able to determine the same character of markings, same width and angle of grooves made by the rifling of the gun, and by such manner determine that the gun, exhibit 17, fired the bullet. He explained that no arms maker makes exactly the same kind of rifling grooves in the barrel of its guns as is made by other manufacturers, and that by reason of the dulling of the tool before the grooves are completed the grooves will show, under the microscope, some difference in the same gun from every other gun of that make and some difference in the grooves in the same gun, and as a bullet fired through that gun takes on markings corresponding to the exact grooves of

the rifling of the barrel, it is shown by the microscope that no gun other than the one from which it was actually fired could make such a mark on the bullet. As to the exhibit in this case, he showed that the cartridge had been loaded with black powder; that the gun, exhibit 17, also showed that it had fired black powder; that the inside of the barrel and cylinder were fouled with black powder; that though this bullet, exhibit 22, was flattened at the end by reason of apparently having struck a bone, his microscopic examination disclosed that the bullet bore rifle marks identical with the bullet which the witness fired from exhibit 17. He also testified that on the end of the bullet were small particles of the materials of which the memorandum book was made, indicating that the bullet had passed through the book. This book, exhibit 24, showed that a bullet had passed through it.[84]

The Court summarized the testimony relating to the cartridge cases as follows:

He explained that under a microscope the imprint of the firing pin of the shot-gun in evidence was unmistakably shown on the caps of the discharged shells found in the automobile. He explained that no two firing pins make exactly the same impression, but that the small rings made in turning out the point of the firing pin are shown under the microscope to vary in each instance, and that the rings formed in the imprint on the cap of the shells fired in this case were identical with the rings shown under the microscope on the firing pins of the shot-gun in evidence.[85]

This statement has not, at the time of writing, been substantiated by research. The same impression of "the small rings made in turning out the point of the firing pin" has been found on cartridge cases fired in two different firearms, as shown in Fig. 132.

The defendants argued that the testimony on firearms identification was novel and should have been excluded; that it was outside the field of expert testimony; that such evidence was not admissible under common law, and that there was no statute in Illinois authorizing its use. The Court in disposing of the defendant's contentions said:

The same objection was raised in People v. Jennings, 252 Ill. 534, to the admission of finger prints as means of identification. So the question was raised when photography was first introduced. (1 Wigmore on Evidence, sec. 795.) Of such evidence it was said in People v. Jennings, supra, that while it may or may not be of independent strength, it is admissible, the same as other proof, as tending to make out a case. The general rule is that

[84] 340 Ill. 216, 238, 239, 172 N.E. 743, 753.
[85] 340 Ill. 216, 242, 172 N.E. 743, 754.

whatever tends to prove any material fact is relevant and competent. (People v. Gray, 251 Ill. 431.) Expert testimony is admissible when the subject matter of the inquiry is of such a character that only persons of skill and experience in it are capable of forming a correct judgment as to any facts connected therewith. (People v. Jennings, supra.) Such evidence is not confined to classified and special professions but is admissible wherever peculiar skill and judgment applied to a particular subject are required to explain results by tracing them to their causes. Such evidence is admissible when the witnesses offered as experts have peculiar knowledge or experience not common to the world, which renders their opinions founded on such knowledge and experience an aid to the court or jury determining the issues. . . . The question of the qualification of an expert rests largely in the discretion of the trial court. Bonato v. Peabody Coal Co. 248 Ill. 422; 3 Wigmore on Evidence, sec. 1923.

In Lyon v. Oliver, 316 Ill. 292, it was pointed out that handwriting, photography of questioned documents and identification of typewriting were subjects for expert testimony. . . . We are of opinion that in this case, where the witness has been able to testify that by the use of magnifying instruments and by reason of his experience and study he has been able to determine the condition of a certain exhibit, which condition he details to the jury, such evidence, while the jury are not bound to accept his conclusions as true, is competent expert testimony on a subject properly one for expert knowledge.[86]

It was also urged that the testimony should have been excluded because the experiments were not performed in the presence of the defendants. The Court dismissed the objection on the ground that there was no rule requiring such experiments to be conducted before the eyes of defendant.

The defense maintained that certain demonstrations made by the expert on a blackboard were incompetent. The Court held that it was not error to permit such demonstrations which had been explained to the jury and which were similar to a plat.

The Court determined that the trial judge had correctly admitted in evidence the caliber .38 bullet taken from the victim's body [87] and all the firearms used by the officers and employees of the

[86] 340 Ill. 216, 239-241, 172 N.E. 743, 753, 754.

[87] The bullet taken from the body of the victim at the hospital had not been marked for identification by the doctor who performed the operation. It was given to the sister in charge of the operating room and she did not testify at the trial. The doctor testified that the bullet exhibited to him was the extracted bullet and his statement was not impeached on cross-examination. The Court held that the bullet had been properly admitted in evidence and the wadding was

bank, on the theory that even though the bullet had not caused a fatal injury it was competent to show that the bullet had not been discharged from the firearm of any one connected with the bank.

The appeal taken from a conviction of manslaughter in State *v.* Campbell (1931) [88] was concentrated upon the testimony of a witness who appeared as an expert on firearms identification. One Harris was slain during the course of a street battle while alighting from his automobile or immediately thereafter. Two shots were fired in the direction of the victim's car by the defendant, who was standing in the entrance to a drug store located on the opposite side of the street. A third shot was fired by the occupant of a passing automobile which had been parked near the drug store until the approximate time of the first shot. The state believed that it was important to prove that the bullet extracted from the body of the victim had been discharged from the defendant's firearm. The revolver taken from Campbell, the two cartridge cases and the four cartridges found therein, the fatal bullet, and a second bullet which had been embedded in the side of the victim's car, were subjected to the necessary comparisons, and subsequently the examiner testified at the trial in an expert capacity.

The witness gave the following testimony with respect to his qualifications as an expert:

Q. What schooling have you had to prepare for that work?
A. I have a Bachelor of Arts Degree from Johns Hopkins Uni-

---

accepted under similar circumstances. Such a flimsy identification is a dangerous practice. In State *v.* Civitan (1932) Hudson County, N. J., a doctor testified that a certain bullet, a caliber .38 S.&W. Sp'l, was the bullet which he had extracted from the body of one Anton Kacya in April 1932. It later developed that this evidence bullet had been taken from the body of one Abraham Fugaro in 1930. The fatal bullet in the Civitan case was a caliber .38 S.&W. The defendant was acquitted. See Chapter I, *supra*, Figs. 131 and 124.

People *v.* Dale (Ill. 1934) 189 N.E. 269, 271. The defendant was convicted of murder. He contended that the Court had erred by admitting in evidence certain bullets found at the scene of the crime by a police officer. It was argued that there was insufficient proof as to the custody of these bullets between the time when they were found and the trial. The court overruled the objection and stated that ''the record shows that the officer who found the bullet put his own distinguishing mark on them at the time they were found and he identified them at the trial. The ballistic expert testified that they were fired from one of the revolvers found in the Dale and Jarman apartment, and we find no error in their admission in evidence.''

[88] 213 Iowa 677, 239 N.W. 715.

versity, and I am a graduate of the Regular Army Medical School in Washington.

Q. What special training have you to prepare you for the special line of work you are doing now? A. Ordnance officer of the army. I have a commission as lieutenant colonel in the ordnance reserve at the present time. I have been at various manufacturing arsenals in this country and taken a special course in the manufacture of rifles and pistols, and rifle and pistol ammunition and powder, and as an individual observer, I have studied the manufacture of small arms and small arms ammunition in practically every American factory, and in factories in England, France, Belgium, Spain, and Germany. For a number of years past I have devoted all my time to the collection of data on arms and ammunition manufacturing, collection of apparatus for use in identification work, and collection of specimens which would give me the reference collection for such work.

Q. Have you made any special study of firearms and ammunition that was used in murder cases? A. Yes, sir.

Q. And what special training have you had along that line? Without saying who you were consulted by, what examination did you make and what cases were you studying? A. I was in the past year,—I worked upon somewhere between one hundred and two hundred murder cases involving the firing of arms, I should imagine.

Q. As a preparation for giving testimony, have you made any special study of any particular cases? A. I have made special study of the subject in all the cases to which I have referred.

Q. Have you made any special study of this kind of work for any particular institution or publication? A. I have made studies for the Department of State and the Treasury Department, and in conjunction with the Department of Commerce, the Bureau of Standards at Washington, D. C., which has established a department,—it might be called "Firearms Identification,"—along lines similar to these that I have in my own office.

Q. Have you been connected with the police organization or law enforcement organizations in your work and study? A. I am a lecturer in the New York Police College on firearms identifications; technical expert to the Pennsylvania State Police on firearms identification.

Q. What apparatus and equipment do you have for carrying on your work in those lines, Professor? A. I have a reference collection of a good many arms, I would say close to one thousand different arms, small arms and many thousands of different types of unfired bullets, empty fired shells and loaded cartridges, of fired bullets fired through known makes of arms of different types, data, covering the manufacture of different types of arms in use in this country today and a great many no longer used, and of a large proportion of those in use overseas. I have instruments which have been largely developed in my own laboratory for the identification of projectiles in shooting cases, that is, including microscopes, photographic apparatus, micrometers, special scales, and so on.

Q. What special study have you made, if any, in regard to microscopes? A. I have been using the microscopes for something over twenty years now on various business.

Q. What kind of microscopes do you use in connection with your work? A. In the firearm work we use what is known as a comparison microscope. It is an instrument made of two single microscopes connected by a cross-arm fitted with two sets of prisms, which enables the user to fuse two images in a single eye-piece and compare these two pieces at the same time.[89]

The Court overruled the defendant's attack upon the qualifications of the expert and in support of his competency it cited with approval the opinion rendered in Jack v. Commonwealth, *supra,* in which the Kentucky Court, rejecting the opinions of two witnesses, referred as authority to an article written by this expert. However forceful the merits of the qualifications may be, the witness had made photographs of the exhibited bullets and by a series of numbers he designated the important markings. When asked to explain how the significant marks were produced, the following testimony on this basic proposition was submitted by the expert:

A. The markings left on the fired bullet consists primarily of the land and groove markings of the barrel. The barrel is rifled with spiral grooves in order to spin the bullet, by reason of which the bullet flies further and truer than if it was not spinning. The lanes are the portion of the bored surface between the grooves. Naturally when the grooves are cut out there is going to be material remain between them elevated with respect to this point which constitutes the grooves. The grooves are cut out by rifling tools which consist of small cutting edges. The secondary cutter as used in the Colt factory, for instance, which cuts the six spiral grooves in the barrel. This cutter appears relatively at right-hand angles to the long axis of the barrel, and cuts a groove exactly its own width, with a sort of hoeing motion. It scrapes first a small section of metal from the barrel where the first groove is to be made. The whole barrel is rotated to a position where the second groove is to be administered, and this scrapes the second groove, and it is rotated on the spiral. The barrel is then rotated again to the position for the third grove, brought into place, and this cutter starts the third groove, and so on it goes, all the way around the barrel, starting as many grooves as the barrel is ordinarily to contain. Having been all the way around the ground, the cutter is automatically lifted, so it starts the second travel, and

[89] 213 Iowa 677, 684-686, 239 N.W. 715, 719, 720.

it scrapes a little more metal, and so on it goes, on around, and scrapes a little more metal, until they vary from the different factories about the five-thousandths of an inch. This cutter edge is theoretically a perfect arc, cutting arc-shaped grooves in its movement. Practically, when metal is drawn against metal no perfect edge exists. And this theoretical perfect arc-shaped edge under the microscope shows a number of fine teeth. These teeth are constantly crumbling by the wear upon them as the cutting edge is drawn through the barrel, and naturally they vary in their arrangement from moment to moment. When the cutter makes its last draw through a certain groove it leaves in the bottom of the groove a series of tiny barrel lines which correspond to the position shown in the microscope of the teeth or edges of the teeth when it is drawn through. When it rotates and cuts the last cut, it leaves another series of tiny barrel lines on the bottom of the groove, different from the ones left by the preceding one by reason of the fact of the wear that it has undergone. The next groove it makes another series, which again vary. The grooves in the same barrel will have the same width, same depth and angle, but they will be unlike each other with reference to the fine barrel ridges left in them by the teeth of the cutting edge. These edges when the bullet is fired through that barrel are covered with powder, and it is these fine barrel ridges that we look for in our pictures and under the microscope to determine whether two bullets have passed through the same irregularities in the same barrel or not.[90]

This testimony is nothing more than a repetition of the subject matter in the article which was discussed by the Supreme Court of Kentucky in Jack *v.* Commonwealth, *supra.* The following questions and answers pertaining to bullets fired from barrels manufactured with the same tools appear in the record:

Q. Suppose that two or more guns were made by the same factory and by the same machinery, would it be possible from your experiments to tell what bullet came from each individual gun? A. Yes, sir.

Q. How do you know that? A. By actual experiment.

Q. And by what experiment? A. By the fact that in an actual experiment six barrels were made at the Springfield Armory, the government armory, where the Sprinfield rifle and the .45 automatic pistol is made, six barrels were rifled consecutively on one machine with no change or adjustment of the cutting apparatus whatsoever, and two of these barrels were then assembled into forms by another officer and bullets were fired through each.[91]

90 213 Iowa 677, 691, 692, 239 N.W. 715, 722, 723.
91 213 Iowa 677, 693, 694, 239 N.W. 715, 723.

The experiment referred to is the one discussed in connection with Evans v. Commonwealth, *supra*, and the same limitations, with respect to its weight as evidence of the reliability of the expert's methods of proof, apply in the instant case.

The Court was confronted with the question raised in Evans v. Commonwealth, *supra*, in so far as the defendant claimed that the expert had invaded the province of the jury. The witness had stated that "the conclusion I reached as a result of a study of these pictures was to the effect that both of these bullets had passed through the same barrel." [92] The Court properly followed the Evans case and pointed out that the fact whether the two bullets had passed through the same barrel was "left for the determination of the jury, taking into consideration the opinion of the witness upon the subject and the credibility which the jury saw fit to give him as a skilled or expert witness." [93] The attitude of the Court was manifested by its belief

[92] 213 Iowa 677, 695, 239 N.W. 715, 724.

[93] 213 Iowa 677, 696, 239 N. W. 715, 724. In Justis v. Union Mut. Casualty Co. (1932) 215 Iowa 109, 114, 244 N. W. 696, 698, the Court said: "Later, in State v. Steffen, 210 Iowa, 196, 230 N. W. 536, 538, 78 A. L. R. 748, in a case involving the science of finger print identification, this Court said: 'We are not disposed to change the rule which has been established in this court for many years to the effect that while an expert may be permitted to express his opinion, or even his belief, he cannot testify to the ultimate fact that must be determined by the jury. Such testimony would invade the province of the jury and determine the very issue which they must decide.' . . . In the case of State v. Campbell (Iowa) 239 N. W. 715, 725 (December 16, 1931) in which case the defendant was charged with murder, this Court, when discussing the admissibility of a ballistic expert's testimony, said: 'The appellant relies upon our pronouncement in State v. Steffen, 210 Iowa, 196, 230 N. W. 536 (78 A.L.R. 748), and cases therein cited; but a careful reading of the Steffen Case shows that the reason for the reversal was because the finger print expert was asked to state the fact instead of his belief, conclusion, opinion, or best judgment. Neither the question asked nor the answer given in the instant case are analogous with those held improper in the Steffen Case.' "

An expert should be allowed to express an opinion on the very question before the jury. See 43 Harv. L. Rev. 497: "The expert, by answering on the very point in issue, does not 'usurp the functions of the jury,' for the latter has still to reach its own conclusion with this aid. Chicago Union Traction Co. v. Roberts, 229 Ill. 481, 82 N. E. 401 (1907); . . . Beaubien v. Cicotte, 12 Mich. 459 (1864). Indeed the expert's opinion is even more useful in such circumstances, for it is then especially important that the jury have all available assistance. And to permit contradictory 'fact' testimony to keep out the expert's opinions would remove the only check on the veracity of these alleged eye witnesses. . . . People v. Storrs, 207 N. Y. 147, 100 N. E. 730 (1912)." In Cropper v. Titanium Pigment Co. (1931) C. C. A. 8th Circ. 47 Fed. (2) 1038, 1043, 78 A. L. R. 737, 744, that "the jury had the benefit of the expert, scientific study, special training, knowledge, skill and experience of the witness, and a full explanation of the exhibits. The theory upon which testimony such as was given by the witness is admissible is that the subject matter of the inquiry is so peculiarly within the range of scientific knowledge or special training and skill that to exclude it would work a denial of the only proof competent to establish the fact."[94]

In support of its belief that the right of the State in the investigation of the truth would be seriously impaired if the opinion testimony of this specially trained witness was rejected, the Court cited the following cases as authorities in which similar testimony had been held competent: Evans v. Commonwealth, *supra;* State v. Boccadoro, *supra;* Galenis v. State, *supra;* People v. Beitzel, *supra;* State v. Clark, *supra;* People v. Weber, *supra;* and State v. Vuckovich, *supra.*

The defendant complained that the photographs of the bullets had been improperly admitted without sufficient verification. They had been taken and developed by a person who was not a witness. The Court held that the objection was without merit since the expert had testified that he had set the two bullets in position and that the photographs were taken and developed under his direction.

In Matthews v. People (1931)[95] the expert on firearms identifica-

---

the Court said: "But if the questions propounded were such that the jury might not be capable of determining from the evidence, then it was proper that they should have the benefit of the opinion of an expert, even though the opinion went to matter directly in issue." See note, 78 A. L. R. 755, on "Testimony of expert witness as to ultimate fact."

It is likewise true that, in ascertaining the truth of a matter, an expert should be permitted to state a fact known to him because of his special knowledge, even though the statement may involve an element of inference. Cropper v. Titanium Pigment Co., *supra.*

See also, State v. Bassone (1932) 109 N. J. L. 176, 184, 160 Atl. 391, 395: "Request No. 27 was properly refused because it asked the judge to declare, in effect, that the testimony of a witness (who happened to be a ballistic expert) 'is not evidence of the facts' to which he testified, entirely apart from his opinion evidence."

[94] 213 Iowa 677, 695, 239 N. W. 715, 724. See note, 47 Harv. L. Rev. 878, on rejection of expert testimony because no opinion formed: "A qualified expert's inability to reach a definite conclusion might in some cases be equivalent to an opinion that there is considerable doubt on the matter, and, particularly in a criminal case, the defendant should be entitled to have brought to the jury's attention any characteristics sufficient to create a reasonable doubt."

[95] 89 Colo. 421, 3 Pac. (2d) 409.

tion testified that the defendant's wife had been killed by a bullet discharged from the defendant's firearm. It appears that aside from mere routine proof the case of the people rested solely upon the testimony of this expert. The Supreme Court of Colorado subjected the testimony to the "most careful scrutiny" in view of the defendant's long residence in the community, his unquestioned character, his perfect alibi, and the absence of any motive. The Court in reversing the conviction of murder discussed the testimony as follows:

> The witness explained to the court how such guns were made and the bullets fired from them were often so marked by the grooves in the rifling as to make identification not only possible but indisputable. He did not, then, rest his conclusion upon his mere opinion that the bullets which caused the death came from defendant's weapon. He made a comparison of the two under a glass, in the presence of the jurors, and pointed out to them the identical markings upon which his opinion was based, so that the jurors were, presumably, as capable of seeing what he saw, and as irresistibly driven to his conclusion therefrom, as the witness himself. The force of his testimony depends, therefore, upon the appearance of the alleged markings on the bullets in evidence. All these bullets are before us. Each of the Justices has examined them under a powerful glass (although not the identical one used at the trial), and has been wholly unable to see anything resembling what the witness says he saw and which he assumed to exhibit to the jurors. The rule applicable here is the identical rule applicable where all the evidence in support of a judgment is presented by deposition. This court is as capable of passing upon such evidence as the jurors who heard it. All the facilities they had we have. If it were alleged, for instance, that a certain assertion appeared in a deposition in such a case and that assertion were indispensable to support the judgment entered, and an examination of the deposition disclosed no such assertion, we would necessarily reverse the judgment. Holbrook Irr. Dist. v. Fort Lyon Canal Co., 84 Col. 174, 193, 269 P. 574.
>
> Here the sole evidence of guilt is the assertion that certain alleged markings appear upon these bullets. We examine them and find nothing of the kind. Hence, the judgment must necessarily be reversed. The thread is entirely too slender to support a sentence of life imprisonment. The evidence is not only weak and uncertain; it is no evidence.[96]

The Court was obviously reluctant to affirm a conviction of murder predicated upon the opinion of the expert in the face of the other evidence which tended to negative the defendant's guilt. The court's

[96] 89 Colo. 421, 424, 425, 3 Pac. (2d) 409, 410, 411.

holding that the identification was not based upon specialized knowledge can be justified unless an intelligent reading of the testimony
clearly discloses the necessity of special skill. If the expert beclouded
the real nature of the sicence of firearms identification and apparently
purported to compare the markings without a skilled interpretation
thereof, the Court correctly excluded the opinion. However, it is a
dangerous practice for any appellate court to conduct an independent
investigation without the assistance of the instruments and witnesses
available at the trial.[97] Clearly, the Court would not have attempted

FIG. 143.

to make a comparison of the markings on the bullets had it been aware
of the requirements necessary to draw a proper inference therefrom.

Testimony on firearms identification, aside from its use as sub-
stantive evidence, has been used to attack the credibility of a witness.
In State *v.* Mangino (1931) [98] the defendant did not deny that the
bullet from his revolver caused the death of his wife. On the witness stand the defendant testified that the shooting was accidental
and that he had thrown his revolver into a river subsequent to the
accident. The prosecution proved that the fatal bullet was fired

[97] 46 Harv. L. Rev. 1168-1169.
[98] 108 N. J. L. 475, 156 Atl. 430.

from a revolver which was retrieved from a vacant lot in the vicinity
of the defendant's home.

Fig. 143 is a photomicrograph of the bullet removed from the
head of Mrs. Mangino, and Figs. 144 and 145 are photomicrographs

of comparisons of the evidence bullet (left) and a test bullet (right).

In People *v.* Meyering (1931)[99] the value of testimony on firearms identification was attacked on the ground that "there was no scientific basis for the conclusions which" the expert "drew from his comparison of the two bullets, and because he refused to take part in an experiment designed to test his ability to identify bullets with the weapons from which they had been fired." The Court ruled that the testimony of the expert was immaterial because the proof on the question of the defendants' possession of the revolvers in evidence was not sufficient. It appears that one Jack Zuta was murdered in

FIG. 144.

the Lake View Hotel near
Delafield, Wisconsin, on August 1, 1930. On September
16, 1930, Smith and Stanton
were arrested in Chicago and
charged with the murder.
When arrested they were driving a Ford Sedan, and a caliber .38 revolver was found in
the pocket of the car door at
Stanton's right. The firearms
expert testified that one of the
bullets extracted from Zuta's
body had been fired from this

FIG. 145.

revolver. On a habeas corpus proceeding against the Sheriff of Cook
County, the Court, in holding that the firearms testimony was immaterial, said:

[99] 345 Ill. 598, 178 N. E. 122.

Admitting that the revolver found in the car at the time of Stanton's arrest on September 16 was the weapon from which Exhibit 11 was fired into the body of Zuta on August 1, this is no evidence that Stanton or Smith was in Wisconsin on August 1, 1930. The car in which the relators were riding at the time of their arrest did not belong to either of them. There is no evidence that the revolver did. It may have belonged to the owner of the car. It was not in the exclusive possession of either Stanton or Smith. Stanton testified he did not know it was there, though the testimony is that the handle of the revolver was sticking out of the pocket in the car door. The exclusive possession of stolen property soon after it was stolen is evidence that the person in possession is a thief and if unexplained may be sufficient to justify a conviction, but to do so it must be not only exclusive but recent. The possession, to be of any value, must be an exclusive, personal possession on the part of the accused. People v. Bullion, 299 Ill. 208, 132 N. E. 577; People v. Kubulis, 298 Ill. 523, 131 N. E. 595. Here the revolver was not shown to have been in the exclusive possession of either Stanton or Smith. The possession, to be of any value as evidence, must be recent. People v. Bullion, *supra;* People v. Kubulis, *supra.*[100]

[100] 345 Ill. 598, 606, 607, 178 N. E. 122, 125, 126. See also, People v. Dale (Ill. 1934) 189 N. E. 269, 271. The defendant was convicted of murder and was sentenced to death. It appears that twelve officers raided an apartment after having received information that the defendants were to be found therein. The defendants were arrested and a package containing four revolvers was found under a bed and neither of the defendants admitted the ownership thereof. The revolvers were produced at the trial in the presence of the jury, and the one which was identified as the firearm employed to fire the fatal shots was admitted in evidence. The defendant contended that it was prejudicial to his rights to have all of the revolvers exhibited to the jury. The court said that "this was not error, as we have repeatedly held that it is competent to prove that a person accused of crime possessed a weapon suitable for the commission of that crime at the time of his arrest even though no claim is made that he actually used it on the occasion in question. People v. Lenhardt, 340 Ill. 538, 173 N. E. 155. . . ."

In People v. Sonoqui (Cal. 1934) 31 Pac. (2nd) 783, the defendants were convicted of murder. The victim was killed by a caliber .22 bullet. The defendants had denied the ownership of any caliber .22 firearm or ammunition therefor. Investigators found caliber .22 firearms and ammunition therefor in the home of one of the defendants. The court held that these firearms and the ammunition therefor were proper evidence to rebut the defendants' contention that they did not possess such articles. The Court said, "being the same size and caliber as those used in perpetrating the crime, they were also competent evidence showing that appellants had the means at hand to commit the offense charged against them. People v. Mooney, 177 Cal. 642, 171 P. 690."

In State v. Banks (Or. 1934) 32 Pac. (2d) 571, the defendant's house was searched at the time he was taken into custody and certain firearms and ammunition were found. Later in the day, after his house had been open and people

In People v. Farrington (1931) [101] the defendant, shortly before the close of the trial, offered in evidence a bullet which was then marked for identification. As the evidence in the case was being closed, the defendant requested that the court permit the bullet to be submitted to an expert in order to determine whether it could have been fired from the pistol in evidence. The request was denied by the trial judge because the case had been set for trial for more than six weeks during which period there was ample opportunity to submit the bullet to an expert. The appellate court affirmed the conviction of murder on the ground that the granting of the request would have unduly delayed the trial and that the trial judge had properly exercised his discretion. The denial of the request was not a prejudicial refusal. In Quinn v. State (1932) [102] the court reversed a conviction of murder and said, in effect, that in the interest of truth and justice the request of the defendant, not objected to by the state, should have been granted, permitting the exhumation of the bodies of the deceased so that the fatal bullets could be recovered and subjected to tests to ascertain whether any of such bullets had been fired from defendant's firearm.

The Supreme Court of Georgia in Williams v. State (1933) [103] approved of testimony on firearms identification. The defendant was convicted of murder. The case presented for the State was based solely upon circumstantial evidence, and the defendant maintained that this evidence was insufficient to support the verdict. An expert

---

had circulated therein, the house was again searched and a pistol and ammunition were found. The court admitted in evidence the pistol and ammunition as evidence of preparedness. The court said that the firearms found in the defendant's house when he was taken into custody were clearly admissible; that the pistol found in the house a few hours later was likewise admissible; the length of the time interval between the apprehension of the defendant and the finding of the pistol, as well as the accessibility of the house to others, merely concerned the probative force to be given this evidence.

[101] 213 Cal. 459, 2 Pac. (2d) 814.

[102] 16 Pac. (2d) 591 (Okla.). In Kent v. State (Tex. 1932) 50 S. W. (2d) 817, 819, 820, the Court affirmed a conviction of murder. A witness at the trial stated that "I have examined this shotgun shell identified as Exhibit 3 (shell found at the scene of the homicide), with the edges rolled down," and he further stated that it was the same kind in all respects as another found in the defendant's gun. The Court said: "The statement of the witness, after describing the shells, that the shell discovered near the body and that found in the gun were the same kind in every respect would seem to have been, in effect, the expression of an opinion that the shells were alike."

[103] 177 Ga. 391, 170 S. E. 281.

on firearms identification testified at the trial, and the Court in affirming the conviction said:

> We have no hesitancy in holding that the evidence was sufficient to support the verdict. Besides the many other convincing circumstances, the testimony of the ballistic expert tended strongly to connect the defendant with the commission of the homicide. In State v. Boccadoro, 105 N. J. Law, 352, 144 A. 612, it was contended that there was no sufficient evidence to identify the defendant as the murderer. Besides proof of some other incriminating circumstances, the state introduced the evidence of ballistic experts whose testimony was similar to that of the expert in the present case. The court held that these various facts, taken together, made the question of the defendant's guilt one for the determination of the jury. See, also, Evans v. Commonwealth, 230 Ky. 411, 19 S. W. (2d) 1091, 66 A. L. R. 360, annotation.[104]

Some time before the murder the defendant borrowed a "pistol," a caliber .38 Smith and Wesson, from a neighbor, saying that he was going away for a short vacation and desired the pistol for protection. The pistol was returned to a kinsman of the lender about two weeks after the murder. Subsequently, the defendant made an inconsistent statement regarding his purpose in borrowing the pistol. The expert on firearms identification testified that in his opinion the two bullets extracted from the body of the deceased had been fired from the borrowed pistol. The Court commented on this evidence as follows:

> The evidence tended to show that the pistol was fired twice while in his possession or custody. The pistol, together with the two bullets which were taken from the dead body, were carried to a ballistic expert in New Orleans, who after experiment and examination attended the trial as a witness and testified that in his opinion these bullets, hereinafter called the evidence bullets, were fired from the identical pistol. The witness further testified that in the manufacture of a pistol the barrel is made first by drilling, and then by reaming, and that in the reaming process the reamer, by wearing, gradually undergoes a change of form and will leave identification marks in each barrel different from those in any other barrel made by the same manufacturer and finished with the same reamer. The result is that each barrel contains its own peculiar lands and grooves, and every bullet fired from the same pistol will bear the imprint of these marks. The witness testified that he had tested the pistol by firing other bullets therefrom, and that by a microscopic comparison of the test bullets with the evidence bullets he found that each bullet carried the same identification

[104] 177 Ga. 391, 398, 170 S. E. 281, 284, 285.

mark. The test bullets and the evidence bullets were all introduced in evidence, and were examined by the jury under a microscope. At the instance of the state, the witness also made tests
and experiments in the presence of the jury. The name of this
witness was Maurice O'Neill and according to his testimony he was
superintendent of the bureau of identification of the police department of New Orleans. He had held this position for about
eight years, and from his own evidence as to his study and experience, together with the demonstrations made by him before the
jury, they were authorized to give credence to his testimony.[105]

This statement evinces an overemphasis of the wearing away of
the reamers. The wearing away of the tools, as a cause of producing
accidental variations upon these surfaces of firearms, sounds extremely
plausible. Tool-marks are produced by the tearing of the metal and
not by the "wearing" of the reamer.[106]

During the week of April 23, 1934, in the circuit court, Wythe
County, held at Wytheville, Virginia, H. F. Bausell and his son,
Bernace, were tried for the murder of Virginia Cornett Bausell, the
estranged wife of the latter. The deceased was struck by a bullet
during an exchange of shots between the Bausells and her father,
T. E. Cornett, who was also killed. By mutual consent the court
deputized G. B. Cassell to take the firearms exhibits to three experts
for examination on behalf of the defense. The exhibits were as
follows:

Revolver 1. A Smith and Wesson revolver, blue finish, chambered for caliber .32 S.&W. long center-fire cartridge, containing
six U.S.C. Co. fired cartridge cases in the chambers of the cylinder,
all of which were of caliber .32 S.&W. cartridges loaded with black
powder.

Revolver 2. A Smith and Wesson revolver, nickel finish, chambered for the caliber .32 S.&W. center-fire cartridge, containing 4
Rem.-U.M.C. Co. fired cartridge cases in the chambers of the cylinder, all of which were of caliber .32 S.&W. cartridges loaded with
smokeless powder.

Revolver 3. A Smith and Wesson revolver, nickel finish, chambered for the caliber .32 Winchester center-fire cartridge, containing three Rem.-U.M.C. Co. and 3 W.R.A. fired cartridge cases in
the chambers of the cylinder, all of which were of caliber .32
W.C.F. cartridges loaded with smokeless powder.

[105] 177 Ga. 391, 396, 397, 170 S. E. 281, 284. See, also, State *v.* Shawley
(1933) S. C. Mo., Div. No. 2, 67 S. W. (2d) 74, 80: "on a comparison of the
evidence bullet with the test bullets he was of the opinion as an expert that it
had been fired through that rifle."

[106] See Chapter I, *supra*, Figs. 27-29.

Bullet "A," Fig. 146. A caliber .32 W.C.F. lead bullet, flat nose, weighing 98 grains and of Rem.-U.M.C. Co. manufacture. The normal weight of this bullet is 100 grains. This bullet was removed from the underclothes of Cornett.

Bullet "B," Fig. 147. A caliber .32 S.&W. lead bullet, round nose weighing 86 grains and of Rem.-U.M.C. Co. manufacture. The normal weight of this bullet is 88 grains. This bullet was recovered from a dresser near Cornett's body.

FIG. 146.—Bullet "A," Bausell Case.   FIG. 147.—Bullet "B," Bausell Case.

Bullet "C," Fig. 148. A caliber .32 S.&W. lead bullet, weighing 82 grains and of U.S.C. Co. manufacture. The normal weight of this bullet is 85 grains. The base of this bullet showed a black powder residue. This bullet was extracted from the skull of Virginia Bausell.

It was admitted by both sides that the three revolvers, together with the discharged cartridge cases in their respective chambers, as submitted to the defense experts, were those used in the shooting, and that the problem was from which of these three revolvers had bullets "A," "B," and "C" been fired. The experts for the defense were not informed as to the ownership of the revolvers. They were told that the most important bullet was bullet "C" as it had been taken from the skull of Virginia Bausell, and the defendants were charged with having fired this fatal bullet. The three experts for the defense independently arrived at the conclusion that bullet "C" was fired from revolver 1, the revolver with the blue finish. As this weapon belonged to Cornett the jury returned a verdict of acquittal.

Essentially the problem, as presented, to determine from which of the three revolvers bullet "C" had been fired was elementary and involved only class characteristics. Bullet "C," a caliber .32 S. & W. lead bullet of U.S.C.Co. manufacture, was fired from the only revolver of the three which contained the U.S.C.Co. fired cartridge cases, that is, revolver 1.

Furthermore, the heavy bullet, bullet "A," could have been fired only from revolver 3, as revolvers 1 and 2 were not adapted to ammu-

FIG. 148.—Bullet "C," Bausell Case.

nition of that size. A comparison of the markings on bullet "C" with the markings on test bullets fired from the three revolvers, under a comparison microscope, demonstrated that bullet "C" was not fired from revolvers 2 or 3, and that it had been fired from revolver 1.

The evidence on firearms identification, owing to a lack of other evidence, was the sole determinant of the guilt or innocence of the Bausells on the charge of shooting Virginia Cornett Bausell. Undoubtedly, if they had been proved guilty the penalty would have been extremely severe. Without testimony on firearms identification it would have been an impossibility to prove, under the circumstances, that Cornett had accidentally killed his daughter. The striking feature of this case is that a tragedy was averted by the success of the defense in presenting competent expert testimony. The prosecution presented a witness, as an expert, who on this simple problem of firearms identification, testified that bullet "C" had been fired from a revolver of one of the defendants, revolver 2. If this testimony had not been intelligently rebutted, it would most certainly have brought about a current miscarriage of justice. The testimony was as follows:

COMMONWEALTH v. BAUSELLS   WITNESS: JOHN H. FOWLER

JOHN H. FOWLER, witness for Commonwealth, sworn:

DIRECT EXAMINATION

BY MR. SHAFFER: Q. What is your name? A. John H. Fowler.

Q. Where do you live, Mr. Fowler? A. Washington, D. C.

Q. Where was your home before you went to Washington? A. Forestville, Maryland.

Q. How old are you, Mr. Fowler? A. 39.

Q. Where did you take your primary education? A. The public schools of Maryland, and the night schools of Washington.

Q. What is your occupation? A. A lieutenant of detectives in the Metropolitan Police Department, Washington, D. C.

Q. How long have you been a member of that department? A. Since June 16, 1919.

Q. What is your official office or title? A. A lieutenant of detectives.

Q. Have you any special assignment, and if so, what is it? A. My assignments consist of firearm identification and forensic ballistics.

Q. How long have you had such assignment? A. About four years.

Q. How long have you been interested in the identification of firearms? A. About ten years.

Q. Have you ever received any instructions with reference to firearm identification, and if so, state from whom? A. My first instruction was received from Sergeant Cornwall, formerly firearm expert of the Metropolitan Police Department. My next instruction was from Major Archer, an expert in small arms in the Ordnance Department. I also received instructions from Major Green, of the Ordnance Department. I have worked with Dr. Sauder at the Bureau of Standards in Washington for around three years, in the research and study of firearm identification.

Q. Is Dr. Sauder the firearm expert of the United States Government? A. He does work for the United States Government, yes, sir.

Q. And what department or bureau is he connected with? A. The Bureau of Standards, Washington, D. C.

Q. Have you received any instructions with reference to firearm identification from any school or university? A. I have, from the Northwestern University, of Chicago.[107]

Q. Have you ever written any papers on the subject of firearm identification? A. I compiled a course of instructions for the Institute of Applied Science in Chicago.

Q. Have you lectured on the study of forensic ballistics? A. I have, before police departments, technical schools, and civic organizations.

Q. Have you ever been in the employ of any manufacturer of firearms? A. I have, for the Remington Arms.

Q. For how long? A. About two years?

Q. Have you studied the manufacture of firearms? A. I was employed in the factory of the Remington Arms.

Q. Have you ever been called upon for opinions by any other jurisdictions outside of Washington, D. C.? A. I have, on several occasions.

Q. Name some of these places, Lieutenant. A. Baltimore, Maryland; Upper Marlboro, Maryland; Brookeville, Maryland; Frederick, Maryland; Hyattsville, Maryland; Richmond, Virginia; Winchester, Virginia; Harrisonburg, Virginia; Fairfax, Virginia, and Sulphur, Virginia; Wentworth, North Carolina; Winston-Salem, North Carolina; Greenville, North Carolina; Kansas City, Missouri.

Q. Have you testified in courts other than the District of Columbia with reference to the identification of firearms? A. I have. I have testified in Wentworth, North Carolina; Winston-Salem, North Carolina; Greenville, North Carolina; Harrisonburg, Virginia; Fairfax, Virginia, and Upper Marlboro, Maryland.

Q. Have you data on firearms and ammunition? A. I do have data on all current ammunition, and a collection of arms—250 different types of firearms.

Q. About how many cases did you handle annually in the District of Columbia, with reference to the comparison and identification of firearms, during the year? A. I would say around a hundred.

Q. What equipment do you have to assist you in your work of identifying firearms? A. I have a comparison microscope, micrometer, a collection of fired bullets fired through different types of firearms, photographic equipment. All of this equipment has been approved by the Bureau of Standards for this type of work.

Q. You say the equipment you use for this type of work has been approved by the Bureau of Standards? A. It has.

Q. You mean the Bureau of Standards of the United States? A. Yes, sir.

Q. I wish you would tell the jury what work the Bureau of Standards is

[107] According to a pamphlet entitled ''Northwestern University Bulletin, Scientific Crime Detection Laboratory, Affiliated with Northwestern University, Outline of Teaching Program for a Course in Methods of Scientific Crime Detection'' (Vol. XXXI, April 6, 1931, No. 32), a course of instruction during a period of four weeks was contemplated in sixty subjects. Of these, there were the following: Firearm Identification; Powder and Cartridge Manufacture (moving picture demonstration); Laboratory Exercises in Bullet and Shell Photography; Pitfalls in Forensic Ballistics; Elementary Ballistics; The Uses of a Microscope.

engaged in. A. It is an institution of several bureaus for research and study in the different commercial fields. They have a bureau for firearm identification, a bureau for handwriting and typewriting identification.

Q. Lieutenant, I suppose you remember an interview that I had with you in the City of Washington with reference to making an investigation in this case, do you not? A. I do.

Q. I wish you would tell the jury whether I, at that time, told you any of the facts of this case at all, or told you the names of any of the parties connected with it. A. No, sir; you did not.

Q. Did you have any identification, whatever, for any of the guns that were presented to you later by the Sheriff of this county? A. No, sir, I did not.

Q. I believe I made arrangements with you, or the commonwealth attorney of this county did, to make an investigation of those guns? A. That is correct.

Q. How are you employed by the Department at Washington? A. On a salary.

Q. How are you employed and paid for your work when you make independent investigations like this? A. I am paid by the state, or the commonwealth.

Q. Did the commonwealth attorney of this county arrange with you to make this investigation? A. Yes, sir; I had a letter from him.

Q. Do you recall the sheriff of this county presenting to you these three guns? A. Yes, sir, I do.

Q. I wish you would tell the jury whether or not the sheriff gave you any information, whatever. A. He did not.

Q. With reference to this case or the facts of this case? A. He did not.

Q. Did he give you any information about any of the parties connected with this case or who may have used either of the guns in question? A. He did not.

Q. Did you get any information from any source? A. I have not.

Q. Before making any of your investigations in this case? A. I have not.

Q. You have not? A. No, sir.

Q. Mr. Fowler, take this gun. I will ask you as to whether or not—I hand you here a bullet—that bullet was presented to you, I believe, was it not, sealed in an envelope? A. It was.

Q. Did you sign that seal? A. I did.

Q. And did you open that envelope in the presence of the sheriff of this county? A. Yes, sir.

Q. I wish you would tell the jury whether or not you made an investigation to see whether or not that bullet was fired from this large gun—this gun? A. I did.

Q. I wish you would tell the jury how you made that investigation and what the result of that investigation was. A. I first made an examination of this gun and the contents of the cylinders. I found the barrel was fouled and all the chambers of the cylinder were fouled, having evidence of being fired since last cleaning; and it contained three Remington U.M.C. empty shells and three Winchester Repeating Arms shells; they were the smokeless powder type. The Remington shells were the 100-grain bullet type, and the Winchester Repeating Arms shells were the 98-grain bullet type. I then fired a .32-caliber Remington U.M.C. cartridge of the smokeless powder 100-grain bullet type through this Smith and Wesson revolver. I recovered the bullet, made a microscopic examination and comparison of the evidence bullet that was presented to me by Sheriff Crockett with the test bullet fired from this Smith and Wesson revolver.

Q. What was the result of that investigation? A. I am of the opinion that that bullet was not fired through that revolver. The type of ammunition of the evidence bullet—this bullet here—is of a different type; that is to say, this bullet, before it was fired, weighed 88 grains; that is a short cartridge. The type of bullet that was in this gun was of the 98 to 100 grain type; that is the long cartridge. This evidence bullet shows many marks and scratches on the side of it, printed there——

By Mr. Campbell: Do I understand you to say the evidence bullet weighs 88 grains? A. It is of the 88-grain type.

By Mr. Campbell: Of course, they are machine-made, and they would vary slightly? A. Yes, sir.

By Mr. Shaffer: Q. In describing it as the "evidence bullet," you mean the bullet that was taken from the skull of Mrs. Bausell? A. The bullet that was presented to me by Sheriff Crockett.

By Mr. Campbell: That is the bullet that was taken from the head of Mrs. Bausell. We admit it. A. This evidence bullet had many marks and scratches on the side which it had received from the gun barrel as it passed through, indicating that the gun barrel was badly fouled and pitted from not being properly cleaned. The test bullet showed that this gun barrel is in comparatively good condition, and the markings and scratchings very much lighter, and I was unable to find any markings and scratchings on the test bullet that corresponded with the markings and scratchings on the evidence bullet.

Q. I wish you would explain to the jury about how the barrel of a pistol is made, and what you would expect in the barrel of a pistol. A. Yes, sir. I have a pistol here that the barrel has been partly cut away, if there will be no objections.

Q. I will ask you to go over and show that to the jury.

By Mr. Campbell: Let me look at that, please. You haven't got it loaded, have you?

By the Witness: No, sir.

By Mr. Campbell: What make is this? It is cut off there.

By the Witness: That is a Spanish gun.

By Mr. Campbell: You are just using this——

By the Witness: Just as an illustration. The gun barrel, at first, is nothing more than a cylindrical block of steel with a hole bored through it; that is the original barrel. After that a reamer goes through it to smooth it up as well as can be done with a machine tool. Later on is what is called the rifling process. Spiral grooves are cut through to give the bullet its spin when it is fired. In making this gun barrel there are a number of tools used. Of course, the edges of no two tools are exactly the same. The teeth of the edges of these tools cause little scratches on the inside of the barrel. Not only that, but after that, if a gun is not properly cleaned there is rust on the inside of the barrel. This spot here is a heavy rust spot. On the second part here there are some tool markings very plain——

By Mr. Campbell: We don't know, if Your Honor please, that the introduction of those marks other than the rifling is material.

By the Court: As I understand, he is testifying as to the process of making a gun, that it will necessarily cause individual markings and scratchings.

By Mr. Campbell: But the rust spots, I don't think we are especially interested in.

By the Witness: I might say that these rust spots are quite characteristic, too, of an individual gun.

By Mr. Campbell: Rust spots don't correspond, either?

By the Witness: Exactly. This first part, in addition to that patch of rust there are several tool marks and scratches; on the back end there the rust spot is missing; however, the tool markings are there. When you turn it over to the opposite side you will notice the difference in the tool markings and scratches to what they are on this side.

By Mr. Shaffer: Q. Mr. Fowler, I wish you would tell the jury if a bullet that is intended for a certain sized gun is of the same size as the bore in the gun. A. It is not. The diameter of the bullet is a slight bit larger than the original bore; the original bore is the barrel before the groove rifling is cut in. There is a difference of about 7 thousandths of an inch. The bullet is forced by the power to shape itself down in these grooves.

Q. Is the bullet of a softer material or a harder material than the gun? A. The bullet is of a softer material. In this particular instance the bullet is of lead—trade lead—the gun barrel being of tempered steel. The bullet passing through it, these elevations and scratches are imprinted on the bullet as it passes through.

Q. I wish you would state whether or not the markings on guns are the same? A. They are not.

Q. I will ask you take this gun, this blue gun, and tell the jury whether or not you made an investigation and a comparison of the evidence bullet with a test bullet fired from this gun? A. I did.

Q. I wish you would tell the jury how you made that investigation and what the result of that examination was. A. I made an examination of the revolver. I found six U.S. cartridge cases of the black powder 85-grain bullet type ammunition; the cylinder was fouled and the barrel was fouled, having evidence of being fired since last cleaned. I then fired a test bullet through this gun, of the black powder 85-grain type bullet.

Q. The same as had been used in the gun? A. Yes. I recovered the bullet and made a microscopic comparison of the two bullets.

Q. What was the result of that investigation and comparison? A. I was unable to find any similarity between the two bullets.

Q. As I understand you, the markings on the bullet that you fired from that blue gun had no similarity to the markings you found on the evidence bullet? A. They had not.

Q. I wish you would explain that to the jury. A. The test bullet showed that the barrel was in first class condition, cutting the bullet clean, leaving no series of lines similar to the lines and markings on the evidence bullet.

Q. What was the condition of the barrel of that pistol? A. Good.

Q. Can you show that to the jury so they can see? A. I will try, sir. If you will hold this rule so it will reflect the light through the barrel, and then look down through the barrel. You have got to get it at a certain angle to reflect the light.

By Mr. CAMPBELL: I don't believe that that is proper evidence, sir. The members of the jury are not experts in the science that the Lieutenant calls forensic ballistics. The condition of the barrel is immaterial.

By THE COURT: The witness testified that the condition of the barrel has something to do with the marks on the bullet fired through that barrel.

By Mr. CAMPBELL: The object of the examination is to let the jury examine the barrel and then let them see if they can determine whether the bullet was fired through it.

By THE COURT: That is not the purpose of it. The purpose of it is to see whether there was any rust in the gun barrel.

By Mr. CAMPBELL: But we think that is unnecessary.

By THE COURT: I don't see anything against it.

To which ruling of the Court the defendants, by counsel, *excepted*.

By Mr. SHAFFER: Q. How many grooves are there in this type of guns? A. Five, with a right twist.

Q. How many grooves are there in a rifle? A. Five. The rifling consists of a number of grooves and lands; the lands are the elevated portions between the grooves, the number of the lands and the number of the grooves being equal. The lands and the grooves make up what is commonly called the rifling.

Q. What is the object of the grooves, Lieutenant? A. To spin the bullet, to stabilize it while in flight.

Q. I wish you would tell the jury whether or not the evidence bullet, the bullet that was taken from this young lady's brain, was fired from that gun. A. It was not.

Q. Now, I hand you this nickel-plated .32, which was presented to you by the sheriff, and I will ask you if you made an investigation of that gun. A. Yes, sir.

Q. Well, tell the jury what the result of that investigation was—first, tell them how you made the investigation and what the result of it was. A. An examination was made of this gun. I found four empty Remington U.M.C. shells of the smokeless powder 88-grain bullet type ammunition. The gun barrel was fouled, the cylinder was fouled, having evidence of being fired since last cleaned; and there was an examination made of the interior of the barrel which showed there was quite an accumulation of rust; it was pitted from being fired and not being properly cleaned out. This rust and pitting has collected over quite a period of time; the length of time I couldn't estimate. A test bullet——

Q. I will ask you to show, before you go further, right at that point, I will ask you to show the condition of that barrel to the gentlemen of the jury. A. I am removing the cylinder so they can see through the barrel easier. I don't think you will need the rule if you will put that up to the light.

By Mr. Campbell: Lieutenant, would shooting that pistol have any effect on its rusted condition?

By the Witness: Yes, sir; if it isn't properly cleaned after.

By Mr. Campbell: You don't know what the condition of this gun was at the time the shot was fired?

By the Witness: Well, from our experience, we know it takes time for rust to develop; a condition like this would require a number of years.

By Mr. Campbell: You don't know that this gun is in the same condition as when it was previously fired?

By the Witness: I have reason to think it was.

By Mr. Campbell: You think it was?

By the Witness: Of course, I cannot say exactly. Of course, every time a gun is fired there is some slight change; some particles that are left in the grooves.

By Mr. Shaffer: Q. I will ask you now to tell the jury how you proceeded to make the investigation. A. I fired a test shot, or a test bullet, with the same type of ammunition, the Remington U.M.C., the smokeless powder 88-grain type bullet. The bullet was recovered and a microscopic examination and comparison was made of the two bullets, that is, the test bullet and the evidence bullet received from Sheriff Crockett.

By Mr. Campbell: What make of bullet was that, Lieutenant?

By the Witness: Remington U.M.C.

By Mr. Shaffer: Q. I wish you would tell the jury whether or not a black powder bullet similar to the one that was fired through the blue gun is a similar bullet to the one that was fired through this one. A. There is a difference in the weight. The U.S. cartridge .32 S. and W. shot weighs 85 grains; the Remington .32 S. and W. shot weighs 88 grains. The U.S. cartridge has the black powder, or this particular cartridge had black powder, and the Remington has smokeless powder.

Q. Is there a difference in the shape of the bullet? A. There is.

Q. I wish you would describe to the jury the difference in the shape of the bullet. A. There is a difference in the shape of the base, there is a difference in the shape of the nose, and there is a difference in the candle works, that is, the wax groove around the bullet.

Q. From your investigation, I wish you would state to the jury what your opinion is with reference to the bullet being fired from that gun. A. The test bullet showed many similarities, in comparisons of the scorings and scratches on the sides of both bullets printed there by the gun barrel as it passed through. The evidence bullet was twisted and distorted; in fact, there is about 6 grains of it knocked off, and I was unable to make a complete examination of the entire bullet. However, due to the many similar scratches and other similarities, I am reasonably sure that the two bullets were both fired from this same gun, the small, break-down Smith and Wesson.

Q. You mean the bullet that came from the brain of Mrs. Bausell— A. The bullet that was delivered to me by Sheriff Crockett.

Q. Was fired through that gun? A. Yes, sir.

Q. How did you make that examination? A. With a comparison microscope.

Q. Did you make any examinations other than with the microscope? A. I made photographic observations.

Q. I would like for you to take the instruments that you have, with which you made this examination, and show the same to the jury. A. I can go ahead and explain this?

Q. Yes, go ahead and explain it. A. This is a comparison microscope, consisting of two single microscopes. One bullet is placed directly under one microscope, say the test bullet, and the evidence bullet is placed directly under this microscope, and through this cross-arm, by the use of prisms, the image of the bullet under the right microscope comes under this single eye-piece, and the image of the bullet under the left microscope also comes under this single eye-piece, making a comparison examination of the two bullets.

Q. Was that done with reference to each of these bullets? A. It was.

Q. I will ask you, in the presence of the jury here in court to fire a bullet from each of those guns and show it to the jury.

Objected to, because experiments before the jury are not proper.

Objection sustained.

Q. I will ask you to place the evidence bullet under the microscope and the test bullets from each of these guns, and let the jury look to see whether or not you have correctly told them about the bullets.

Objection.

BY THE COURT: I think, gentlemen, this expert on the science of ballistics, if he is an expert—I suppose he is an expert—may testify as to what he finds, and that is as far as he can go.

BY MR. SHAFFER: Q. I believe you told the jury that that instrument has been approved by the Bureau of Standards? A. It has, yes, sir.

Q. And you used that instrument in determining from what gun this bullet was fired? A. Yes, sir.

Q. Did you make any further examination to determine which gun this bullet was fired from? A. I did not.

Q. Did you make a photograph of it? A. I did.

Q. I wish you would tell the jury how you photographed this bullet in order to determine which gun the bullet was fired from. A. I made arrangements for a special camera, a camera that fits down over this single eye-piece and operates the same as any professional camera.

Q. Have you those camera photographs present? A. I have.

Q. I wish you would produce those photographs.

(Witness produces three photographs.)

Q. I wish you would show those and explain them to the jury.

BY MR. CAMPBELL: May we see them first.

BY THE WITNESS: I have some more.

BY MR. CAMPBELL: Let us have those, too, if you will let us have them. I notice a separation here; is that a photograph of two bullets?

BY THE WITNESS: Yes, sir.

Counsel for defendants objected to the introduction of these photographs, and the objection was sustained by the Court.

BY MR. SHAFFER: Q. Mr. Fowler, I wish you would explain to the jury how you fire and get the evidence bullet—I mean the test bullet. A. Fire it into cotton waste contained in that box.

Q. I wish you would show them how that is done. A. It is just ordinary cotton waste. The bullet is fired into it, and recovered after being fired.

Q. And that was done with each of these test bullets? A. It was.

Q. I want to ask you whether or not the evidence bullets—that is, the bullet taken from the brain—I believe you said that there was five grooves in these guns? A. That is correct, sir.

Q. And would that make five sections on the bullet? A. That would make five lands and five grooves on the bullet. The lands on the bullet come from the grooves in the gun, and the grooves on the bullet come from the lands in the gun; in other words, it is just vice versa on the bullet from what the barrel is.

Q. How many sections of the evidence bullet did you find in condition to examine? A. I found three sections; they were found in different places, that showed evidences of marks being printed on the side of the bullet by the barrel it passed through.

Q. How many of those sections compared with the test bullet from this little blue-steel gun? A. Three.

Q. Three? A. This is not a blue-steel; this is the nickeled.

Q. I mean the nickeled gun; three of them? A. Yes, sir.

Q. And they were in a condition to examine? A. That is correct, sir.

Q. The other two sections were mutilated so that you could not make a comparison; is that correct? A. Well, I could find many places which I could examine. You see, we examine a very very small part at a time. Due to its condition, I could only compare three portions.

Q. And the three portions which you examined, did those three portions compare identically with the test bullet from the nickel-plated gun? A. They did not compare identically. We don't find any two bullets that do compare identically. However, there were a great number of similarities.

Q. I wish you would state whether or not any markings on the evidence bullet compared with the markings on the test bullet which was taken from the blue-steel gun. A. There was not.

Q. No comparisons whatever? A. There was not.

Q. And I believe you stated positively that the evidence bullet was not fired from the blue-steel gun? A. I am of the positive opinion that it was not.

Q. I wish you would take that evidence bullet and the test bullet that was fired from the blue-steel gun and show them approximately the sections that you compared together. A. I am unable to do that without the aid of the microscope.

By THE COURT: He has stated how he made the comparison, and has given his opinion on that.

By MR. SHAFFER: Q. I believe you stated the weight of the bullet from the big gun was 98 grains; is that correct? A. There were two types of cartridges. One was 98 grains and the other 100 grains; that is, the Remington shells were the 100-grain bullet type, and the Winchester Repeating Arms shells were the 98-grain bullet type.

Q. And what was the weight of the bullet that was fired from this gun? A. That was the U.S., of the 85-grain bullet type.

Q. That was fired from the blue gun? A. Yes, sir.

Q. And then the other gun, what was the weight of the bullet that was fired from the nickeled gun? A. That was the Remington 88-grain, smokeless powder type.

Q. What did the evidence bullet weigh? A. 82 grains.

Q. Was any sections or part of that bullet gone? A. There appears to be, yes, sir.

Q. I wish you would tell the jury whether or not, in firing bullets of that kind, which are mutilated, as to whether a part of the bullet is lost as a usual thing. A. Under the ordinary circumstances there is very, very little lost in firing a bullet, unless it comes in contact with some hard substance which will tear part of it away.

Q. Did you examine those barrels by any other instruments, other than what you have here? A. I used a helixometer at the Firearms Identification Bureau in the Department of Justice. That is an instrument used to examine the interior of gun barrels.

Q. What did that examination disclose? A. This small, nickel-plated gun, that is, the interior of the barrel was in a very rusted and pitted condition. The large Smith and Wesson revolver showed some evidence of rust and pitting, but not so much as in the small gun. The blue-steel Colt revolver showed the condition of the barrel to be in very good condition.

Q. You mean Colt? A. Or Smith and Wesson.

Q. You mean by the Department of Justice, that is, in Washington? A. Yes, sir. The reason I used their instrument, I don't have such.

CROSS-EXAMINATION

BY MR. CAMPBELL: Q. Lieutenant, I understood you to say you had worked in an arms factory? A. Yes, sir.

Q. What factory was that? A. The Remington.

Q. And when was it you worked there? A. In 1915 and '16, if my memory serves me correct—or '16 and '17.

Q. And then you went with the Police Department in 1919? A. No, sir. Oh, I went with the Police Department in 1919, yes, sir.

Q. Where was that Remington plant? A. Chester, Pennsylvania.

Q. You say that you had taken instruction under two majors in the Ordnance Department? A. Yes, sir, from time to time, as problems would arise.

Q. The Ordnance Department is that part of the military establishment that has firearms and ammunition as its only business? A. That is correct, sir.

Q. Furnish all the ammunition and firearms for the Army and Navy. Lieutenant, who is the present Chief of Ordnance? A. I do not know, sir.

Q. You are well acquainted around Washington, aren't you? A. I have been acquainted around Washington all my life with the exception of about six years.

Q. You have told us of your work around Washington. Do you know Major Wilhelm of the Ordnance Department? A. I do not.

Q. You have heard of him? A. I have not.

Q. Did you ever hear of General Tasker H. Bliss? A. I have not.

Q. Did you ever hear of Mr. August Vollmer? A. I have, sir.

Q. Are you familiar with —— of the Institute of Criminology, published at Northwestern University? A. I am.

Q. That is regarded as an authoritative magazine on police investigation, is it not? A. That is correct, sir.

Q. And the Northwestern Institute of Criminology and Criminal Law is one of the leading institutions of its kind in the United States, is it not, sir? A. That is my opinion, sir.

Q. I believe you went there, yourself, did you not? A. Yes, sir.

Q. That is called the scientific crime detection laboratory, is it not? A. Affiliated with the Northwestern University.

Q. Did you know Lieutenant-Colonel Goddard there? A. Very well, sir.

Q. Did you also know Lieutenant-Colonel C. O. Gunther, of the Stevens Institute of Technology? A. I did not know him, sir.

Q. Did you know of him? A. Yes, sir.

Q. What is his reputation with reference to forensic ballistics? A. Not so good, sir.[108]

Q. Is that his reputation as you know it? A. Yes, sir.

Q. I hand you here a letter from the Department of Ordnance and ask you if Lieutenant-Colonel Gunther's name is not there?

Objection.

Q. I will ask you to read that letter and see if your name is there among the ones recommended.

Objection.

BY THE COURT: Wait just a minute. You are going into the reputation of some man.

BY MR. CAMPBELL: He has testified as to his qualifications. We want to show that the Department of Ordnance did not recommend him.

Objection sustained. Defendants, by counsel, *except*.

BY THE COURT: I don't want any misunderstanding about it. The question you asked was about some lieutenant-colonel. I sustained the objection to that question.

BY MR. CAMPBELL: I will withdraw it. I am trying to show his own qualifications now.

BY THE COURT: You may show that.

BY MR. CAMPBELL: He is a man engaged in this business in Washington, D. C., and claims that he took his instructions, part of them, in the Ordnance Department, or the Bureau of Ordnance——

BY MR. SHAFFER: We object to those statements before the jury.

BY THE COURT: I wouldn't allow the introduction of this letter at all.

BY MR. CAMPBELL: The whole letter is perhaps not altogether admissible.

BY THE COURT: Well, any part of it. Objection sustained.

---

[108] It is interesting to note that Lt. Col. Gunther was the first expert on firearms identification presented by the defense. The state's witness Fowler heard the testimony. After the direct examination of this defense expert, Fowler and the prosecuting attorneys conferred. Subsequently the prosecutors failed to cross-examine Lt. Col. Gunther and they did not recall their own witness Fowler.

By MR. CAMPBELL: Q. Now, let's see, Lieutenant; I believe you said that you did know Mr. Vollmer, or knew him by reputation? A. I know him by reputation.

Q. The Encyclopedia Britannica is regarded as a standard work of reference, is it not? A. I have consulted it on several occasions.

Q. And have you ever consulted in that the article by August Vollmer? Objection.

Q. August Vollmer is regarded, is he not, as an outstanding authority in forensic ballistics? Objection.

By THE COURT: Who is regarded as an outstanding authority?

By MR. CAMPBELL: August Vollmer?

By THE COURT: What was the question?

By MR. CAMPBELL: If he is not an outstanding authority on forensic ballistics.

By THE WITNESS: I have never heard Mr. Vollmer's name mentioned in connection with forensic ballistics. I have never read any papers in connection with forensic ballistics where Mr. Vollmer's name was mentioned.

Q. And you never have read his article in the Encyclopaedia Britannica on this subject, have you? A. I have not, sir.

Q. You said you had never heard of General Tasker H. Bliss? A. I have not, sir.

Q. You did not know that he was at one time Chief of Ordnance? A. I have never heard of him, sir.

Q. And I imagine you have never heard of General Teskemovitch? A. I never heard of him, sir.

Objection.

By THE COURT: I haven't seen anything yet to show the purpose of this. I wanted to see if he would get any place.

By MR. CAMPBELL: This man has said he is an expert. We think we have a right to test him to see if he is familiar with textbooks and writings of authorities on the subject. I imagine he is one of the few people in the courtroom who never heard of General Bliss, for instance.

Q. You are not familiar with the "Theory of Ballistics" published by Captain James Currans, are you? A. No, sir.

Q. Or any of the things contained in that work. You are not employed by the United States Government in any capacity, are you? A. No, sir; by the District Government.

Q. By the District of Columbia as a lieutenant of detectives in the police department? A. Yes, sir.

Q. In Washington, D. C.? A. Yes, sir.

Q. How long have you been a lieutenant of detectives? A. Since 1930.

Q. Did you ever work for the Smith and Wesson Arms people? A. No, sir.

Q. They made the three pistols that are here in question, did they not? A. I believe so, yes, sir.

Q. When you worked for the Remington people, what was your line of work there? A. I worked in the barrel department, filing and assembling departments.

Q. Is there any difference in the number of grooves and lands in the Smith and Wesson and Remington products? A. There is a difference in the Remington revolver and the Smith and Wesson.

Q. Could you give us the approximate age of any of these pistols? A. I could not.

Q. You don't know how long they have been manufactured? A. No, sir.

Q. Or anything about them? Don't answer this until the gentlemen on the other side have an opportunity to object. I read here from an article by Chief Vollmer in the Encyclopaedia Britannica which deals with forensic ballistics and fire arms——

Objection.

BY MR. SHAFFER: We want to read that first.

BY MR. CAMPBELL: Q. The serial number of a pistol indicates approximately the date of its manufacture, does it not? A. I do not know, sir.

Q. You haven't sufficient familiarity with them to know that? A. I do not know.

BY MR. SHAFFER: I don't think it is proper to read to a jury——

BY MR. CAMPBELL: He claims he is an expert, and we claim he ought to know what an expert can do.

BY MR. PARSON: By what authority?

BY MR. CAMPBELL: By the Encyclopaedia Britannica; it is generally accepted as being about the last word on anything.

After a conference in chambers, the Court permitted two questions dictated by Mr. Campbell to be read to the witness, and upon resumption of trial the first question was read by the reporter as follows:

Q. Is it not a fact that experts in forensic ballistics should be able, after examining a bullet, to state the make, caliber, type, approximate date of manufacture, and approximate serial number of the gun from which it came? A. I only interest myself in one question: was the given bullet fired from the given gun?

The second question was then read by reporter as follows:

Q. I hand you the Encyclopedia Britannica, 14th Edition, and refer you to page 561, article on criminal investigation, subdivision, firearms and forensic ballistics, by August Vollmer, Chief of Police of Berkeley, California, and will ask you to please state whether or not this authority does not lay down the proposition that an expert should be able to determine these matters. A. I am only interested in one thing in the examination of firearms and bullets, and that is, if the evidence or fatal bullet was fired from the suspected weapon.

BY MR. CAMPBELL: That is not the question I asked you. I will ask you to listen while the stenographer reads the question, again, and ask you to answer yes or no.

(Question read again by reporter)

A. I have not read this article.

Q. Will you please read it.

BY THE COURT: Just read the article——

BY THE WITNESS: "Forensic Ballistics"——

BY MR. SHAFFER: Just read it to yourself and then answer the question. A. I don't fully appreciate the question. In fact, I don't understand just what you want. Can I repeat the way I understand you?

By Mr. Campbell: Q. I will try to clear it for you. Isn't it a fact that the Encyclopaedia Britannica, to which I have referred you, states that experts in forensic ballistics, after examining a bullet, should be able to state the make, caliber, type, approximate date of manufacture and approximate serial number of the gun from which it came? A. That does not assist in the identification.

Q. I am asking you if the Encyclopaedia Britannica does not state that. A. It does.

Redirect Examination

By Mr. Shaffer: Q. In your investigation of this case you are seeking, as I understand, to find which of these guns fired the fatal bullet, instead of seeking to find the date of manufacture of the gun in question; is that correct? A. That is correct, sir, and I never do try to seek the date of a gun in any of my investigations.

Q. Would the serial number or date of a gun, or knowing the date of manufacture of a gun or the serial number of a gun aid you in any degree in determining which of the guns a bullet was fired from? A. It would determine the type of firearm. That is to say, each firearm manufacturer has his own standards and specifications that he makes his rifling, by, the rifling consisting of the land the groove. We use those methods when a bullet is brought in without a suspected weapon, in an effort to inform the officers of the type of weapon to look for.[109]

Conclusions

The analysis of the authorities on firearms identification from ammunition fired therein, illuminated by the scientific principles expounded in Chapter I, leads to the inevitable conclusion that juries are not qualified to make intelligent determinations without the assistance of proper expert testimony. Even if the issues arising in particular cases appear to involve only conspicuous characteristics, such as the different sizes and shapes of ammunition, it will be advanta-

---

[109] Stuart B. Campbell, of counsel for the defendants, sent Lt. Col. Gunther this testimony with the comment that "this transcript was written up hurriedly and does not set out the cross-examination as accurately as might have been done. There were some proper names that the reporter did not get, and I haven't the book before me now, so I can't give them to you. They were not important . . . ." There is no official citation for this case.

In Gibson v. State (Ga. 1934) 174 S. E. 354, 356, the Court used the following language: "shells 10 and 11 'were fired out of the shotgun that we got from Raider Davis' . . . the shell marked No. 1 was found by the officers at the scene of the homicide, and that the plunger mark on that shell corresponded with the plunger mark on the shells 10 and 11, there was no error in admission of the evidence over the objection stated."

geous, as a precautionary measure, to consult experts, because they alone are equipped to construct the inferences ascertainable by examinations of the firearms exhibits. In many cases the experts will be able to draw helpful inferences which would not be otherwise available for the courts.

The authorities exhibit the confusion and injustice which arises when the courts, laboring under misconceptions as to the real nature of firearms identification from ammunition fired therein, admit in evidence incompetent expert testimony. A great deal of the difficulty in successfully excluding such testimony is that a large part thereof appears most attractive and plausible to the courts untrained in the field of firearms identification. And once the courts approve of the experts' qualifications, the weight to be given their testimony, under our scheme of justice, is for the determination of the jury, and as the juries have no special knowledge in this field of identification they are likely to allocate undue weight to the apparent or attractive value of unreliable testimony, unless the deficiencies thereof are intelligently brought to their attention. Consequently, in order to prevent the juries from being misled by positive opinions predicated upon fallacious foundations, it is of the utmost importance that they should evaluate only competent testimony, and that they should confine their determinations to questions arising from conflicting items of proper expert testimony. However, under the present condition of the law the courts will not take judicial notice of the fact that a particular witness is not a proper expert. On the contrary, the tendency is to allow experts to qualify easily upon a relatively insignificant representation of specialized experience, such as mere experience and familiarity beyond that of the ordinary juror. And once the court has admitted in evidence the opinions of the witness, his entire testimony, however incompetent, is likely to carry some weight with the jury because an air of authority often surrounds a witness who has qualified as an expert. It is error to submit such testimony to the jurors for what it is worth, unless the jurors have an intelligent basis for determining its value. Of course, the deficiencies of testimony may be brought to the attention of the jury through the exercise of an intelligent cross-examination and by the understandable testimony of rebutting experts. But sadly enough the trial lawyers, lacking superior knowledge in the field of firearms identification, are not always able to bring out effectively the limitations in the expert testimony on cross-examination. Similarly, the trial lawyers have not always been successful in obtaining the services of competent experts.

Therefore, in the future, the courts should raise the standards of qualification and make it more difficult for a witness to qualify as an expert.

In connection with the question of qualifications the courts should keep in mind that the art of identifying firearms from ammunition fired therein is a science based upon principles which have been developed by constructive research, and that the only way by which a working knowledge can be acquired at present is by individual research along scientific lines, under the direction of a trained investigator. The vital element in this field of identification is *scientific judgment* which can be gained only through the performance of actual experiments. The training of the experts must include experiments with ammunition fired from weapons collected at random, such as the firearms confiscated by the police departments, ammunition fired from firearms manufactured under approximately the same conditions at a particular factory, and ammunition fired from a single firearm under various conditions. It is clear that neither the mere reading of material on firearms identification nor the execution of careless research constitute the necessary experience of an expert.

The expert must be skilled in the use of the comparison microscope. It is obvious from the methods used in firearms identification that an expert cannot successfully employ a microscopist, untrained in the science of firearms identification, to perform the necessary examinations with the microscopes. In the case of photography, however, the expert may employ a skilled photographer to make photographs and photomicrographs under his personal supervision. Likewise, it is unnecessary for an expert on firearms identification to be a gunsmith. All parts of a firearm with which the ammunition may come in contact are accessible for visual inspection without the need of disassembling the firearm. The expert in firearms identification must confine himself to a microscopic examination of the markings upon the ammunition fired in a firearm, and all his conclusions must be based thereon, just as in identifying a typewriter all conclusions are reached and predicated upon an examination of typewritten material. The identification of firearms rests upon a comparison of the effects produced upon ammunition fired therein, and there is no more occasion to examine the interior of a barrel than there is to examine personally the writer when identifying his handwriting. An inspection of two bores might reveal apparently similar visual conditions in each, and yet the effect of these conditions upon the surfaces of the bullets might be quite different. Surely a competent expert in firearms identification will not have a great deal of time to devote to other fields.

By a proper use of instruments the experts can ascertain the facts pertinent to the identifications. In many instances the experts will not agree as to the inferences to be drawn therefrom, and unless they disagree on a mooted technical question about which there can be an honest difference of opinion, one of the experts must be wrong. In explaining conclusions the experts should submit only statements which are supported by controlled research, eliminating all material which is based upon mere hypothesis. Any part of expert testimony which is not supported by research substantially affects the weight to be given the entire testimony. For example, it would not have been a difficult matter to ascertain that the wearing away of the tools was of minor importance as a cause contributing to the individuality of the surfaces of firearms at the time of manufacture. The neglect to perform this basic research necessarily casts doubt upon the performance of other research.

The authorities clearly illustrate the prevailing attitude of the courts towards scientific questions. The expert witnesses are now permitted to appeal to the intelligence of the juries by demonstrations and by pictorial evidence. By the same token, the juries are no longer forced to rely upon the mere assertions of the experts. They look primarily to the explanations of the data upon which the opinions are predicated for proof that the conclusions are warranted. By means of these comprehensive explanations, the juries are equipped to consider more intelligently a single opinion on its own merits, and to evaluate more intelligently the merits of conflicting opinions. The disagreement between experts is becoming a disagreement as to the reasons for opinions rather than a mere conflict of opinions.

Despite the liberality of the courts in permitting scientific explanations of scientific questions, the personal integrity of the expert witnesses continues to be a very important consideration. A great deal of the subject matter of firearms identification is highly technical and it involves the application of the sciences of microscopy and photography. It may not always be possible for the juries to understand fully in detail the scientific explanations of competent experts, and to this extent, the juries are forced to rely upon the assertions of the experts even though the evidence is properly handled by capable counsel. Likewise, the jury must rely upon the assertions of the experts as to the results of research. The experts should express opinions, predicated upon thorough examinations of the data, which concisely reveal their personal convictions. In some cases the insufficiency or condition of the data may preclude any reliable conclusion. In others, the observable data may only support the conclusion that the bullet or cartridge case might have been fired in a particular firearm. The experts who set forth carefully and frankly the limitations inherent in the data of particular cases will soon earn meritorious reputations. The integrity of the experts will undoubtedly continue to be tested by the vicious practice of zealous attorneys and officials who present the problems of identification in a partisan manner in an effort to secure favorable testimony. An advance knowledge of the circumstances of a case will be a source of embarrassment to the conscientious witness seeking to render unbiased opinions. The deficient or corrupt witness will materially benefit from the advance information that all the circumstances create the probability that a given weapon had been employed in the crime. The obvious danger in this practice is that a reliable identification of firearms may indisputably show that all the other circumstances lead to an erroneous inference as to the guilt or innocence of a particular defendant (State v. Bausells, *supra*). The opportunity for deception also arises in the use of microscopes and photomicrographs. A truthful representation of the marks appearing upon the surfaces of bullets and cartridge cases can be prevented by a manipulation of the lighting effects.[110] However, with the comparison microscope and the firearms exhibits in court, a careful scrutiny of the expert's methods will, for the most part, be an adequate check upon the data which are purported to be the basis of a given conclusion.

PRESENTATION OF EVIDENCE

The authorities show that regardless of its accuracy the evidence on the identification of firearms from ammunition fired therein has

[110] See Chapter I, *supra*, Figs. 89, 90.

An article has appeared containing an illustration purporting to show the corresponding markings on a test and fatal bullet as viewed under the comparison microscope. An examination of this illustration leads to the inevitable conclusion that it is a photograph of a model of a bullet, and that the marks were made with a tool to simulate the markings made on the surface of a bullet as it passes through the bore of a firearm.

A general objection may be made to the illustrations which have appeared in publications for the reason that they do not depict conditions which confront the investigator in his examination of bullets and cartridge cases which have actually been recovered in connection with homicides.

not been presented in the most effective manner. In this connection a logical procedure should be followed.[111]

The qualifications of a competent expert should always appear on the record for the purposes of appeal, even if the other side admits the competency of the witness. The qualifications can be presented most effectively in answers to properly framed questions which develop the essential points of experience and training in a clear and concise manner. When an expert is asked to state his qualifications in response to a general question it often appears to the jurors that the witness is manifesting an unnecessary degree of conceit.

After the witness has been qualified, he should be asked to state whether or not he has made the necessary comparison of the signatures to be identified with the signatures used as the standards of comparison. If the witness has made the comparison, he should be asked to state whether or not he was able to reach definite conclusions with respect thereto, and if so, what his conclusions are.

It is at this stage in the presentation of expert testimony that the greatest of care must be exercised. The most important function of the expert is to present in an intelligible fashion the reasons for his opinions. "The range and force of expert testimony is to explain to the jury things not capable of being understood by the average person. His special or technical knowledge license him to express an opinion of cause and effect of things with which he is familiar. The weight of such an opinion necessarily rests upon the reasons or explanations given by such witnesses in support of the opinion." [112] Mere opinions may neutralize each other, but this is seldom true of two reasons. The honest witness is anxious to support his opinions with reasons; he does not wish to force any ready-made conclusions upon the jury; his aim is to assist the jury in reaching a correct conclusion predicated upon a proper interpretation of the data. The competent expert realizes that by this approach the jury will be in a position to discount incompetent and corrupt testimony. *Therefore great emphasis should be focused upon reasons for opinions, and the importance of mere assertive opinions should be minimized.*

The two principal phases in presenting opinion testimony on firearms identification are: first, to point out, describe, and illustrate the

[111] Invaluable assistance may be obtained from Osborn, ''Questioned Documents,'' and Osborn, ''The Problem of Proof.'' The procedure to be followed here is similar to that suggested by Mr. Osborn in the field of disputed documents.

[112] Gulf, C. & S. F. Ry. Co. v. Downs (Tex. 1934) 70 S. W. (2d) 318, 322.

data which the witness has observed through his specially trained faculties; and second, to give a definite interpretation of the data. And it is of the utmost importance that the observed data are not confused with the interpretations thereof; testimony as to facts must be distinguished from testimony relating to the inferences which the witness has drawn from the facts.

The reasons for opinions should be developed for the jury in a convincing manner. This can be done with the greatest efficacy by the use of intelligently phrased questions designed to develop the various points logically, with the proper emphasis; and in this connection it is essential that a distinction be made between class and accidental characteristics. The witness should avoid giving an extended technical lecture. He should always apply the principles of the science of firearms identification directly to the case at bar. Any unnecessary use of complicated technical terms should be condemned; it is far more effective to use language readily understood by the jurors. When it is found necessary to use technical terms they should be clearly and precisely defined for the jury. The witness should carefully state his beliefs, and in so doing he must avoid being either dogmatic or weak.

A competent expert is aware that any problem of identification involves reasoning with respect to similarities and differences, and that he must point out the significance and the comparative value of all the elements relative to the identification. He will therefore, as fits the necessities of particular cases, clearly interpret the differences found in the signatures of the same firearm and the similarities found in the signatures of different firearms. He will never base a conclusion upon general appearances; he will never minimize the importance of either differences or similarities; and he will never form an opinion without giving a proper consideration to both.

In enabling the jury to reach a conclusion instead of accepting a ready-made conclusion from the witness, pictorial and demonstrative evidence must be effectively presented. Photomicrographs are invaluable because they afford to the jury a permanent record of the markings observed through the comparison microscope. They can be arranged to afford the greatest clarity for the jury, and it is best to produce copies for each member of the jury as well as the judge and counsel. (Photomicrographs can be admitted in evidence if they are made under the personal supervision of the expert, or if there is testimony to show that they are accurate. The proof of genuineness may be made by the expert who has by an examination ascertained that the negatives reproduce the originals and that the prints reproduce the negatives.) On many occasions it will be possible to demonstrate a particular fact to the jury conclusively; for example, it might be pertinent to prove that a certain cartridge could not be fired in a given firearm.

When the reasons for opinions are presented in response to a proper series of questions it may happen that the expert will be precluded from conveniently explaining some important points. Therefore it will be advisable to submit a final comprehensive question, giving the witness a great deal of latitude in explaining matters which would have been unresponsive to the former and more specific questions. The witness may be asked a general question, permitting him to give his conclusions, detail his examination and all matters affecting his opinion, and in connection therewith to make use of the exhibits and photomicrographs.

CROSS-EXAMINATION. QUESTIONS

Generally speaking, a competent and reliable expert will not be cross-examined. On the other hand, the right of cross-examination is the most effective weapon which can be employed to expose successfully the deficiencies in incompetent testimony, and the misrepresentations in corrupt testimony. It is clear that the importance of the proper exercise of the right of cross-examination can not be exaggerated. A proper cross-examination directed along intelligent lines will effectively expose the weakness of testimony without any unnecessary abuse of the witness.

It will be found that some of the witnesses presented as experts are wholly unable to reason logically. Others will be found to be incompetent because they have not made the necessary study; they do not possess the qualifying experience and training. In some cases it will be found that a witness has made a mistake either honestly or otherwise.

With the exception, perhaps, of a totally unqualified witness, the trial lawyer should hesitate to undertake any extended cross-examination unless he is substantially well informed as to the basic principles of the science of firearms identification from ammunition fired therein. It is essential that in a proper cross-examination the trial lawyer should not lose sight of the particular points to be developed; he must not allow himself to be sidetracked by the digressions of the witness. Whenever possible direct questions should be given calling for a yes or no answer, the theory of which is to build up a series of admissions which lead to an ultimate distressing conclusion. Likewise, after a yes or no answer the witness may be contradicted by reference to some authoritative book on the subject.

In analyzing the testimony given by the witness on direct examination, certain items require particular attention. If the witness has identified two signatures as having been made by the same firearm, it is important to note whether the witness has ignored the differences therein, and if he has mentioned them, whether he has reconciled them by an accurate interpretation. If the witness has stated that two signatures have not been made by the same firearms, it should be noted whether he has ignored the similarities existing therein, and if he has mentioned them, whether he has successfully interpreted them. It is likewise important to notice whether the witness has based his opinion upon general appearances. A most effective line of cross-examination arises out of the failure of the witness to point out properly and interpret particular similarities and differences.

If a witness on direct examination has insisted upon using complicated technical terms not readily understood by the jurors, a few typical sentences should be quoted to him on cross-examination. He should then be interrogated as to whether he believed that his testimony would assist the jury, and on the other hand, whether it was not his purpose merely to create an impression upon the jurors? It is usually found that such a witness is too incompetent to be able to discuss the subject intelligently; persons who are well versed in their specialties usually are able to discuss matters pertaining thereto in simple language readily understood by the layman.

The following groups of questions are considered to be pertinent, and they have been selected as relating to the fundamental principles of firearms identification from ammunition fired therein. The questions are designed to test the ability of the witness and to bring out his familiarity with and knowledge in this field. They must, of course, be intelligently adapted to the necessities of particular cases.

GROUP I. QUESTIONS RELATING TO THE TRAINING AND EXPERIENCE OF THE
WITNESS

Q. Where did you receive your experience for this work?
Q. Did you equip yourself for this work under the direction of a properly trained investigator?
Q. How much time have you spent in actual research in this field?
Q. What instruments do you use?

Q. What instruments do you own?

Q. What material have you read in this field?

Q. Have you read Gunther on "Identification of Firearms"?

Q. Have you read any material relating to the general problem of identification?

Q. How much time did you spend in making the necessary comparisons in this case?

GROUP II. QUESTIONS DESIGNED TO EXPOSE THE UTTERLY DEFICIENT AND UNQUALIFIED WITNESS

Q. Do you realize that you are testifying as an expert in this case?

Q. What does the word expert mean to you?

Q. Are you so confident of your ability that you would be willing to have a person sentenced to death on the basis of your opinion?

Q. Would you appear for the defense in a capital case if the defendant were a close relative?

Q. Knowing your relative to be innocent, would you feel that your relative's chances of acquittal would be better if the services of a more widely known expert could be obtained?

Q. Are you a ballistics expert?

Q. What is ballistics?

Q. Do you distinguish between skilled observation and skilled interpretation?

Q. What is skilled observation?

Q. What is skilled interpretation?

Q. What is your understanding of the word inference?

Q. What is "reasoning"?

Q. Is your testimony based upon scientific principles?

Q. What is a scientific principle?

Q. What is the application of the law of probability in the science of identification?

Q. What is the application of the law of probability in the science of identifying firearms from ammunition fired therein?

Q. How do you identify a particular firearm from all those in existence when your research has been confined to an examination of markings on ammunition fired from a relatively small number of firearms?

Q. How do you distinguish between class and accidental characteristics?

Q. Do you always examine the surfaces of the firearm with which the ammunition fired therein comes in contact?

Q. You do not confine your examination solely to markings on ammunition, do you?

(If the witness places any emphasis upon his examinations of the surfaces of firearms, the following questions should be asked.)

Q. If you were given three bullets involved in three different crimes, would you be able to determine whether all three had been fired from the same firearm if no firearm had been recovered?

(The totally unqualified witness is quite likely to state that he has heard of fictitious publications and experts in this field, if the fictitious titles and names are employed in conjunction with the actual titles of publications and the actual names of persons outstanding in the field of firearms identification.)

GROUP III. QUESTIONS RELATING TO MICROSCOPY

Q. Do you use a microscope?
Q. Do you use a comparison microscope?
Q. What is a comparison microscope?
Q. What is a photomicrograph?
Q. What is a composite photograph?
Q. What is magnification?
Q. What is enlargement?
Q. What power of magnification do you use?
Q. How long have you been using a comparison microscope?
Q. How long have you been using any microscope?
Q. What have you read on microscopy?
Q. Have you read Gage on "The Microscope"?
Q. Do you think it possible to make a reliable comparison under a single microscope?
Q. Is it not a dangerous practice to predicate opinions upon a comparison of two signatures under a single microscope?
Q. Is it not true that a witness in order to qualify as an expert should be able to use the comparison microscope intelligently?

GROUP IV. QUESTIONS RELATING GENERALLY TO SIMILARITIES AND DIFFERENCES IN THE SIGNATURES OF FIREARMS

Q. Do similarities exist in the signatures of different firearms?
Q. Give some examples of such similarities.
Q. Do differences exist in the signatures of the same firearm?
Q. Give some examples of such differences.
Q. If you are making a comparison of two signatures, assuming common class characteristics, and you find that, of the effects of eight accidental characteristics appearing in one signature, only five are found in the other signature, what would your conclusion be as to whether the two signatures were made by the same firearm?
Q. Is it not dangerous to base an identification upon the general appearances of two signatures?
Q. Is it not necessary to have the effects of some outstanding characteristics upon which to base your opinion?
Q. Did you ever give an opinion when there were no outstanding characteristics?
Q. What is an outstanding characteristic?
Q. What is an individual peculiarity?
Q. How do you know whether or not a particular combination of characteristics is an individual peculiarity?
Q. You have not examined ammunition fired from all the firearms in the world, have you?

Q. Are you not identifying this firearm from all the other firearms in the world?

Q. Is it not true that you have examined ammunition fired from but a small number of firearms?

Q. How many test bullets did you fire?

Q. How many test cartridge cases have you?

Q. You did not consider all these test bullets, did you?

Q. You did not consider all these test cartridge cases, did you?

Q. You discard particular test bullets if they do not show the outstanding markings which you find on other test bullets, don't you?

Q. You discard test cartridge cases if they do not show the outstanding markings which you find on other test cartridge cases, don't you?

GROUP V. QUESTIONS RELATING TO THE MARKINGS ON BULLETS

Q. Has every firearm permanent characteristics?

Q. What are the permanent characteristics?

Q. Is it not true that powder fouling will produce changes in the signature of a firearm?

Q. Can you identify the signature on a bullet fired through a clean barrel with the signature on a bullet fired through the same barrel, after ten black powder cartridges have been fired therein?

Q. How many black powder cartridges can be fired before it is impossible to identify the signature on the first bullet with the signature on the last bullet?

Q. Can you always distinguish between the differences in the signatures of the same firearm produced by powder fouling, corrosion and erosion, and the dissimilarities found in the signatures of two different firearms of the same class characteristics?

Q. Is it true that the interior surfaces of no two barrels are identical?

Q. Is it not true that the groove engravings may be reproduced by two different barrels of the same class characteristics?

Q. Is it true that the entire tool-marked pattern of a groove does not produce an effect upon a bullet?

Q. Is the groove engraving the cumulative effect of the bullet's contact with the groove?

Q. Is it not true that the individuality of the tool-mark pattern of the groove depends upon the wearing away of the rifling cutter?

Q. The cutting of the groove is a shearing operation, is it not?

Q. You do not think that the cutting of the groove is a tearing operation, do you?

Q. The cutting of the grooves is of great significance in this field of identification, is it not?

Q. Would you be able to tell whether a bullet struck the forcing cone with its axis inclined to the axis of the bore because the chamber of the cylinder was out of line, or because there was a defect in the crimp of the cartridge case.

Q. Can the individual peculiarity of a particular firearm be established by the marks on a bullet caused by the bullet entering the forcing cone with its axis inclined to that of the bore?

Q. Is not a double land impression caused by the failure of the bullet to take up the motion of rotation immediately upon entering the bore?

Q. Is it not true that the so-called "stripping marks" are brought about by the failure of the bullet to enter the bore with its axis parallel to the axis of the bore?

Q. Are fine striae of any significance in themselves?

Q. Are they misleading?

Q. What is their comparative value?

Q. What consideration do you give them?

Q. Will all the differences in the surfaces of two bores be perceptible from an examination of bullets fired therein?

GROUP VI. QUESTION RELATING TO THE MARKINGS ON CARTRIDGE CASES

Q. Is it not a fact that a defect on a breeching face may produce an effect on the head of a cartridge case which may vary in all proportions from shot to shot?

Q. How do you distinguish between the differences in the effects produced by a particular defect on a breeching face, and the dissimilar effects produced by defects of various proportions on the breeching faces of different firearms?

Q. Is it not true that the firing-pin indentation is of primary importance?

Q. Is it not true that the effect of the firing pin upon the primer is an individual peculiarity?

Q. What is the part played by the relative hardness of the metal in the primer cup?

Q. Does the firing pin always strike the center of the primer?

Q. Is the firing-pin indentation in the primer exactly the same in the signatures of any two firearms?

GROUP VII. QUESTIONS RELATING TO THE DIFFERENCES IN AMMUNITION

Q. Do differences in ammunition cause differences in the signatures of a firearm?

Q. You did not use the same ammunition as the fatal bullet, did you?

Q. You did not use the same ammunition as the cartridge cases picked up at the scene of the crime, did you?

Q. Are you sure that the differences in the markings on the test and fatal bullets has not been caused by the differences in ammunition?

GROUP VIII. QUESTIONS RELATING TO MEASUREMENTS

Q. Is it not true that all variations in the dimentions of a bore can be ascertained by a measurement of the markings on bullets fired therein?

Q. When the measurements of the land and groove widths on the bullet exactly agree with the dimensions of the groove and land widths of the bore of a particular barrel, is it not conclusive that the bullet passed through the barrel?

Q. There will be no variations in the dimensions of the effect of the accidental characteristics of a bore from shot to shot, will there?

Q. Is it not a simple matter to measure the markings on the surfaces of bullets?

Q. What instruments do you use in making precision measurements?

Q. How long have you used the instruments?

Q. What training have you had in the use of these instruments?

Q. The surface of a bullet is curved, is it not?

Q. Most of the bullets located at the scenes of crimes are deformed, are they not?

Q. Isn't it true that it is a difficult matter to measure markings appearing upon the curved surface of a distorted bullet?

Q. Is it true, then, that measurements are of but little value in the identification of firearms from ammunition fired therein?

GROUP IX. QUESTIONS RELATING TO A CORRUPT WITNESS WHO HAS TESTIFIED AS TO AN ERRONEOUS CONCLUSION

Q. Did you have any hesitancy in arriving at your opinion?

Q. How long did it take you to arrive at your opinion?

Q. Did you find anything unusual in the markings on these bullets (or cartridge cases) when you made your comparison?

Q. Would you say that these markings are typical of the usual case of two bullets (or cartridge cases) fired from the same firearm (or different firearms)?

Q. As this is an especially difficult problem of firearms identification, will you admit that it is possible that you are mistaken?

Q. Do you always willingly admit any mistake that you have made?

Q. What information have you received with respect to this firearm, this evidence bullet, and the defendant.

FUTURE DEVELOPMENTS

The future development of the science of firearms identification from ammunition fired therein depends upon exhaustive and constructive research. Data must be collected concerning as many of the various kinds and types of firearms and ammunition as possible. Comprehensive experiments, conducted under many sets of conditions, must be performed to determine further the nature, causes, and extent of the similarities found in the markings on ammunition fired in different firearms, and to determine further the nature, causes, and extent of the dissimilarities which appear in the markings on ammunition fired in the same firearm. In this manner only can the scope of firearms identification be substantially enlarged.

The collection of the vast amount of required data could be accomplished most successfully through a central collection agency which would possess more adequate resources than the most enterprising of individuals. Likewise, the necessary experimental work is beyond the scope of individual facilities and, consequently, it could be more thoroughly and efficiently performed at a central bureau. Furthermore, this central bureau of firearms identification, not being devoted to the selfish interests of any individual, would serve as a common source of information; all the collected data and results of research could be filed and made available to interested parties actuated by proper motives. The bureau could be made largely self-supporting by the publication of a journal containing information of vital importance to the police, expert witnesses, lawyers, and the courts. In addition there are numerous other ways by which the bureau could render an efficient, invaluable, and continuous service to the police in their daily work of detecting criminals. The police departments could contribute to the maintenance of the bureau in consideration of this service.

The establishment of a central bureau of firearms identification would encourage and expedite a more widespread use of the fruits of the science of firearms identification. The dissemination of proper and helpful information would raise the standards of expert testimony and make it more uniform, thereby precluding the present pernicious evil of incompetent expert testimony being admitted in evidence.

Unquestionably, it is a simple matter to advocate the establishment of a central bureau of firearms identification. The great difficulty lies in the administration of the bureau. If improperly handled it will be but a continuance of the charlatanism which has been manifested in the expert testimony. And to those who believe that the present expert testimony is reliable, attention is called, for example, to the Bausell case (1934) *supra*.[113] By necessity, then, the bureau must be

---

[113] Sadly enough there are current publications in this field which are highly unscientific and unsupported by actual research. For example, it is stated that an examination of the markings on a bullet will reveal the approximate date of manufacture and the approximate serial number of the firearm from which it was fired. Likewise, a recent writer contended that in the comparison of the markings on cartridge cases an ordinary single microscope was more effective than the comparison microscope; he did not exhibit an understanding of the principles involved in tool operations, and he was not able to explain the principles which govern the engraving of the surfaces of bullets. Consistently enough, he advocated that test bullets should be discarded when they did not appear to be properly engraved under the circumstances. Surely, a true scientist will not destroy pertinent data. The ability to reason between similarities and differences depends upon an accurate interpretation of all relevant data.

placed under the direction of a competent scientist personally experienced in meritorious research in this field of identification. Such a scientist, having actually performed constructive research, would be familiar with the important lines of research to be followed as well as the uselessness of other lines. The energies of the bureau would be directed to the most productive channels, and under this proper leadership the bureau would soon become of paramount importance. Furthermore, only a real scientist who has already demonstrated his ability in this field is competent to organize the bureau and to make it function most effectively through an intelligent classification of the accumulated data. It will be worse than useless to put the bureau into incompetent hands because the science of firearms identification is one of ascertaining the truth, and *Magna est veritas et praevalet.*

The science of firearms identification, properly and intelligently applied, not only will substantially contribute to a more accurate administration of justice, but it also will be an important aspect of firearms control. Clearly, then, the science of firearms identification is invaluable to social welfare, and its further development must be stimulated and made possible under the most advantageous conditions.

Fiat justitia, ruat caelum.

BIBLIOGRAPHY

EDWARD S. FARROW, American Small Arms. The Bradford Co., New York, 1904.

SIMON HENRY GAGE, The Microscope. Fifteenth Edition. The Comstock Publishing Co., Ithaca, N. Y., 1932.

WIRT GERRARE, A Bibliography of Guns and Shooting, being a list of ancient and modern English and Foreign Books relating to Firearms, etc. Roxburghe Press, London, 1895.

DR. HANS GROSS, Criminal Investigation. Translated by J. and J. C. Adam, Madras, 1906.

JULIAN S. HATCHER, Pistols and Revolvers and Their Use. Small Arms Technical Publishing Co., Marshallton, Del., 1927.

A. L. A. HIMMELWRIGHT, Pistol and Revolver Shooting. The Macmillan Co., New York, 1928.

JAMES V. HOWE, The Modern Gunsmith. Funk & Wagnalls Co., New York, 1934.

WILLIAM JAMES, Selected Papers on Philosophy (Everyman's Library). E. P. Dutton & Co., New York, 1917.

A. LUCAS, Forensic Chemistry. Edward Arnold & Co., London, 1931.

F. W. MANN, B.S., M.D., The Bullet's Flight from Powder to Target. Munn & Co., New York, 1909.

JOHN A. MARSHALL, The Manufacture and Testing of Military Explosives. McGraw-Hill Book Co., New York, 1919.

L. C. MARTIN, Optical Measuring Instruments.

EARL MCFARLAND, Textbook of Ordnance and Gunnery. John Wiley & Sons, Inc., New York, 1929.

JOHN STUART MILL, A System of Logic. Eighth Edition. Longmans, Green and Co., London, 1872.

ALBERT S. OSBORN, The Problem of Proof. Second Edition. The Essex Press, Newark, N. J., 1926.

ALBERT S. OSBORN, Questioned Documents. Second Edition. The Boyd Printing Co., Albany, N. Y.

CHAS. W. SAWYER, Firearms in American History. Cornhill Publishing Co., Boston.

LLOYD I. SNODGRASS, The Science and Practice of Photographic Printing. Third Edition, Revised. Falk Publishing Co., Inc., New York, 1931.

WILLIAM H. TSCHAPPAT, Text-Book of Ordnance and Gunnery. John Wiley & Sons, Inc., New York, 1917.

E. M. WEAVER, Notes on Military Explosives. Fourth Edition. John Wiley & Sons, Inc., New York, 1917.

JOHN H. WIGMORE, The Principles of Judicial Proof. Second Edition. Little, Brown & Co., Boston, 1931.

Photography as a Scientific Implement. A collective work. Second Impression. Blackie & Son Limited, London, 1924.

Textbook for Small Arms. His Majesty's Stationery Office, London, 1929.

INDEX